For Shade and For Comfort

For Shade and For Comfort

Democratizing Horticulture in the Nineteenth-Century Midwest

By Cheryl Lyon-Jenness

Purdue University Press
West Lafayette, Indiana

For Shade and For Comfort
by Cheryl Lyon-Jenness

© 2004 Purdue University
All rights reserved

A portion of chapter 5 appeared under the title "A Telling Tirade: What Was the Controversy Surrounding Nineteenth-Century Midwestern Tree Agents Really All About?" in *Agricultural History* © 1998 by the Society for the Agricultural History Society. Reprinted from *Agricultural History* vol 72:4, by permission of the University of California Press.

A portion of chapter 7 appeared under the title "Bergamot Balm and Verbenas: The Public and Private Meaning of Ornamental Plants in the Mid-Nineteenth-Century Midwest" in *Agricultural History* © 1999 by the Society for the Agricultural History Society. Reprinted from *Agricultural History* vol 73:2, by permission of the University of California Press.

MANUFACTURED IN THE UNITED STATES OF AMERICA

Library of Congress Cataloging-in-Publication Data

Lyon-Jenness, Cheryl, 1948–
    For shade and for comfort : democratizing horticulture in the nineteenth-century midwest / by Cheryl Lyon-Jenness.
    p. cm.
Includes bibliographical references and index.
    ISBN 1-55753-28699
1. Horticulture—Middle West—History—19th century. I. Title.
SB319.2.M54L96 2003
635'.0977'09034—dc21
                                                  2003009396

Contents

Acknowledgments   vii

Introduction   ix

One
Embellished Landscapes
1

Two
A Reason for Planting
25

Three
Hands-on Horticulture
53

Four
Commercial Realities: Supplying the Needs
87

Five
Commercial Realities: Promoting Demand
115

Six
A Neighborly Nudge
147

Seven
Bergamot Balm and Verbenas
179

Eight
Conclusion
215

APPENDIX A	223
APPENDIX B	226
NOTES	231
BIBLIOGRAPHY	295
INDEX	325

Acknowledgments

Many family members and friends have expressed interest as I completed this manuscript, and I am grateful for their support. This study grew out of a dissertation written while a doctoral student at Western Michigan University. My dissertation committee, including Dr. Linda Borish, Chair, Dr. Peter Schmitt, Dr. Michael Nassaney, and Dr. Thomas Schlereth, contributed useful comments, good direction, and unfailing optimism as the initial work progressed. Dr. Thomas Schlereth was particularly helpful as I contemplated turning the dissertation into a book manuscript. At that early stage, Dr. Janet Coryell kindly offered to read the manuscript and made many pertinent suggestions. Many colleagues at Western Michigan including Dr. Michael Chiarappa, Dr. Robert Galler, Dr. Marion Gray, Dr. Bruce Haight, Dr. Barbara Havira, Dr. Judy Stone, and Dr. Kristin Szylvian have consistently given me sound advice and strong encouragement.

A Macmillan Grant from the History Department at Western Michigan University provided financial support for obtaining illustrations, and for that I am grateful. Friend and photographer Robert Havira was generous with his time and expertise as we mulled over which images to include. His photographic and technical expertise made the preparation of illustrations for publication a real pleasure. Many librarians provided research assistance, but Judy Garrison and her staff at the Western Michigan University Resource Sharing Center deserve grateful acknowledgment for all their efforts in obtaining resource materials. Colleagues at the Western Michigan University Archives and Regional History Collections also provided helpful assistance. First Karl Kabelac, and then Melissa Mead at the Rush Rhees Library, University of Rochester offered important insight into the horticulture industry of that region and access to a wealth of

primary materials. The competent staff at the National Agricultural Library in Beltsville, Maryland, and the Bentley Historical Library in Ann Arbor, Michigan, also provided invaluable assistance. Mr. Stephan Swanson, director of The Grove National Historic Landmark, generously supplied information and access to the letters of Illinois nurseryman John Kennicott. Evelyn Leasher of the Clarke Historical Library, Central Michigan University, guided me to many Midwestern horticultural resources, and Catherine Larson of the Kalamazoo Public Library offered ready access to local horticultural materials.

Margaret Hunt, Purdue University Press managing editor, has also provided a good deal of support and helpful direction. She patiently fielded many telephone calls and frequent questions, but was always gracious and willing to talk over the issue at hand. Donna VanLeer, Jennifer Tyrrell, and others at Purdue University Press have also contributed time and expertise in seeing this project through to completion, and I thank them for both their competence and their support. In addition, the comments of anonymous readers were most helpful as I worked to refine the manuscript.

Finally, both my husband, Mark Jenness, and my father deserve a very special thank you. Mark extended kindness, patience, and a most valuable practical perspective through many rough moments. He had an unfailing faith in my progress as this project moved from dissertation to book, and his encouragement and support over the years have meant a great deal to me. My father, Ronald Lyon, instilled in me a love for Midwestern landscapes. Although he did not live to see this book completed, he would have delighted in a ride through the countryside to admire the sights and to discuss the meaning of those towering Norway spruce and carefully planted sugar maples. It is to honor his fine intelligence, his lively curiosity, and his great love that I dedicate this book.

—Cheryl Lyon-Jenness

Introduction

Mature sugar maples lining Midwestern country roads or city streets, Norway spruce towering over farm yards, and lilacs and spireas marking abandoned home sites are silent testimony to a common theme. Nineteenth-century residents of the Old Northwest used ornamental plants to define the landscapes of their daily lives. Trees, shrubs, and flowers, like parlor furnishings, kitchen stoves, fences, or farm implements, were a part of their material culture, and, like those manufactured items, both shaped and reflected the values, beliefs, and concerns of the people who promoted and nurtured them.

Despite their seeming permanence in contemporary Midwestern landscapes, shade trees and flowering shrubs have not always graced dooryards and roadsides. In the early decades of the nineteenth century, New England agricultural and horticultural commentators, as well as moral reformers, frequently noted that domestic landscapes nationwide lacked even the simplest of horticultural refinements. They painted a dismal picture of village and city homegrounds or country dooryards bereft of trees and flowers, and complained that if by some chance a garden existed, it was usually "miserably overrun with weeds."[1] According to reformers, the problem was widespread, and plagued the homes of the well-to-do along with the more modest dwellings of farmers, mechanics, and village shopkeepers. Long-settled regions had few horticultural amenities to brag about, but the lack of ornamental plants was particularly notable in newly settled areas like the Old Northwest where, as one Ohio commentator remarked, "the useful and needful must precede for a long time the ornamental."[2] In an 1839 New York state report on horticulture and home embellishments, agricultural reformer Alexander Walsh summarized what many others repeatedly expressed. Ignoring their importance as a "beau-

tifying and moralizing branch of social economy," Americans simply had no time for horticultural embellishments.[3]

This grim portrayal aside, America was not totally devoid of horticultural interest in the late eighteenth and early nineteenth centuries. Some wealthy Americans enjoyed horticultural literature from abroad, and occasionally visited notable European gardens before landscaping their own luxurious grounds. Botanic gardens and horticultural collections, established in the early decades of the nineteenth century, encouraged the systematic study of botany along with an appreciation for horticultural diversity and beauty. Dr. David Hosack, for example, developed the Elgin Botanic Garden on Manhattan Island in 1801 as an offshoot of his medical practice, and in the process, assembled over two thousand plant species for comparative study. In addition, several universities, including Yale, Harvard, and the University of South Carolina, established horticultural collections in the early nineteenth century. Shortly after the Revolution, the Baltimore area boasted a number of privately operated "pleasure gardens" that provided visitors with ornamented grounds for strolling, and a few cities, such as Philadelphia, incorporated horticultural embellishments into public parks as early as 1815. Associations, like the New York Horticultural Society founded in 1818, the Pennsylvania Horticultural Society established in 1827, and the Massachusetts Horticultural Society organized in 1829, also provided both professional and amateur horticulturists with a forum for exploring their interests and disseminating information.[4]

While they viewed any horticultural advancement as better than none, the nineteenth-century commentators who so regularly complained about the dearth of trees and flowers were not satisfied with tastes and trends that influenced only the wealthy or well-educated. They thought middle- and lower-rank farmers, mechanics, and urban workers should all enjoy the simple beauty and moral uplift so readily available through ornamental plants, and argued that those pleasures were most influential when enjoyed at home. As the century progressed, some observers pointed to evidence for a broadening of horticultural interest. In 1828, Jesse Buel, a well-known agricultural adviser and author, noted that American nurseries were growing in "number, respectability, and patronage." Charles M. Hovey, a nurseryman and the editor of the *Magazine of Horticulture*, offered a similar assessment in 1845. Since the mid-1830s, Hovey claimed, "Nursery collections [were] rapidly springing up in various parts of the country," horticultural societies were "steadily on the increase," and more and more Americans landscaped the grounds around their newly built homes.⁵

By the mid-1850s, the cautious optimism of an occasional horticultural observer had given way to an enthusiastic chorus of approval. No longer the realm of professional gardeners or wealthy patrons, commentators claimed, the culture of trees, shrubs, and flowers had transformed farm and village dooryards across the nation. In 1860, the widely circulated *American Agriculturist* typified the glowing appraisal by reporting that "the farmer, the mechanic, the doctor, and the clergyman—indeed all classes and professions" had developed a "generous enthusiasm" for planting trees, establishing lawns, and cultivating gardens. Several years later, *Moore's Rural New Yorker* voiced a similar opinion. "In no country in the world has there been such a general improvement in gardening taste during the past ten or fifteen years, as is in our own land," the *Rural New Yorker* claimed, and then added, "Trees, shrubs, plants and seeds of the rarest kinds, are purchased and planted freely, and nurserymen and seedsmen find it difficult to keep up with the popular demand."⁶

According to commentators from the Old Northwest, the growing enthusiasm for horticulture extended into their own region. In 1858, the *Michigan Farmer* proclaimed, "There has sprung up amongst the [citizens] a desire for horticultural embellishment and there are many private residences now adorned with conservatories and fruit and flower gardens, where there had been nothing of the kind." In the *Prairie Farmer* of Illinois, Chicago horticulturist Edgar Sanders pointed out that the newfound interest in plants crossed both geographic and economic borders. Reviewing an 1861 article that described the huge increase in varieties of flower seeds, Sanders asked rhetorically, "Who buys all the flower seeds?" He responded to his own question with a sweeping "everybody," and then went on to suggest that along with wealthy patrons, "the poor man must have his packages of Balsams, Asters and Scarlet Runners, and the man of moderate means sends his one or two dollars to the seedsmen."⁷ From across the nation, commentators in religious journals, general interest magazines, and local newspapers joined their brethren in the horticultural and agricultural press in proclaiming a democratization of horticultural interest. Americans of all ranks and circumstances had taken up their shovels and hoes, observers claimed, and they unanimously applauded the resulting advancement in taste and refinement.

These widespread, enthusiastic descriptions created a whole new image of American domestic landscapes, and at the same time raise a number of

provocative and persistent questions. How valid were these claims for advancing horticultural refinement? What prompted and promoted this expanding interest and new-found enthusiasm? And what, after all, did those shade trees and flowering shrubs mean to the people who promoted their culture, or to the farm residents and city dwellers who enjoyed their shade and beauty? Careful consideration of these questions reveals an interpretive dilemma. Mid-nineteenth-century horticultural advocacy was a nationwide phenomenon, but it was accepted or rejected at a very local level. A myriad of voices urged the planting of trees and flowers, but real people in real places made the horticultural decisions that actually shaped their lives and their landscapes in traditional or more progressive ways. To integrate both the broad national perspective and the specific regional detail needed to understand horticultural improvement and its meaning, this book turns to the states carved out of the Old Northwest, and explores how residents of Michigan, Indiana, Illinois, Ohio, and Wisconsin came to accept ornamental plants as important components of their "Midwestern" material culture.[8] Focusing on the heyday of horticultural enthusiasm between 1850 and 1880, this study evaluates the use of ornamental plants in and around Midwestern homes, and then analyzes why, at mid-century, horticulture became such a popular means of home improvement. Finally, the book investigates the cultural significance of ornamental plants, and argues that this democratization of horticultural interest was closely linked to the possibilities and problems of a rapidly modernizing nation.[9]

The temporal focus of this study and its regional base were selected for several reasons. As historians have pointed out, American horticulture experienced a kind of "golden age" between 1850 and 1880. In those decades, horticultural literature proliferated, the commercial seed and nursery industry boomed, and worldwide plant exploration introduced new plant varieties that excited amateur gardeners and professional horticulturists alike. By the early 1880s, however, this passion for horticulture was slowly ebbing as many of the early horticultural proponents passed from the scene. Interest in ornamental plants certainly continued into the late nineteenth century, but as garden historian Charles Van Ravenswaay has noted, much of the "creativity and excitement" that characterized home-ground embellishment and its promotion in the middle decades was gone.[10]

These "golden" years of horticultural innovation and advocacy coincided with sweeping changes in many other aspects of American culture. In the middle decades of the nineteenth century, American industry expanded, large cities proliferated, and newly established communication and transportation networks linked much of the nation. A market economy driven by national or even international demands altered traditional systems of local exchange in many regions, and touched the lives of farm families as well as urban dwellers. Historians often characterize these changes as typical of a modernizing society, and point out that social transformations accompanied long-term economic and political trends. At midcentury, Americans were increasingly mobile, cosmopolitan in outlook, and clearly tied to a capitalist system that provided for and encouraged consumption. Those who embraced this changing social order appeared to value "secular-rational" ideas, appreciated innovation, and understood a competitive spirit as key to worldly success.[11]

For many, development of the Old Northwest offered an opportunity to test the possibilities of this new social and economic order, and with it, the promise of American democracy. Settlers moved into the Northwest Territory in the early decades of the nineteenth century, establishing the states of Ohio, Indiana, Illinois, Michigan, and Wisconsin between 1803 and 1848. With the political framework of the Northwest Ordinances guiding settlement patterns, promoting education, and denying slavery a foothold, the region seemed rooted in egalitarian principles. Over much of the area, rich, relatively inexpensive farm land, along with a diversity of other natural resources, promised a widely distributed future prosperity. The developing economy was firmly rooted in agriculture, but soon thriving villages, towns, and cities served as commercial centers and home to at least some industry.[12]

Over time, a pluralistic society also developed in the region. Indian removal and racial barriers limited access for Native Americans and former slaves, but others streamed into the Old Northwest.[13] Westward migration from the Mid-Atlantic and the Upper South, but especially from western New York, Pennsylvania, and New England, provided the backbone of settlement, and in many ways dominated the cultural climate of the region.[14] Immigrants from Germany, Scandinavia, and the British Isles added to the ethnic diversity of the newly settled states. As historian Jan Gjerde has noted, by the 1880s, the proportion of "foreign-born" citizens in the Midwest rivaled that of urban locales along the eastern seaboard.[15] Though varying in source, these migrant and immigrant streams peopled the region with many "ambitious, hopeful, and diligent" settlers who, drawing on their varied religious backgrounds or cultural proclivities, valued hard work and believed in its rewards.[16]

Many of these Midwesterners understood the advantages of the new economy and promoted internal improvements like railroads and canals to enhance their access to national markets and progressive trends. With the help of advancing transportation and communication facilities, residents quickly transformed frontier communities into well-established farms, villages, and cities. By the 1860s, their diligent efforts and demand for the region's agricultural products brought unprecedented prosperity, firmly tied the Midwest to a national marketplace and culture, and established the region as "one of the most important centers of commercial agriculture in the world."[17] As historians Andrew Cayton and Susan Gray have noted, the story of the emerging Midwest was about "near total identification" with both the market and the nation. At mid-century, they maintain, the Midwest became the "place that best exemplified the United States." For their part, residents of the Old Northwest often portrayed the history of the region as a simple tale of wilderness conquered, rapid progress, and advancing prosperity.[18]

While at first glance, nineteenth-century Midwesterners seemed to whole-heartedly embrace modernizing trends, the reality was, in fact, more complex. Historian Sally McMurry, in her study of the Oneida County, New York, dairy industry, emphasizes that rural people combined "modern" innovation with traditional practices in ways that make it "extremely difficult to characterize the outcomes as one or the other."[19] In a similar fashion, Midwesterners tried to control just how much and in what ways distant processes or new ideas changed their lives. They supported certain innovations and rejected others; they valued local traditions and customs; and they sometimes carried on practices that enhanced community bonds and neighborly interaction without seeming regard to modern values like profitability or efficiency. Recognizing these inconsistencies, historians increasingly argue that modernization was at best a fitful process, its effects varying by region or locale, and its outcomes tempered by the response of local communities.[20]

This confluence of expanding horticultural interest, broad cultural change, and a prosperous, newly settled region receptive to new ideas makes the Midwest a particularly useful "setting" to consider the mid-nineteenth-century democratization of American horticulture.[21] As horticultural advocacy burst upon the national scene, Midwesterners were in the process of shaping new domestic landscapes, they had access to markets, and in many cases, they enjoyed the means to cultivate at least some level of refinement. Like others across the nation, Midwesterners experienced the challenges and changes inherent in a rapidly modernizing society, yet seemed willing to embrace new ideas and innovative practices. Just as horticultural proponents found fertile territory for their ideology and their wares in the region, the dooryards of the Old Northwest offer us a palimpsest for understanding social change, the communication of broad cultural trends, and the translation of widespread ideas into the concrete reality of local landscapes and everyday lives.

To unravel the story of Midwestern horticultural interest, this study draws on a variety of documentary and visual resources. Chapter 1 develops a descriptive portrait of actual ornamental plant use with the help of a richly detailed, but seldom used nineteenth-century visual resource called "combination atlas maps." Published primarily between 1870 and 1880, and usually focused on individual Midwestern counties, these large, lavishly bound books were an intermediate stage between illustrated county wall maps, popular prior to the Civil War, and the more text-oriented county histories so commonly produced in the last decades of the nineteenth century. Atlas maps typically featured a brief county history, biographies of prominent county residents, cadastral maps for each township, and most importantly, hundreds of lithographs of farm, village, and city homes and their domestic landscapes.[22]

Prosperous, rapidly developing Midwestern counties with a population base of at least ten thousand residents provided a fruitful market for atlas map companies, and a number of firms entered the business during its brief heyday. A Chicago publishing firm led by L. H. Everts was among the most successful and Mr. Everts produced, with various partners, atlas maps for counties in Michigan, Iowa, Illinois, Wisconsin, and Ohio, as well as several New York, Pennsylvania, and New Jersey counties.[23] Atlas map production was a model of both effective promotional practices and efficient preparation techniques. After canvassers had secured sufficient interest among subscribers and patrons to make the atlas maps economically viable, a crew of map makers, surveyors, writers, and artists scoured the selected county, gathering information and developing maps. If residents agreed to purchase a lithograph of their home or business, they could expect a visit from an atlas map artist, who, in consultation with the family members, selected a favorable perspective and set to work drawing the premises.[24]

Atlas map images usually focused on the home and its immediate surroundings, and artists pictured details of Midwestern material culture like architectural refinements, streetscapes and sidewalk designs, outbuild-

ings, wagons, buggies and agricultural machinery, fences, windmills and wells, orchards, field crops, and even livestock. Many views also included some evidence of human activity such as farmers working the fields, women gardening or feeding chickens, city dwellers walking on sidewalks, children and pets running about in the yard, groups playing croquet, or occasionally individuals seated in the yard or on porches. Almost without exception, homeground illustrations also detailed horticultural elements like trees and shrubs, potted plants, or garden beds.[25] Scholars have long debated the reliability of these tidy, seemingly idealized views, but recent work with the genre suggests that they accurately document both homeground layouts and the presence of specific domestic landscape features.[26]

In April 1873, Mr. Everts and his partner D. J. Stewart moved their crew of canvassers, artists, and writers into Michigan. They began their work in Lenawee County in southeastern Michigan, then moved to Washtenaw County just to the north, and by early 1874 were hard at work in adjoining Jackson County.[27] By the time they completed their efforts in all three counties, Everts and Stewart agents had persuaded 731 families to support the project by purchasing lithographs of their homes or properties. Many others residents thought the opportunity to memorialize the region's progress and prosperity was well worth the $9 price, and agreed to purchase a copy of the atlas map. Sales in these three Michigan counties came very close to the industry average of 5 percent of a county's population.[28]

Atlas map firms left behind an unprecedented record of ornamental plant use within a defined, yet diverse geographic area. Using a methodology termed content analysis, the number, type, and arrangement of specific horticultural elements were determined for all 731 illustrated homegrounds.[29] In addition, the 1870 Federal Population and Agricultural Censuses provided demographic information for most of the families with homes pictured in the atlas maps, and added economic, social, and cultural dimensions to the descriptive portrayal. The combination of visual data and demographic information produced a broad composite portrait of widespread ornamental plant use, and confirmed that Midwesterners of varying ranks, occupations, and personal circumstances had taken an interest in homeground embellishment. These three adjacent southeastern Michigan counties typified other thriving Midwestern agricultural locales, and the "case study" that emerges from this methodology has significance for the entire region.[30]

Introduction xvii

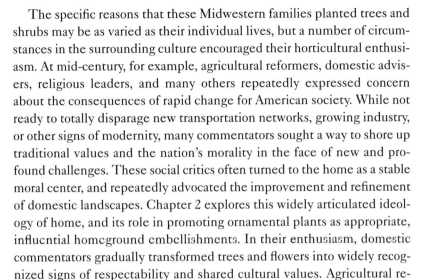

The specific reasons that these Midwestern families planted trees and shrubs may be as varied as their individual lives, but a number of circumstances in the surrounding culture encouraged their horticultural enthusiasm. At mid-century, for example, agricultural reformers, domestic advisers, religious leaders, and many others repeatedly expressed concern about the consequences of rapid change for American society. While not ready to totally disparage new transportation networks, growing industry, or other signs of modernity, many commentators sought a way to shore up traditional values and the nation's morality in the face of new and profound challenges. These social critics often turned to the home as a stable moral center, and repeatedly advocated the improvement and refinement of domestic landscapes. Chapter 2 explores this widely articulated ideology of home, and its role in promoting ornamental plants as appropriate, influential homeground embellishments. In their enthusiasm, domestic commentators gradually transformed trees and flowers into widely recognized signs of respectability and shared cultural values. Agricultural reformers also attached meaning to ornamental plant culture, frequently insisting that the presence of dooryard embellishments signified a progressive outlook and a rational, scientific approach to home and farm management.

While domestic reformers articulated important reasons for planting trees or cultivating flowers, an expanding array of horticultural professionals offered further encouragement and practical information. Chapter 3 examines the horticultural advice literature available to Midwesterners at mid-century and then, drawing on resources ranging from specialized journals like the *Gardener's Monthly* to horticulture columns published in local newspapers, analyzes the recommendations that commentators routinely promoted. These experts provided information for beginners as well as experienced gardeners, but consistently emphasized modest, attainable standards of ornamentation. Everyone, horticultural professionals claimed, could enjoy the benefits and beauties of ornamental plant culture even though they had limited means or little space.

For all this advice to bring about a real transformation, Midwesterners needed access to ornamental plants, and by the mid-nineteenth century, a burgeoning American nursery and seed industry offered a vast selection to meet the growing demand. With the help of advertisements, seed and nursery catalogues, and business records, chapters 4 and 5 consider the commercial realities of home embellishment, describe where and how Midwesterners obtained ornamental plants, and explore the horticultural

industry's role in developing new markets and attaching new meanings to home ornamentation. Nurserymen and seed dealers built their clientele through innovative promotional techniques, and as they did so, always stressed how easy and inexpensive it was for all Americans to obtain and culture trees and flowers. To promote sales, horticultural entrepreneurs played off the cultural values that reformers had attached to plants, but also added their own attributes of novelty, fashion, and good taste at a reasonable price.

Although nationwide trends encouraged horticultural interest, not all Midwesterners subscribed to horticultural journals, read agricultural newspapers, or sent for beautifully illustrated seed catalogues. Many men and women decided to take up horticulture by simply looking around them. Using evidence drawn from local and regional newspapers, county fair records, Grange reports, and personal accounts, chapter 6 investigates the role of local models or "cultural mediators" in promoting ornamental plant culture. These progressive individuals and local organizations translated prescriptive literature and commercial promotions into a concrete reality of trees and flowers for others to see and emulate. Many hoped their actions would be contagious, and that with a little local nudging, their neighbors would clean up the dooryard, mow the lawn, and plant a tree or two.[31]

As they went about their varied tasks, all of these horticultural advocates enhanced their cause by imbuing ornamental plants and their culture with social significance. Dooryard trees and flowers emerged from these efforts as multifaceted but widely recognized symbols of personal virtue, respectability, good taste, or even a progressive, up-to-date outlook.[32] Although a "public" meaning was frequently articulated, the significance of trees and flowers for the Midwesterners who actually selected and nurtured them was not so clear. To develop a more comprehensive understanding of how individuals integrated these public symbols into private lives, chapter 7 draws upon diaries from members of three closely intertwined southern Michigan farm families and, with the help of other local resources, explores the meaning of ornamental plants for members of a rural Midwestern community. These families read horticultural and domestic advice literature, planted trees by the roadside, and ornamented their schoolyard, but also traded floral cuttings with neighbors, treasured plants passed down from mother to daughter, and wept when carefully packaged flower "roots," meant to grace a new home in Michigan, were jostled from a wagon and lost. In the minds of those who planted and nurtured them, ornamental plants did express public meaning, but more importantly reflected their own lives and memories.³³

The shaded rural roads or towering Norway spruce of contemporary Midwestern landscapes are merely teasers, emblems of a deeply layered past. The pages that follow attempt to trace their story and to document and interpret the unprecedented expansion of horticultural interest in the mid-nineteenth century and the transformation of Midwestern domestic landscapes that it engendered. As the study explores the ideological, professional, commercial, and personal roots of that burgeoning interest, a pattern emerges. Americans forged a link between ornamental plants and the modernizing society around them. On the one hand, advocates considered horticultural amenities an emblem of progress and used all the modern resources at their disposal to promote ornamental plant culture. Yet, at the same time, they and many of the Midwesterners who heeded their advice understood trees and flowers as a sign of traditional virtues and stable values, an antidote for the social ills and personal upheavals that accompanied rapid, undirected change. Proponents frequently advocated ornamental plants as the most democratic of refinements, but then imbued trees and flowers with a particular set of values that reflected their own vision of a proper society. In their quiet beauty and in their seeming permanence in the landscape, ornamental plants took on these often contradictory layers of meaning without shedding their naturalness or their power as cultural symbols. In the middle decades of the nineteenth century, trees and flowers added a very welcome touch of shade and beauty to many lives, but roadside sugar maples or lilacs at the doorstep also became what some have termed a "sociotechnic tool," helping horticultural advocates, Midwestern families, and their counterparts across the nation negotiate a path through the turmoil and uncertainty of nineteenth-century American society.³⁴

For Shade and For Comfort

One

Embellished Landscapes
The Image from Southeast Michigan

In 1879, the Annual Report of the Michigan State Pomological Society contained a "portfolio" on landscape gardening and arboriculture, and tucked amidst articles on lawn grass and yellowwood trees was a commentary on the embellishment of Michigan farms. The author, S. Q. Lent, reviewed the common complaints about farmers' practicality, lack of refinement, and insensitivity to beauty, but then observed, "As one travels over the railways of Michigan and takes notes of the farm-house surroundings, it is a rare exception to find a single farm on which something has not been done toward the ornamenting of the premises." Mr. Lent suggested that the level of ornamentation varied a good deal, ranging from lines of dooryard trees to "shrubs, evergreens and flowers scattered about in profusion." Sometimes no ornamental plants were evident, but weathervanes atop the barn attested to the family's interest in refined appearances.[1]

Mr. Lent's assessment, along with that of other nineteenth-century observers, creates an expectation of trees and flowers adorning Midwestern farm and village dooryards, but their enthusiasm deserves a judicious appraisal. Does other evidence corroborate ornamental plant use in Mid-

western domestic landscapes? If Midwesterners were actually improving their homegrounds, what types of horticultural refinements did they select and how did they arrange them? And finally, how widespread were these horticultural embellishments? Did all manner of Midwesterners, no matter what their economic rank, personal circumstances, or where they lived really jump on the horticultural bandwagon, as commentators claimed?

We cannot retrace Mr. Lent's journey to verify his comments, but we can reconstruct at least a portion of what he observed with the help of atlas map lithographs. The work of Everts and Stewart and their crew of mapmakers, writers, and artists left behind a detailed accounting of material cultural and ordinary landscapes in southeastern Michigan. Careful scrutiny of each Jackson, Lenawee, and Washtenaw County atlas map image offers a series of individual homeground "snapshots" that are rich in horticultural detail and varied in content. When tree and shrub plantings, garden beds or other horticultural amenities pictured in each image are quantified and their frequency within the entire sample of 731 homegrounds calculated, a broad sense of horticultural proclivities and ornamental plant preferences emerges. Mr. Lent observed the external signs of horticultural interest, but if we are to evaluate its pervasiveness, we also need to know something about the region and the local families who selected trees and shrubs for their dooryards. The pages that follow first describe this corner of southeastern Michigan, and place it within the bustling agricultural economy so typical of the Upper Midwest at mid-century. Then, using the 731 families with pictured homes as a base and drawing on the 1870 Federal Population and Agricultural Censuses for information, the study weaves demographic characteristics like family composition, economic rank, occupation, and ethnic origins into the tapestry of horticultural choices revealed in homeground lithographs. In many ways, the portrait of ornamental plant use that emerges from this place, these people, and their pictured homes confirms Mr. Lent's very simple observation—a passenger riding Michigan railways in the 1870s was likely to see many signs of horticultural interest. The reality, however, was more complex, and while patterns of embellishment did emerge, this Midwestern horticultural landscape remained a mosaic shaped by individual choices and personal realities.

Like atlas map publishers, who usually preferred prosperous, largely agricultural counties for their work, any mid-nineteenth-century rail tour of Michigan would certainly pass through Jackson, Lenawee, and Washt-

Figure 1.1. A prosperous agricultural region in the mid-nineteenth-century, Lenawee, Jackson, and Washtenaw Counties are clustered in southeastern Michigan. From State of Michigan Specialized Technology/Mapping Unit, Lansing, Michigan.

enaw Counties. These southeastern Michigan counties had, in fact, ranked among Michigan's most populous and productive agricultural regions since the early days of statehood. When a steady stream of settlers began to flow into Michigan in the late 1820s and early 1830s, many entered the territory from the east, using the Erie Canal and Great Lakes steamboats to transport their family and household goods. This transportation pattern favored southeastern Michigan, and placed Lenawee, Jackson, and Washtenaw Counties in line for earliest settlement. In expectation of population growth, Congress "set off" Lenawee and Washte-

naw as counties in 1822, and followed with Jackson and eleven other Michigan counties in 1829.[2] (See Figure 1.1)

Settlers in the region encountered a relatively flat or gently rolling countryside dotted with lakes, bogs, and marshes. Several major rivers, including the Huron in Washtenaw County, the River Raisin in Lenawee, and both the Grand and the Kalamazoo Rivers in Jackson County, linked the area to the Great Lakes, and eventually provided power for mills and manufactories. Deciduous woodlands covered much of southern Michigan as early settlers moved into the region. Forests of oak and hickory dominated the well-drained upland areas, while moister soils supported extensive stands of American beech and sugar maples, along with other deciduous trees. Conifers, like tamarack, black spruce, and white cedar, normally occurred only in scattered swamps and bogs. Due to the range and types of glacial deposits, soils in these three counties were extremely variable, but generally fertile, and responded well to agricultural practices. The climate, while also subject to great variation, provided adequate rainfall and a sufficiently long growing season to accommodate most crops.[3]

Agricultural potential and convenient location within the path of antebellum westward migration promoted rapid population growth in the three-county area. At the time of Michigan statehood in 1837, Washtenaw County was second only to Wayne County in population, and Jackson was the most populous county west of Washtenaw, while Lenawee contained the most settlers in the southern tier of Michigan counties. In all three counties, towns and villages sprang up along the principal roads, which in turn linked villagers to other parts of the Midwest. The Chicago Road, one of the earliest and most important, extended from Detroit to Ypsilanti in Washtenaw County, into Lenawee County and then due west, linking the forts and eventually the cities of Detroit and Chicago. Slightly to the north, the Territorial Road ran from Detroit to Ypsilanti and then through Ann Arbor, Jackson, and west to the mouth of the St. Joseph River at Lake Michigan. Authorized in 1829, the Territorial Road supported stagecoach traffic by the mid-1830s. Along with Ann Arbor, the Washtenaw County villages of Dexter and Chelsea grew up along its route, while Saline in Washtenaw County and several Lenawee County villages benefited from the traffic and convenience of the Chicago Road. Located at the point where these two major roads diverged to go west, Ypsilanti also enjoyed early growth due to its favorable location.[4]

Beginning in the 1830s and continuing through much of the mid-nineteenth century, railroad construction provided access to markets, frequent passenger service, and encouraged the growth of towns and villages along rail lines. In 1836, due to the aggressive efforts of local residents, rail service linked Adrian in Lenawee County with Toledo, Ohio. The Michigan Southern Railroad ran track from Detroit to Adrian in 1840, and several years later pushed into the Lenawee County village of Hudson and adjoining Hillsdale County. By 1874, three branches of the Lake Shore and Michigan Southern served Lenawee County, with several other lines proposed or in partial operation. The state began construction of a railroad from Detroit to St. Joseph, passing through the second tier of Michigan counties in 1837. By 1839, workmen completed track to Ann Arbor and construction began to Jackson. After considerable financial difficulty, the railroad extended west to Kalamazoo in 1846. Eventually, branches of the Michigan Southern also served towns and villages in Washtenaw and Jackson Counties. As railroad construction continued, the city of Jackson gained increasing stature as a rail center. In 1871, with six different lines passing through, the Michigan Central established its manufacturing and repair shops at Jackson, creating hundreds of jobs and adding to the hustle and bustle of the city. At that point, Jackson also topped the state in rail passenger traffic and was second only to Detroit in amount of freight shipped by rail.[5]

Access to transportation networks certainly enhanced future prospects, but the citizens of Jackson, Lenawee, and Washtenaw Counties also jockeyed for other political and cultural advantages. The state legislature designated Tecumseh as the county seat of Lenawee County in the 1820s, but in 1837 agreed to move it to the rapidly developing city of Adrian. Legislators selected Ann Arbor as the county seat of Washtenaw County in 1824, and made Jackson the county seat of Jackson County in 1833. For all three towns, the presence of public institutions also assured future growth and prosperity. In 1837, Ann Arbor landed the new state university, soon to be the University of Michigan. Despite competition from Detroit and nearby Napoleon, Jackson became the site of the first state prison. Selected largely because nearby stone quarries could provide good building materials, the Jackson facility accepted its first prisoners in 1839. Adrian, in turn, supported the Raisin Institute, a Quaker school, and by the 1850s the Methodist-affiliated Adrian College. Legislators established the state normal school at Ypsilanti in 1849, and by the 1870s the school had trained more than four hundred teachers.[6]

At mid-century, a number of local accounts suggested that southern Michigan was coming into its own as an agricultural center, and that Lenawee, Jackson, and Washtenaw Counties shared in the general prosperity and growth. *The Michigan Farmer* pointed out that the state had emerged from a position of "entire dependence for most of the necessaries of life upon other states," to one of abundance, and could now boast of selling its surplus to Ohio, New York, and "other portions of the world."[7] Local residents also noted the change in their circumstances, and proudly described their progress. In 1850, William S. Maynard, a Washtenaw County farmer, spoke for many others when he declared, "I think there is no better country in the world, all things considered, for the farmer," than Washtenaw County. "Our soil is good," he continued, "land as easily tilled as it ought to be to prevent idleness; climate mild and healthy; sufficient quantity of wood, and good water; land rolling, but not hilly or mountainous." And, Mr. Maynard concluded, rapidly improving transportation facilities gave Washtenaw County residents access to markets "almost equal to a day's ride of Eastern cities."[8]

By the 1860s, corn, wheat, livestock, and dairy production, as well as wool and fruit, dominated the agricultural economy of Lenawee, Washtenaw, and Jackson Counties. Manufacturing and commercial enterprises also flourished in the three-county area. In Lenawee County, the railroad yards at Adrian employed many citizens. The city of Jackson emerged as a center for agricultural tools and wagon production, while firms in both Adrian and Jackson manufactured railroad equipment. In Washtenaw County, industries like the Ypsilanti paper mills and carriage factories developed in the mid-nineteenth century, enhancing the general prosperity and boosting population levels. Throughout the region, millers, merchants, furniture manufacturers, and a host of other small businessmen met the growing needs of an expanding population and a booming economy.[9]

The three counties enjoyed a thriving cultural and social life as well. Jackson County residents, for example, could choose from five local newspapers, Lenawee supported eight, and Washtenaw County residents could read any of nine local papers. Churches of all denominations flourished, but the Methodists led in terms of membership in all three counties, with the Baptist Church running a rather distant second. County residents supported a number of voluntary organizations, educational institutions, and a variety of political activities. Each county boasted a countywide agricultural society along with several smaller local agricultural organizations. Adrian had a long-established and very active horticultural society, and by the 1870s, some residents from the three counties belonged to the State Pomological Society. The Patrons of Husbandry, an important new nationwide rural reform organization founded in 1867, also drew county residents in the mid-1870s.[10]

By 1870, Lenawee's population reached nearly 46,000, Washtenaw boasted slightly over 41,000 residents, and Jackson supported over 36,000 citizens. The city of Jackson had grown to be the third largest in the state, trailing only Detroit and Grand Rapids in terms of population. Adrian and Tecumseh in Lenawee County and Ann Arbor and Ypsilanti in Washtenaw County remained bustling commercial, cultural, and manufacturing centers, but were exceeded in size by other Michigan cities like East Saginaw and Kalamazoo.[11]

While Lenawee, Jackson, and Washtenaw Counties exemplified Midwestern agricultural prosperity and bustling commercial enterprise, the counties remained a mosaic of individuals drawn together by geographic proximity and some common experiences, but also varying widely in many of their economic, cultural, and personal circumstances.[12] The 731 families in the atlas map sample reflected that diversity, differing, first of all, in where they made their homes. The majority, 87 percent, lived on farms sprinkled throughout the region, but ninety-seven families lived in one of twenty-nine villages, towns, or cities in the three-county area. Of those, twenty-seven families or just over 4 percent lived in one of the three largest cities, either Ann Arbor with an 1870 population of 7,363 citizens, Jackson with 11,947, or Adrian with 8,438 residents. Others lived in tiny villages like Grass Lake in Jackson County or in small towns such as Hudson in Lenawee or Dexter in Washtenaw County. Most of these families obviously made their living as farmers, but atlas map images also pictured the homes of dentists, doctors, lawyers, newspaper editors, merchants, bankers, millers, lumber dealers, and manufacturers. In addition, saloon and hotel proprietors, blacksmiths, carpenters, and a number of other craftsmen purchased lithographic images of their homes and businesses.[13]

Families with pictured homes also represented considerable variation in economic rank. Nearly 60 percent had a combined worth of real and personal property ranging from $5,000 to $15,000 in 1870. An additional 142 families or 21 percent had assets valued in a range from $15,001 to $25,000. Sixty-one residents had a combined economic worth under $5,000, with several having less than $1,000 in assets. At the other end of

the spectrum, nearly 10 percent or sixty-five of the pictured families reported combined assets of more than $25,000. Several lawyers, mill owners, merchants, and a few farmers were among the wealthiest residents in this group. Of the 663 families for whom the 1870 economic rank was available, the combined value of real and personal property averaged $14,644.[14]

Men headed most households pictured in the combination atlas maps (98 percent), but women appeared to either own the home or head the family in sixteen instances. Most of the men were well-established in their careers by 1870. The average age was forty-eight, with only twenty-four of the pictured households headed by someone under thirty. Most families (86 percent) had children, but varied considerably in how many still resided at home in 1870. Frequently, other family members, boarders, or hired help expanded the household size even further. Domestic servants, for example, assisted with household chores in 36 percent of the households, while 63 percent of the families received assistance from at least one farm laborer.[15]

These Jackson, Lenawee, and Washtenaw County residents came from many parts of the United States and Europe. The majority (86 percent) of heads-of-households were born in the United States, and of those 65 percent hailed from New York state. A few (11 percent) were native Michiganders, the second generation of early settlers in the region. Others moved to the region from Vermont, Massachusetts, Maine, New Hampshire, New Jersey, Ohio, Pennsylvania, and several other states. In 6 percent of the illustrated households, the head of the family was born in England, and in 4 percent, the head of the family immigrated from Germany. In addition, a few families from Canada, Ireland, Holland, Scotland, and Wales made their homes in the region. English immigrants were quite evenly distributed throughout the area, but nearly two-thirds of the German immigrants made their homes in Washtenaw County.[16] (See Appendix A for a more complete breakdown of demographic information.)

These southeast Michigan residents were indeed a varied lot, and they lived in a newly settled, thriving agricultural region closely tied to national markets and trends. What, then, can atlas map images tell us of their horticultural interests?[17] At the most basic level, Jackson, Lenawee, and Washtenaw County atlas map lithographs documented a phenomenon that Mr. Lent must have noted, but failed to mention in his Pomological Society report. In a region filled with thriving agricultural entrepreneurs and enterprising businessmen, families willingly set aside valuable land around their homes for ornamentation rather than production. All 731 homegrounds pictured in the atlas maps had at least some ornamental space adjoining the house, but varied significantly in the amount and its configuration. Eighty-four families (11 percent) put aside a "front yard" for ornamental purposes. Typically, this small but clearly defined area extended out from the front of the house and was bordered by fences or hedges. Other county residents (183 families or 25 percent) demarcated the area in front and to one or both sides of the home for embellishment, and like their neighbors, used fences or hedges and the house itself to define this relatively limited ornamental area. In both homeground layouts, no obvious utilitarian components such as roadways or farm outbuildings interrupted the ornamental space. In contrast to these clearly segregated ornamental areas, most county residents (464 families) seemed to integrate ornamental components into the productive areas of their homegrounds. In many cases, fencing or hedges marked the distinguishable boundaries of these extended yard areas, setting this "mixed use" space off from adjoining cropland or neighboring property. Atlas map images also revealed that some residents, particularly those with smaller ornamental areas limited to the front or front/side yards, complemented that ornamental space with an additional fenced area. Located on the side or occasionally behind the home, residents dedicated at least a portion of these areas to ornamental plant culture and sometimes displayed deciduous or coniferous trees, shrubs, or a large-bed garden in the area.[18]

Most county residents put their ornamental space to good use. In his assessment of Michigan farms, Mr. Lent reported the widespread planting of trees, and atlas map images corroborate his observations.[19] Although slightly more likely to plant deciduous trees rather than coniferous or evergreen varieties, the vast majority of residents (703 or 96 percent) cultivated one or the other tree type in their yard. Even more astonishing, over two-thirds of families in the study sample selected both shade trees and evergreens like Norway spruce as horticultural adornments. When examined from the perspective of specific demographic groups, the frequency of coniferous and deciduous trees remained uniformly high. Only households headed by women and German immigrants revealed percentages slightly below the norm for coniferous trees, and only the homes of English immigrants showed a preference for conifers much above the frequency for all images. Most demographic groups seemed to enjoy deciduous trees at a nearly equivalent rate.[20]

Figure 1.2. Mrs. A. W. Palmer embellished her Jackson County home with a weeping tree, several shade and coniferous trees, and an ornamental hedge. From *Combination Atlas Map of Jackson County, Michigan* (Chicago: Everts and Stewart, 1874), 92.

While these "upright" deciduous and coniferous trees could arguably provide shade, wind protection, and beauty all at once, a few county residents (41 families or slightly over 5 percent) selected purely ornamental "weeping" trees to embellish their homegrounds. Distinguished by gracefully drooping branches, weeping trees were most striking when planted as a single specimen in an open stretch of lawn. Residents seemed to recognize these ornamental qualities, and often planted their prized trees in a conspicuous portion of the yard. A Washtenaw County farm family, for example, planted a weeping tree in the center of the walkway leading from the road to their front porch, and encircled it with a cultivated garden bed. The walkway diverged to pass on each side of the tree, further emphasizing its distinctive shape and refined appearance. Several demographic groups including women, English immigrants, and wealthier residents planted weeping trees with a slightly greater frequency than other county residents.[21] (See Figure 1.2.)

A nineteenth-century observer like Mr. Lent, with knowledge of the region and a thoughtful bent, might have noticed a curious paradox when it came to Midwestern tree planting. Most of the trees pictured in atlas map images were relatively small in relation to the height of the home, probably indicating they were young or recently planted. Tall, obviously mature deciduous trees appeared singly or in clusters about the home in only 41 percent of atlas map images. These large trees may have represented uncut original growth, small trees simply allowed to mature in place after the home was built, or fast growing deciduous trees that residents planted very early in the history of the homestead.[22] Since deciduous forests dominated the pre-settlement vegetation of southern Michigan, the preponderance of small trees in ornamental plantings suggests a change of heart. Midwesterners who had once busily cleared space around their homes were just as busily, at mid-century, planting new trees. Even residents who had seemingly preserved native trees showed an interest in additional plantings. In a few cases, only large deciduous trees were evident in the homeground illustration, but more frequently, residents combined one or more original trees with much smaller specimens, or other types of ornamental plantings.[23] Farm yards were more likely to boast of these large original trees than were the more limited town and city homegrounds. Families at the low end of the economic scale and German immigrants enjoyed original trees even less frequently than urban dwellers.[24]

In addition to trees, some atlas map illustrations pictured shrubs scattered about the ornamental grounds, but their numbers (37 percent of the homegrounds) suggest they were generally a less popular form of homeground ornamentation.[25] Most families cultivated five or fewer specimens, and except for one instance, always nurtured shrubs along with trees. A few county residents, however, seemed to emphasize shrubs in their landscape plans. Lenawee County farmer Nelson Kinney lined the front walk of his home with nineteen shrubs, while neighboring farmer Horatio Hicks used shrubs to border a walkway and the front of his house. Generally these cultivated shrubs appeared to be deciduous species, but a few illustrations pictured low-growing evergreens.[26] While shrub plantings were fairly consistent among most demographic categories, residents in the lowest economic group and urban families selected them less frequently than others in the region.[27]

Although often described in horticultural literature as a particularly pleasant and simple way to embellish the home, ornamental vines appeared in only 17 percent of the homeground illustrations. Some families supported their vines with a trellis against the house while others planted vines in the yard and trained them to grow on free-standing vertical or horizontal trellises or low arbors. Usually, residents grew just one or two vines, but several lithographs revealed three or four trellises spaced about the homegrounds. In one instance, Washtenaw County farmer Charles H.

Wines displayed four trellises in his front yard, two placed symmetrically on each side of the front walk, one as an arch at the side of the house, and one shading the side of the front porch.[28] Occasionally, a long horizontal trellis supported a number of vines.[29] Families in the lowest economic rank and households headed by German immigrants planted vines less frequently than other groups, while city residents revealed a higher preference for vining plants.[30]

Finally, eighty-six county residents (12 percent) planted and pruned trees or shrubs to create hedges around their homes, but used the hedging for varying purposes.[31] At the Washtenaw County farm home of Heman N. Hicks, a low deciduous hedge bordered a circular drive in front of the home, and lined the interior of a fence delimiting ornamental space to the side of the house. In another practical use of hedging, the Hendersons of Washtenaw County planted a coniferous tree hedge as a windbreak for a series of large-bed gardens to the side and rear of their home.[32] Hedges appeared to be most common among village, town, and city families, who frequently used them as a way to mark property lines.

In addition to their selection of plantings, families varied their domestic landscapes by arranging trees and shrubs in noticeably different ways. Most appeared to plant trees and shrubs randomly, without landscape plan or pattern guiding their placement. Other residents sought order and balance in their dooryards, placing pairs of trees or shrubs symmetrically, using them as borders for roadways or paths, or, infrequently, planting a line of trees or shrubs across their ornamental grounds with no discernible relation to other landscape features. Still other families clustered their trees and shrubs in diverse plantings that appeared to reflect the irregularity of natural growth. Finally, a few families selected foundation plantings, arranging several trees, a few shrubs, or a combination of the two next to their homes so that the vegetation acted as a kind of natural frame for the house itself.[33]

Some residents (43 families or 6 percent) in the three-county area were so proud of their tree plantings that they circled favored specimens with garden beds. These "tree beds" provided a cultivated space at the base of trees or shrubs, but normally contained no additional plantings. Occasionally tree beds encircled all the trees in the dooryard, but more frequently accentuated only one or two selected specimens. Stones or other bordering materials defined the perimeter of at least some of the illustrated tree beds. The use of tree beds was fairly consistent among demographic groups, but women and English immigrants selected this

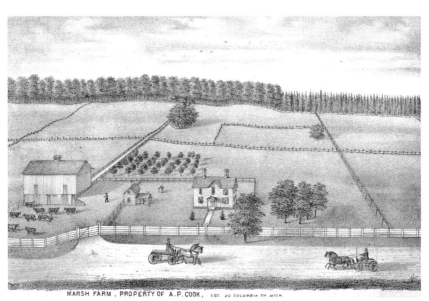

Figure 1.3. The Cook family of Jackson County proudly planted their two prized evergreens on each side of the front walkway. From *Combination Atlas Map of Jackson County, Michigan* (Chicago: Everts and Stewart, 1874), 65.

horticultural amenity a bit more frequently than their neighbors.[34] (See Figure 1.3.)

While happy to see an expanding interest in tree and shrub culture as he traveled about the state, Mr. Lent was particularly delighted with the occasional dooryard that had "flowers scattered about in profusion."[35] Although far less common than tree plantings, images do portray a variety of garden types and arrangements. Of the families who established gardens, most (148 families or 20 percent) cultivated "large-bed" gardens, an often extensive rectangular or square area, in which they routinely planted vegetables and small fruits along with flowers. These large garden spaces occasionally occupied ornamental areas near the front of the house, but more often, residents relegated them to the side or rear of the property or to an adjacent fenced area, probably reflecting their combined utilitarian and ornamental purposes.

Usually a single large-bed garden served the family's needs, but in a few instances, pathways divided an unusually large area into smaller spaces, facilitating access to all parts of the garden. The Anson Updike family of Jackson County, for instance, divided their large-bed garden into ten smaller gardens, and cultivated several additional large-bed gardens

Figure 1.4. The Quigley family of Jackson County divided their large-bed garden with pathways to provide easier access. From *Combination Atlas Map of Jackson County, Michigan* (Chicago: Everts and Stewart, 1874), 57.

on the other side of their farm house. On another Jackson County property, farmer A. A. Quigley cultivated a huge large-bed garden occupying the entire adjacent fenced area to one side of the front yard. Pathways divided this unusually large garden space into smaller rectangular beds.[36]

In a few cases, residents developed large-bed gardens solely for ornament, and used pathways within the garden space to create geometric patterns. E. M. Cole, a Washtenaw County widower, cultivated the most elaborate example of this design form in the study sample. In his extensive large-bed garden located to the side and slightly to the rear of the house, Mr. Cole established a pathway system that encircled and joined smaller circular or oval garden beds, creating a geometric design within the large-bed space.[37] A few additional families created geometric designs in large-bed gardens by varying their planting patterns, or by planting flowers of contrasting colors in adjoining sections of the garden, perhaps experimenting with fashionable parterres or ribbon beds occasionally described in horticultural literature.[38] These geometric garden designs provided a highly ornamental embellishment, but only nine families pictured in the atlas map sample seemed willing to undertake the labor-intensive

Figure 1.5. The Chandler family of Lenawee County accentuated their cut-in flower gardens by placing them in a line leading up to the front of their home. From *Combination Atlas Map of Lenawee County, Michigan* (Chicago: Everts and Stewart, 1874), 61.

project. Both very simple and the more complex large-bed gardens appeared in rural domestic landscapes much more commonly than in more urban areas. English immigrants also revealed a greater propensity for large-bed gardens than their neighbors from other regions.[39]

The combination of plantings that typified most large-bed gardens offered residents a kind of transitional space. They could, within its boundaries, produce fruits and vegetables, and at the same time satisfy their desire for beauty. Ninety-eight families (13 percent), however, chose another path, and established compact, often round gardens cut into an open section of lawn. Mid-nineteenth-century horticultural advisers labeled these beds "cut-in" gardens, recommended them for ornamental use, and noted that they were most effective when planted with brightly colored bedding plants.[40] Some families placed these purely ornamental gardens in a conspicuous position in front of the house while others

arranged them symmetrically on each side of the front walkway or used them to mark opposite corners of the front yard. In a rather uncommon arrangement, Lenawee County farmer Thomas Chandler used three round cut-in gardens positioned in a line to accentuate the front of his home.[41]

Some county residents experimented with the shape of their cut-in gardens. The home of William Preston featured a heart-shaped cut-in garden ornamenting the center of the front lawn; the Davis family of Lenawee County cultivated a long, narrow cut-in garden; and Washtenaw County farmer John M. Braun created a triangular cut-in garden.[42] Occasionally, these ornamental gardens appeared as raised beds, bringing the planting surface several feet above the surrounding lawn. Some cut-in gardens accentuated a planting or special horticultural embellishment in the center of the bed, and a few residents, like Washtenaw County farmer G. W. Gale, bordered their cut-in gardens with decorative stones.[43] Households headed by women, English immigrants, and families living in Adrian, Ann Arbor, and Jackson selected that type of garden space with greater frequency than their neighbors. Town and village residents did not share the preferences of their city neighbors, however, for only 4 percent of those families chose a cut-in garden for their yards. Interest was also low among German immigrants, with only 3 percent of their pictured homes showing cut-in gardens.[44]

Cut-in gardens were notable for their isolation. Their prominent position within an open space accentuated the color and bounty of the floral display. Some homeowners (45 families or 6 percent), however, selected a more diversified ornamental garden arrangement. Usually designed as a long, narrow rectangle, these gardens bordered some other landscape feature such as a walkway, fence, or drive. Like cut-in gardens, residents probably used border gardens solely for flower culture, but according to horticultural advisers were more likely to include herbaceous perennials along with annuals and bedding plants in this garden layout.

A few families turned to more specialized garden designs. Five atlas map images pictured rockeries, a garden in which residents arranged decorative rocks, lined the interstices between boulders with soil, and then placed a variety of plants in the cracks. Although horticultural advisers usually described rockeries as best situated on natural outcroppings in secluded or at least sheltered circumstances, families in this sample arranged their rockeries in flat, very open front yards without any addi-

Embellished Landscapes 17

Figure 1.6. The Hitchcocks were one of the few families in the region to culture plants in a cold frame. From *Combination Atlas Map of Jackson County, Michigan* (Chicago: Everts and Stewart, 1874), 83.

tional protective plantings. In a typical example, John Tallman of Lenawee County placed stones in a cascading form facing the roadway with plants positioned on the gradually sloping surface. A large pyramid-shaped rockery adorned one side of the front yard in Woodland Owen's highly embellished Adrian homegrounds.[45]

A number of other horticultural accessories graced the lawns of Lenawee, Washtenaw, and Jackson County residents, but were selected far less often than trees or even flower gardens. Only six families tended greenhouses, slat houses, or coldframes. (See Figure 1.6.) A few county residents (eleven homes or 1.5 percent) ornamented their yards with architectural structures like gazebos, summer houses, or arbors.[46] Chairs and benches or other types of yard furniture appeared on the porches or in the yards of only thirty-seven county residents. In some cases, a low bench encircled a well-established tree, and in other instances, residents placed a free-standing bench invitingly in a corner of the yard. In most instances, however, family members restricted lawn furniture to a single chair temporarily placed in the yard. Town, village, and city residents along with the wealthiest residents selected yard furniture somewhat more frequently than did others in the sample.[47] Hanging baskets and ornamental vases filled with plants also embellished a few homes in the three-county area. In one instance, Lenawee County farmer Ezra Cole was so proud of his hanging baskets that he placed them in a tree near the road. Although not readily visible from the house, passersby could enjoy his good taste.[48]

Atlas map illustrations also revealed some evidence of indoor ornamental plant culture. Images of twenty-one homes or 3 percent of the sample pictured potted plants at windows, displayed on porches, or placed in the yards of county residences. In addition, seventy-one illustrated homes or nearly 10 percent had bay windows. Mid-nineteenth-century homeowners considered bay windows to be a fashionable architectural detail, but horticultural advisers often promoted them as an ideal way to raise indoor plants, and their presence offered an additional clue to ornamental plant cultivation. Several groups, including women, families from England, and city dwellers, seemed to favor indoor gardening more than their neighbors.[49] (See Appendix B for the complete breakdown of the frequency of selected horticultural embellishments by demographic categories.)

With their renderings of tree-embellished yards, a variety of garden beds, or an occasional ornamental vase or hanging basket, images from the Lenawee, Jackson, and Washtenaw County atlas maps confirmed Mr. Lent's observations about the pervasiveness of homeground embellishment. In the three-county area, all families with illustrated homes set aside at least some land for ornamental purposes. More importantly, with the exception of eight households whose yards exhibited no sign of embellishment, every family cultivated ornamental plants to at least some

Embellished Landscapes 19

Figure 1.7. The Adrian home of Mrs. Dorcas Whitney boasted many horticultural amenities. Ornamental vases graced the front yard, while potted plants lined windows at the rear of the home. Mrs. Whitney also enhanced her ornamental garden design with pathways. From *Combination Atlas Map of Lenawee County, Michigan* (Chicago: Everts and Stewart, 1874), 72.

degree. Most residents selected simple ornaments like trees as the primary means of embellishment, but others planted shrubs and vines, established flower gardens, or enjoyed a variety of horticultural accessories. (See Figure 1.7.)

County residents varied not only in the type of embellishments they selected, but also in the degree of ornamentation they incorporated into their domestic landscapes. Again, most chose a fairly simple landscape design, with only a few horticultural elements in the ornamental space surrounding their homes. Although 224 families or nearly 31 percent used trees as the sole means of embellishment, most families typically planted coniferous or deciduous trees, and then added some other type of plant, a garden space, and one or two other decorative elements to the landscape plan. Only a very few homegrounds, like that of Adrian dentist Woodland Owen, featured more than ten ornamental plant components.[50] Some families created highly embellished homegrounds not by increasing the types of ornamentation, but by developing or refining a particular aspect of the design. Several county residents, for example, cultivated very extensive and ornate large-bed gardens, enjoyed multiple cut-in gardens, or

nurtured diverse types of hedging within the ornamental space, but had few other types of horticultural ornamentation.

The more detailed examination of ornamental plant use by demographic categories emphasized the ubiquity of at least some horticultural embellishments, but at the same time revealed diversity and variation in the selection of landscape design elements. Urban dwellers and women, for example, tended to display a higher level of certain kinds of embellishments like vases or hanging baskets. English immigrants were often far ahead of other ethnic groups in their selection of weeping trees, ornamental shrubs, or other elements of design. German immigrants, in contrast, showed little interest in horticultural ornamentation like vases, potted plants, hanging baskets, hedges, or vining plants, but tended to select simpler embellishments such as deciduous trees and shrubs at the same rate as their neighbors. Farm families, in turn, exhibited greater interest in large-bed gardens or arranged their trees and shrubs in symmetrical patterns, lines, and borders far more frequently than did town, village, and city residents. All of these patterns, however, were indistinct and often interwoven. At best, the varying percentages revealed a somewhat greater interest in certain kinds of embellishments among certain groups, but did not preclude many exceptions and richly varied combinations of horticultural choices that crossed economic, gender, occupational, and ethnic lines.

In some measure, the eight families whose homeground illustrations revealed no horticultural embellishment exemplified the individual nature of these landscape decisions. Among those residents, no common demographic patterns emerged, and hence no apparent answer as to why they chose not to embellish their homes. One of those families belonged in the lowest economic rank, but one was a miller with economic assets valued at $75,000. Four lived on farms, but four others lived in small towns or in larger cities like Manchester and Ann Arbor in Washtenaw County. An immigrant from Germany headed one of the families, but the rest came to Michigan from New York State or other parts of the United States. Their family size also varied. Some had the assistance of farm laborers or domestic servants while others did not. Five of the men ranked below the average age of forty-eight years for heads-of-household, but the other three were well above that age. The reasons for their choices remain obscure, but the message was clear. No matter how widespread horticultural embellishments had become in Midwestern domestic landscapes, the decision to plant trees or cultivate flowers remained up to

individual families and involved a very private appraisal of their circumstances and preferences. Although most families in this sampling selected at least some level of embellishment, these eight weighed the options and made other choices.[51]

While the literal presence of trees, shrubs, and flowers around southeastern Michigan homes signaled interest in embellishment, an economic analysis of plant culture confirms that residents were willing to invest in refined appearances. Many families made their first economic commitment to horticulture when they reallocated productive land for new and essentially nonproductive use. Although the issue was perhaps more direct for farm families, who lost potential crop land, it also had economic significance for village and city dwellers, where their already diminished homegrounds afforded even less opportunity to grow fruit and vegetables, raise chickens, or keep a cow. Without exception, families in this sample were willing to make the sacrifice.

The loss of productive space had long-term economic implications, but other aspects of ornamental plant culture required outright expenditures. Many of the deciduous trees, flowering shrubs, vines, and even some herbaceous plants cultivated in gardens may have been native species, transplanted to the homegrounds at little or no cost. Several plant types, however, although routinely visible in atlas map illustrations, were not available in the native flora, and probably indicated purchase from commercial nurserymen. A few species of coniferous trees grew in the wetlands and occasionally in the forests of southern Michigan, but were uncommon when compared to deciduous varieties and difficult to transplant. Underscoring the problems with native species, horticultural advisers frequently promoted easily transplanted non-native trees like the Norway spruce. These comments and the environmental realities of southern Michigan suggest that county residents probably purchased the coniferous trees so commonly featured in their homeground layouts.[52]

Weeping trees also represented a financial investment for the homeowner. Most were literally the creation of nurserymen who grafted trees with a drooping habit onto a rootstock that they could easily propagate. A newly grafted tree required careful nurturing, adding to its cost, but by mid-century, commercial nurseries routinely included a number of attractive weeping trees among their ornamental offerings.[53] Hedge plantings were also costly. Horticultural advisers occasionally recommended some native species like white cedar for hedging, but preferred non-native plants like privet or Norway spruce.[54]

A comparison of identifiable plant types with contemporary nursery catalogues underscores the economic implications of many horticultural embellishments. In 1859, well-known Illinois nurseryman F. K. Phoenix offered coniferous trees, planted by 78 percent of the families in this sample, at prices ranging from $.25 to $1.50 per tree. Ten years later, Rochester, New York, nurserymen Ellwanger and Barry charged between $.60 and $2.00 for conifers, with the widely planted and most popular Norway spruce selling at the low end, and uncommon varieties like a dwarf Norway spruce selling for the highest rate.[55] County residents seldom planted a single specimen, so the cost of coniferous tree culture could quickly add up. Other types of woody plants presented a similar picture. In 1870, Ellwanger and Barry priced deciduous and coniferous shrubs from $.50 to $1.00. Hedge plants, in turn, routinely ranged in price from $4.00 per hundred for privet to $25.00 per hundred for a dwarf arborvitae.[56]

Flower culture, too, could be costly. Bedding plants like verbenas or geraniums, typically planted in cut-in gardens, ranged in price from $.10 to $.50 per plant according to Detroit nurseryman William Adair's 1872 catalogue. Special varieties often cost even more. Hubbard and Davis, another Detroit nursery, offered several geraniums for $1.00 and bouvardias for $1.50.[57] Individual bedding plants were relatively expensive when compared to the price of trees, but more significantly, cut-in gardens often required hundreds of these plants. In addition, most bedding plants could not withstand Midwestern winters and required replanting each season. Commonly planted flowers like dahlias and tuberoses cost $.25 to $.30 per tuber or bulb, while herbaceous perennials ranged from $.25 to several dollars per plant. In 1872, B. K. Bliss, a major supplier of the popular gladiolus, listed over two hundred varieties priced from $.10 to $4.00 per corm.[58] Many nurserymen and florists also offered indoor plants for sale at varying prices. In 1872 William Adair listed hundreds of available greenhouse plants ranging in price from $.25 to $1.00. Horticultural accessories like hanging baskets or plant vases required an even greater investment. In 1872, B. K. Bliss advertised a variety of hanging baskets priced from $.50 up to $2.00. Both cast-iron and terra cotta plant vases, although widely available, were quite costly, ranging in price from $5 to $50.[59]

Catalogue prices underscore the economic realities of plant culture, but actual orders to nurserymen provide a more comprehensive view of what it cost to ornament a dooryard. In the mid-nineteenth century,

Rochester, New York, nurserymen George Ellwanger and Patrick Barry operated the largest and one of the most complete nurseries in the country. Between 1852 and 1867, at least forty-six Jackson, Lenawee, and Washtenaw County residents placed ornamental plant orders with the firm. The names of most county residents appeared in the Ellwanger and Barry order book only once, suggesting that they may have turned to this well-established firm as they initially planned and planted their homegrounds. Orders from southeast Michigan averaged $21.00, with individual bills ranging from $1 to $103.[60]

A few southeastern Michigan residents, like Artemus J. Dean of Lenawee County, placed several orders with Ellwanger and Barry. In 1850, the Dean family lived on a farm in Raisin Township, Lenawee County, but sometime between 1850 and 1860, sold the farm and moved to Adrian, where Mr. Dean operated a livery. A long-time member of the State Agricultural Society and the Adrian Horticultural Society, Mr. Dean held a number of offices in both organizations. Mrs. Dean, in turn, regularly exhibited her roses and perennial flowers at the Lenawee County Fair. Placing one Ellwanger and Barry order in the fall of 1854, and a second in the spring of 1855, the Deans selected a wide variety of trees, shrubs, flowers, and hedge plants, totaling $36.36. Both the variety of plants ordered and the fact that the family moved during the period suggested that the Deans were establishing new homegrounds in Adrian. The $36.36 investment may well have represented the total cost of nursery stock for a family of modest means but with strong interest in horticulture. Given their horticultural bent, the Dean family probably made an additional investment each year as they purchased flower seeds, bulbs, or bedding plants.[61]

The fact that some horticultural amenities required an expenditure seemed to make little difference in this sampling of Lenawee, Jackson, and Washtenaw County families, for the vast majority purchased at least some of their embellishments. If plants that required purchase are combined with other horticultural elements like vases, baskets, lawn furniture, gazebos, fountains, and greenhouses that also involved a financial outlay, at least 596 or nearly 82 percent of the sample made some financial commitment to ornamental plant culture. Although large, this percentage of homeowners probably represents a conservative estimate, for many families may have also purchased at least some of their deciduous trees, vines, shrubs, and flowers, not included in the 82 percent.

In addition to their economic implications, certain horticultural embellishments demanded time and labor if county residents were to properly maintain their homegrounds. At a most basic level, householders took on a new task as they initially set aside ornamental space and established a lawn. Depending on the method and frequency of mowing, families had additional maintenance chores throughout the growing season. The planting of ornamental trees and shrubs was also labor intensive, but after they were established, these woody plants needed, and probably received, little care. In contrast, families who cultivated gardens or maintained hedges spent many hours throughout each growing season planting, pruning, and weeding. Horticultural amenities like greenhouses, hot beds, and cold frames, indoor plants, hanging baskets, and ornamental vases needed special care, and often involved additional tasks like potting and careful watering that continued throughout the growing season.

If one considers the initial work involved in planting trees and shrubs, nearly 99 percent of the families pictured in the atlas maps devoted at least some labor and time to horticultural embellishment. More significantly, 327 families or 45 percent made an additional and on-going commitment as they cultivated garden beds, watered house plants, or tended their greenhouses. Comparison of that number with those families who purchased at least some of the horticultural embellishments around their homes suggests that when they considered how to express their horticultural interest, county residents may well have preferred a financial investment to the prospect of long hours of horticultural labor. The beautification of their domestic landscapes was important, but it clearly had limits.

Based on the number and variety of horticultural elements pictured, or on an analysis of the cost and labor they represented, atlas map images from Lenawee, Jackson, and Washtenaw Counties confirmed what Mr. Lent and many others asserted. These Midwesterners were indeed on the horticultural bandwagon. The illustrations cannot explain, however, why this varied group of county residents so willingly took to planting trees and flowers, or what a well-tended, shady lawn meant to the rural and urban families who dedicated time, labor, and money to its culture. To place this sampling of Midwestern ornamentation into a broader context, it is necessary to explore the social and cultural circumstances surrounding both horticultural proponents and practitioners in mid-nineteenth-century American society.

Two

A Reason for Planting

Ornamental Plants and the Ideology of Home

If the residents of Lenawee, Washtenaw, and Jackson Counties and their Midwestern counterparts needed a reason to plant trees and tend flowers, they did not have far to look. Beginning in the antebellum era and extending through the 1880s, a plethora of domestic reformers, religious writers, rural advocates, and other social commentators argued that proper American homes offered a stable moral center for a nation beset by rapid and often worrisome change. Their descriptions of ideal homes, surrounded by shady lawns and flower beds, poured from the pages of domestic, architectural, and agricultural advice books, general interest magazines, religious journals, and women's periodicals. Even agricultural newspapers, horticultural magazines, the reports of agricultural and horticultural societies, and local newspapers proffered opinions on the ideal home, its importance, and its influence. This carefully constructed, yet complex, ideology often focused on the home's appearance. Reformers urged their readers to establish neatly ordered domestic landscapes while emphasizing the importance of ornamentation in creating a desirable

home environment. Although they recognized a number of ways to refine and beautify the home, many advisers insisted that ornamental plants provided a particularly effective and influential addition to the home scene.

This interest in the home, its appearance, and influence emerged against the backdrop of a rapidly modernizing American society. As mid-nineteenth-century reformers and social commentators looked about them, they saw a nation on the move, and a bewildering array of unfamiliar events and new circumstances. The unprecedented growth of American cities, increased industrialization, and expanded communication and transportation networks linking previously isolated rural regions to national markets and new ideas signaled progress to some, but others worried about the consequences of rapid change and seemingly constant turmoil. The dedicated observer could find threats to social stability almost anywhere. Issues like crowded city living conditions, the rise of a new and perhaps unrestrained working class, the drain of rural youth from their farm homes to the city, and the seemingly unending influx of new immigrants gave the scene a particular urgency. Political controversy bubbled through the 1840s and 1850s as the nation struggled with the issue of slavery and the acquisition of new territory. Capped by the Civil War, the political turmoil extended into the corruption and scandal of the postwar years, adding to the nation's sense of unease and social concern. Even seeming advantages like increasing prosperity and more readily available consumer goods posed a problem, for throughout the period, many reformers feared that rampant materialism and pursuit of wealth would soon shoulder aside other more beneficent values.[1]

All of these challenges, coupled with what many perceived as the declining effectiveness of traditional sources of social stability like the church or tightly knit communities, prompted a good deal of soul searching and hand wringing among social reformers and citizens alike. As historian John Higham has argued, American society at mid-century moved from a spirit of "boundlessness" and infinite possibility to one of "consolidation" where the perils of uncertainty and change often fostered the desire for a "more stable and cohesive society."[2] Clearly, Americans needed a solid moral center to form and guide their lives. Many reformers not only voiced concerns, but cast about for mechanisms to build an "enduring moral order."[3] Despite diverse perspectives, some of these critics looked homeward, and insisted that individual homes provided an effective, available, and essentially portable means to establish character and inculcate appropriate social values, while offering individuals the flexibility and inner resources to properly negotiate the changes, opportunities, and temptations of mid-nineteenth-century society.[4]

The ideology of home had a number of distinct yet closely interwoven themes that repeatedly appeared in advice and reform literature. Some advisers emphasized the importance of family relationships, and argued that women, as wives and mothers, had a particularly significant role in making the home a pleasant and potent influence on other family members.[5] Others understood home as an increasingly vital source of Christian precepts and values, and promoted home influences as a way to establish and maintain Christian character. Beginning in the late 1840s with the pronouncements of Hartford, Connecticut, minister Horace Bushnell, the idea of "home religion" surfaced frequently in the reform literature. In his influential yet controversial "Discourses on Christian Nurture," Bushnell argued that every child should grow up a Christian, never knowing any other influence. A loving Christian home, Bushnell insisted, was central to the process. As other thinkers added their voices to the argument, home religion eventually defined new roles for family members, and often gave physical aspects of home a religious significance, if not a church-like appearance.[6]

A third very important and persistent theme emphasized home as "place." Mid-nineteenth-century reformers repeatedly argued that a proper, neatly ornamented home environment set the physical stage for a decent family life, and was so critical that few could overcome the negative influence of truly degraded surroundings. More importantly, the home environment actively shaped household members, exerting a quiet yet meaningful influence on daily life, individual conduct, and moral sensibilities. It was within the context of home as place that ornamental plants and other embellishments received their most significant endorsement, and gradually emerged as important visible reminders of proper attitudes and values.[7] To understand the influence of these widely touted ideas on Midwestern horticultural interest, this chapter first documents the sources of domestic ideology available to people of the region, then considers why reformers promoted ornamental plants as particularly effective additions to the home scene, and finally explores how horticultural advocates transformed trees and flowers into powerful and very eloquent cultural symbols.

The residents of Lenawee, Jackson, and Washtenaw Counties, like other Midwesterners, encountered the precepts of this place-oriented domestic ideology frequently, for proponents derived at least some of their

power by spreading their message far and wide. Boosted in part by technological advances in printing and expanded transportation and communication networks, and often fired by social concerns and a reform spirit, America, at mid-century, witnessed a flowering of books, advice manuals, association reports, and other print resources. Also between 1850 and 1880, tempered only briefly by several economic downturns and the Civil War, the number of journals and newspapers steadily increased, offering citizens an amazingly diverse selection of periodical literature. Although certainly not a center for the publishing industry, Michigan experienced a growth in periodicals that reflected national patterns. In 1850, the state boasted 58 magazines and newspapers published within its boundaries. By 1860, the number jumped to 118, and in 1870, totaled 211. Other states registered similar growth, causing one pundit to note in the 1870s that the nation was experiencing a regular "mania of magazine-starting."[8]

Ranging from nationally circulated journals voicing a largely urban point of view to newspapers and reports published within the region, the residents of Lenawee, Jackson, and Washtenaw Counties could learn about the ideal home and the importance of ornamental plants in a myriad of sources and from a variety of perspectives. In terms of nationwide coverage, literary or general interest magazines like the *North American Review*, the *Atlantic Monthly*, *Harper's*, and *Hours at Home* were an important source of domestic ideology, and subscribers read about the topic right along with articles on politics, economics, and American culture. General interest and literary magazines with a more Midwestern slant also discussed domestic ideology. The *Western Journal of Agriculture, Manufactures, Mechanic Arts, Internal Improvement, Commerce, and General Literature*, published in St. Louis, routinely featured articles ranging from the moral uses of plants to commentary on the planning and adornment of "proper" homes. Women's literary magazines like *Godey's Lady's Book* offered its many readers a menu of fiction and fashion information, but also published some articles on the influence of homes and home gardening.[9]

In the mid-nineteenth century, religious journals emerged as both a popular genre and an important source of articles on domestic ideology and home embellishments. The Unitarian *Christian Examiner*, for instance, aimed at "wide coverage and broad readership," and regularly published articles on gardening, landscape concerns, and their importance to a proper home. A second Unitarian journal, the *Monthly Religious Magazine*, founded in Boston in 1844, had a similar goal, and featured a number of articles on "home influence," proper home life, and the importance of making home attractive. The *Methodist Quarterly Review*, published in New York City from 1841 until 1881, also had a readership from across the nation, and like the Unitarian journals, discussed a variety of issues including landscape gardening and its importance in enriching the home environment.[10] Potentially these articles reached a large audience, for by 1870, over four million subscribers enjoyed the more than four hundred religious journals published in the United States. Their circulation outdistanced that of general interest, literary magazines by several hundred thousand readers.[11]

The agricultural and horticultural press also functioned as very important conduits for domestic ideology, and by the mid-nineteenth century, both journals and advice books abounded. Agricultural historian Albert L. Demaree has estimated that over four hundred agricultural journals appeared in the United States between 1819 and 1860, and although many were short-lived, those that prospered profoundly influenced American agricultural practices and social perspectives.[12] By 1870, U. S. census figures recorded ninety-three agricultural or horticultural periodicals in active circulation, and that number nearly doubled between 1870 and 1880.[13] Among the most important of regional publications, newspapers like the *Michigan Farmer*, first published in Jackson, Michigan in 1843, or Chicago's *Prairie Farmer*, also beginning in the early 1840s, provided their Midwestern readers with both access to nationwide points of view, and Midwestern interpretations. Many residents of southeastern Michigan who hailed from New York State also subscribed to *Moore's Rural New Yorker* published in Rochester, New York. Of the national agricultural publications, the New York City–based *American Agriculturist* enjoyed the most prominent reputation and the widest circulation. All of these journals and newspapers regularly published articles on the importance of home, its ideal characteristics, and the desirability of ornamental plant embellishments, frequently linking domestic ideology with precepts of progressive farm management practices. Echoing the perspective so routinely voiced in agricultural journals and newspapers, essays and speeches published in agricultural and horticultural society reports often described innovative farming methods and domestic ideology as part of the same progressive outlook.

Several indicators suggest that Jackson, Lenawee, and Washtenaw County residents had ready access to these publications. Between 1853 and 1865, the postmaster in South Lyon, Michigan, a village in Oakland County near the northeastern boundary of Washtenaw County, recorded

all the magazines and newspapers that came through his post office, and noted which residents of this largely agricultural region received them. Not surprisingly, more local families subscribed to the *Michigan Farmer* than any other journal or newspaper. These Midwesterners were by no means parochial in their reading habits, however, for in addition, over sixty other journals and newspapers circulated in the area. Some residents subscribed to regional agricultural publications such as *Moore's Rural New Yorker* or the *Farmer's Companion*. Other local families read general interest publications like *Harper's* or the *Western Literary Cabinet*, published in Detroit, or journals with a scientific focus such as the *Scientific American*. Some subscribed to urban newspapers like the *New York Mercury*, *New York Evening Post* and the *New York Tribune*, or the *Detroit Free Press* and the *Detroit Advertiser*. A number of religious magazines including the *Christian Advocate*, the *Presbyterian Banner*, and the *Methodist Quarterly Review* also had readers in the region. Some residents enjoyed journals with a focus on women like the *Ladies' Wreath*, *Godey's*, *Mother's Magazine*, and *Hearth and Home*. A few residents even read journals with a very specific reform agenda like the *Water Cure Journal*.[14]

Even if Midwesterners read only their hometown newspaper or the *Michigan Farmer*, they still had many opportunities to encounter national issues or perspectives from other regions. Due to a liberal policy of exchange among editors, articles initially published in one journal often appeared in other publications across the nation. Sometimes editors copied the article in its entirety, and at other times, simply drew their readers' attention to what the article or journal had to say. In 1863, for example, the *Tecumseh Herald*, published in Lenawee County, notified its readers that it had just received the *Atlantic Monthly*. The newspaper then went on to describe the "varied table of contents" and the many distinguished contributors, including Ralph Waldo Emerson. The *Herald* capped its endorsement by suggesting that the *Atlantic*, although "a partisan publication," ranked "high in the literary world" and was well worth the price. Finally, the *Tecumseh Herald* noted that readers could purchase the journal through a local merchant.[15]

Clearly, interested Midwesterners had many opportunities to read about the home and it importance, but while pervasiveness lent power to the message, domestic commentators relied on their carefully crafted arguments and descriptions of ideal homes to convince readers and promote action. Many began their discussions by explaining how the home environment actually functioned to guide and stabilize society. Some commentators simply emphasized that a proper home provided an ordered landscape in which to live, and that a decent environment could counteract the unseemly influence of gambling halls or the corner saloon. If pleasantly and appropriately ornamented, they argued, homes offered individuals satisfaction, mental stimulation, and interesting diversions, and they would have no need to seek entertainment and excitement in morally dangerous circumstances. In a typical appraisal of the issue, members of the Illinois State Horticultural Society learned that "whatever makes home pleasant and attractive lessens the temptation to stray into paths of evil. Tippling houses, gambling hells and dens of darker deeds do not draw their victims from congenial, happy homes."[16]

Worried about a drain of young talent to the city, agricultural advisers like Lewis Allen promoted a pleasant farm home as a way to attach rural youth to their situation in the country and to agriculture as a profession. But Allen warned, if the sons and daughters of farm families lived in a "squalid, miserable tenement" they would, without question, soon look for more enriching, stimulating surroundings far away from their rural roots.[17] The failure to provide a decent home environment had serious consequences for not only farm families, but the nation as a whole. In 1852, Rev. H. L. Baugher told readers of the *Evangelical Review* that if individuals lacked a proper home, and therefore had "nothing to bind the soul to the order and proprieties," the nation should expect many new candidates for "houses of correction and penitentiaries," or even worse, mobs, riots, and social upheaval.[18]

Other social critics complained about what well-known landscape gardener, horticultural journal editor, and nurseryman Andrew Jackson Downing termed the "feverish unrest" of American life, and offered home as a reason for Americans to slow down and stay put.[19] Some critics linked constant movement with moral lassitude, and worried that individuals, unencumbered with attachments to place and the social restraints nourished by familiarity, had no reason to behave themselves. Anonymity was dangerous, these reformers concluded, and they persistently lamented the loss of the family ties and community approbation that served as a kind of watchful eye to guide and perhaps restrain individual actions. In answer to the problem, advisers suggested that love and affection for home and homegrounds would temper all the seeking and searching that seemed to pervade so many American lives. In 1864, the *American Agriculturist* stated the argument quite plainly when it urged readers to "get a home and keep it." The "unsettled morals" so clearly linked with the

American tendency to be "too roving in their habits," would stabilize once families invested in a home and again knew their neighbors, the *Agriculturist* insisted.[20]

For many domestic commentators, these concerns for social stability boiled down to an issue of class, and they worried about the "working-class" citizens who increasingly flocked to town and cities, or who made their living as tenant farmers or renters in the rural countryside. Reflecting a persistent emphasis among many social critics, Unitarian minister E. C. Guild pointed out that home ownership made working-class citizens "house proud," and gave them concrete reasons to pursue sobriety, thrift, and hard work. A well-loved home, Guild continued, would hold people, vulnerable to "careless and dangerous" ways of living, in check, and would elevate them "to a better standard by the desire to own house and land." Knowing that drunkenness and debt could lead to the loss of their home, working-class homeowners were sure to "exert self-control," reformers like Rev. Guild argued.[21]

While they agreed that a pleasant home offered a simple alternative to bad habits and evil influences, some commentators went a step further, and insisted that a proper home played an active role in shaping character and establishing appropriate values and habits. Landscape architect Horace William Shaller Cleveland described the concept in an 1856 advice book entitled *The Requirements of American Village Homes*. Home, Cleveland wrote, offered "the means by which men are to be transferred from the government of Sense and Passion, to that of Reason and the Affections." According to Cleveland, the home taught lessons of order and neatness, instilled concern for comfort and decorum, and established a love of parents and of place. In sum, homes were the "nurseries of filial and fraternal affection," and the "earliest and best schools of obedience and duty, of patriotism and piety."[22] The argument and exhortations went on for decades. In 1874, Mrs. Jeremiah Brown, an amateur horticulturist and resident of Battle Creek told members of the Michigan State Pomological Society, "the subject ... of making homes attractive and educative is fraught with vital importance to every individual, every community, every country; most especially a Republican Country." From the sacredness of home, Mrs. Brown continued, radiated "all that is known of honor, morality, patriotism, and all the higher principles which endow a free people with wisdom for self-government."[23]

The physical circumstances of home influenced the behavior of the whole family, but was particularly significant as it shaped the perceptions,

memory, and emotions of children. Mid-nineteenth-century commentators termed the process "association," and insisted that children linked the moral and behavioral lessons taught in the home with their surroundings.[24] In his essays on "home feeling," published in the mid-1860s, domestic adviser and minister Henry Harbaugh explained that young children experienced an exchange between their inner and outer lives in which images and experiences formed and conditioned the soul. "Scenery on which our eyes have rested in childhood's years, while our minds and affections were peculiarly plastic and growing, transfers its own images into the eye and spirit, leaving there a bent and a bias which remain part of our inmost selves," Harbaugh told his readers. Because of this link between early impressions and memory, the physical circumstances of the childhood home followed individuals into adulthood and led Harbaugh to define home as a "continuous location of our thoughts, feelings, impressions, and memories." This "mysterious consciousness" of home was a part of human nature, and could be expected to exist in every "uncorrupted bosom."[25]

Proponents of association insisted that once connected with particular objects or scenes, the meaning and messages learned in the home environment became deeply ingrained in the mind of the child, and persisted in the memory. When stimulated by events, experiences, or the physical details of adult life, these "associations" or memories of home reemerged to guide behavior in an appropriate and hopefully moral way. Needless to say, the formative power of these early circumstances proved immensely important, and Harbaugh told his readers that people who experienced "perverted" home influences or whose "feelings and memories are not pleasant as life itself [were] not to be trusted in family, church or state."[26] In 1848, the *Western Journal of Agriculture, Manufactures, Mechanic Arts, Internal Improvement, Commerce, and General Literature* described the concept quite clearly. For an adult with a conscience "not quite at ease," early "emblems" of childhood like well-loved ornamental plants would stir memory and in the process, purify the heart, revise the affections, and refresh the mind. No job was too big for these "silent yet eloquent monitors," the *Western Journal* declared, for even a man "beguiled" by passion, "lured by thirst of gain," or "goaded by ambition" would turn aside from evil, when an appropriate association from his childhood came to the rescue and reminded him of the proper course of action.[27]

Once they had established and confirmed that proper homes could influence character and guide conduct, mid-nineteenth-century reformers

turned their attention to stimulating commitment and action on the part of their readers. For the nation to reap the obvious social benefits, Americans had to not only "get a home and keep it," as the *American Agriculturist* urged in 1864, but must maximize the positive influences of their own home environments.[28] To help their readers, domestic commentators offered a myriad of suggestions aimed at defining the proper home and the most influential improvements to select. Though varied, these descriptions or visual images yielded some consistent themes, and eventually established an idealized domestic landscape against which readers might measure their own homes and dooryards.

With the concepts of association firmly in mind, many domestic advisers began their descriptions by emphasizing beauty and refinement, and the need for some level of ornamentation in an appropriate and appealing home environment. Henry Harbaugh minced no words when he told readers that "home feeling" and the process of association did not thrive on "naked walls, floors and ceilings."[29] Other commentators also pointed out that effective home influences required objects to stimulate the imagination and form early impressions. In 1853, the *Mother's Magazine and Family Monitor*, a women's magazine published in New York City, described the life-long relationship between things and a moral sensibility, telling readers that "the heart instinctively clings to home," and noted that even as adults, each room of the childhood home was "endeared by its peculiar associations, the furniture, the pictures upon the wall ... have become in some way linked with interesting reminiscences."[30] A *Potter's American Monthly* article published in 1881 summarized thirty years of prescriptive literature on the issue of home ornamentation. Homes, the author Marian Ford asserted, were becoming more than a place to "hang up the family portraits ... where the beefsteak is broiled, the children are spanked, the newspapers read, and the wife reproved for her lack of economy." Instead, homes should be full of things, appropriate objects "which attract eyes and minds of the children as they grow up."[31]

Many types of ornamentation could strengthen home ties and encourage appropriate associations, and advisers frequently described the value of books, pictures, or music in enhancing the efficacy of home. In a number of portrayals, however, ornamental plants emerged as the most obvious and effective means of embellishment. In a widely read account of her travels through America and of American homes published in 1853, Swedish authoress Fredrika Bremer provided a typical example as she described a modest, but tasteful and convenient dwelling. The piazza of

this ideal home, according to Bremer, was "formed of lovely trellis-work up which clambered vines." Clematis, roses, and honeysuckle surrounded the house, while beautiful trees, "the natives of all zones" and a "great number of odoriferous flowers" helped to complete the perfect picture.[32] Nearly twenty years later, Charles W. Garfield, a well-known horticulturist from Grand Rapids offered a similar, but more modest portrayal of the ideal Midwestern farm home. According to Garfield, proper rural families had thrifty, productive households and farms with "convenient arrangements and industrious occupants." A "neat lawn embellished with trees and the flower garden teeming with floral beauty" was, according to Garfield, the crowning glory of the progressive household.[33] In many descriptions, ornamental plants played an important role in the ideal home's interior as well. The *Ladies' Floral Cabinet and Pictorial Home Companion* offered a typical perspective in a revealing comparison between keeping house and making a home. The ideal home, the journal insisted, made room for "choice plants and thrifty hanging-baskets," a Wardian case filled with ferns, and vines and ivies encircling pictures and windows. A house, in contrast, lacked those embellishments and the power to influence its inhabitants as a true home should.[34]

Reformers considered a number of issues when they promoted plants as a favored mode of embellishment. Some simply suggested that ornamental plants encouraged good habits or appropriate values. In 1843, for instance, the *Michigan Farmer* argued that a proper home provided a daily "association with interesting plants and flowers," and that, in turn, "exert[ed] a salutary influence on the human character." The *Farmer* went on to point out that gardeners were seldom rude or uncivil, and normally exhibited both refinement and intelligence. In addition, a tidy yet prolific garden fostered thrift, and frequently established love of order and neatness among those who tended it. To top it off, the care of homegrounds and the nurture of ornamental plants provided important lessons in the value and meaning of labor. Families quickly saw the positive results of work as they planted, mowed, and weeded. This industriousness, the *Farmer* asserted without blinking an eye, was sure to lead to a sweet, cheerful disposition.[35]

Other commentators grappled with concerns that home ornamentation might lead to excessive materialism. Many resolved the issue by promoting a restrained and purposeful materialism that could "minister to the growth of heart and mind" as it improved and refined the home. Of the many objects that might embellish the home, commentators frequently

asserted that ornamental plants were least likely to lead to inappropriate display. Most based their recommendations on the "naturalness" of plants, and argued that as a part of the natural world, they were freely available to all. This free access to plants, so the argument went, meant that trees and flowers provided beauty without the negative associations of cost or pretension. Assuming at least some proximity to the natural world, they insisted that families unable to purchase plants could always transplant trees and shrubs from nearby fields and forests. If that source was unavailable, aspiring horticulturists could grow flowers and other plants from seed, or solicit cuttings from friends and neighbors.[36]

Even the desire for ornamental plants had an air of naturalness about it, and because their use was so simple and innocent, many reformers considered them the most appropriate way to indulge a love of beauty without "any corruption of the heart." In a typical appraisal, the *Methodist Quarterly Review* maintained that flowers and trees would never inspire greed for more or competition among neighboring families to have the best and most beautiful. Rochester, New York, horticulturist and seedsman James Vick brought the message close to home for Michigan residents in 1872, when he told members of the Michigan State Pomological Society that gardening had God's blessing. If God had created a garden as humankind's first home, Mr. Vick argued, then the acquisition of trees, shrubs, and flowers must certainly have His blessing as well.[37]

Eventually, horticultural proponents came to argue that a love of nature was deeply ingrained in the human spirit, and that everyone admired beautiful trees and colorful flowers. Wealth, education, occupation, age, or gender simply had no bearing when it came to nature's appeal. In 1848, the *Michigan Farmer* declared that the cultivation of ornamental plants was "open to the pursuit of high and low, rich and poor, the over-toiled man of business and the industrious mechanic." Both men and women enjoyed gardening, the *Farmer* added, and embellishments ranging from a "single flower pot or ornamental border, to the princely green-house or parterre" brought pleasure.[38] Nearly thirty years later, J. J. Thomas expressed a similar sentiment when he told members of the Michigan State Pomological Society that "horticulture is adapted to every dweller in this broad country, who either owns or occupies a square rod of ground." Low cost and universal appeal pointed many reformers to an obvious conclusion—ornamental plants were the embellishment for the "million," and eminently suited to adorn the dooryards of a democratic nation.[39]

The universal appeal of ornamental plants was especially compelling to reformers for another reason. For many nineteenth-century advocates of home, the natural world reflected underlying moral laws and revealed divine purpose, and its messages were both inevitable and inviolable.[40] In some situations, ornamental plants strengthened or perhaps provided the only link to these important natural influences. Emphasizing their ubiquity and adaptability, reformers often pointed out that trees and flowers could grow almost anywhere, and in their versatility offered everyone, whether busy farm residents or inhabitants of a crowded city, an immediate and very personal contact with the natural world. In contrast to less cultivated forms of nature, many proponents perceived ornamental plants as a kind of condensed and refined natural expression that clearly indicated God's presence in the world. In 1848, the *Western Journal of Agriculture* promoted the idea by suggesting that with a shady lawn or a brilliant flower garden as a daily reminder, homeowners were sure to feel "love and gratitude to the Creator for his benevolence" as a part of their day-to-day experience. Nearly twenty-five years later, Mrs. Jeremiah Brown of Battle Creek voiced the same concepts before the Michigan Pomological Society. A properly ornamented home graced with shade trees, ornamental shrubs, and luxuriant vines provided a very simple, obvious, and readily available invitation for residents to "look from Nature to Nature's God," Mrs. Brown declared.[41]

Once they had firmly established the suitability of plants as ornamentation for the ideal home, many domestic reformers turned their attention to the home's physical setting, and in the process, provided a very obvious place to cultivate trees and flowers. According to many commentators, the ideal dwelling should stand "alone, roomy [and] convenient." In her commentary on American homes, Fredrika Bremer noted that emphasis on space, and remarked that the dwellings she had observed "hasten to get apart from each other." Even when built in rows, homes in America "soon make open spaces and surround themselves with a greens-sward, and trees, and flowers," Bremer insisted. Andrew Jackson Downing, in his influential study of country homes, explained that spatially separated, individual homes fostered the "distinct family," and that institution, in turn, was the best social form for American life. He, like many others, understood the family as a well-balanced tool for inculcating social values such as civic responsibility and self-restraint, but emphasized that it could function effectively only within a proper physical envi-

ronment. More importantly, Downing thought that widely spaced homes provided both the solitude and freedom necessary to produce independent citizens, and to encourage the "highest genius and finest character" so critical for the "purity of the nation."[42]

While physically separated from other dwellings, many reformers also insisted, the ideal home should occupy a particular geographical location. Both urban reformers, concerned about the fast pace, increasing anonymity, and growing materialism of city life, and rural advisers, worried about the drain of rural residents to the city, promoted the charms of country living, and placed their ideal homes within a clearly rural setting. In a typical example, landscape gardener H. W. Cleveland portrayed a home in the countryside as free from the "baits and haunts of vice" so abundant in the burgeoning cities. Parental authority was "less counteracted," family discipline maintained, and the benefits of home "more effectively secured" within rural households. The purported tendency for solitary work and leisure activities in a rural setting, Cleveland insisted, encouraged thoughtfulness, a "marked individuality in character," and offered a pleasing contrast to city dwellers, who were "all moulded into one form by the surrounding pressure."[43]

Commentators routinely decried city homes, but they also recognized that contact with urban life was increasingly unavoidable, and in fact, beneficial for many Americans. The city, they were forced to admit, provided jobs, prosperity, and at its best, cultural and educational advantages that could enhance home life. Total isolation from those influences was no more desirable than complete immersion in urban culture. To solve the dilemma, many reformers performed a delicate balancing act, positioning their ideal home in what some historians have termed a "middle landscape." Even though in a rural setting, this new kind of space emerged untainted from the drudgery and cultural isolation long associated with farm life, and yet offered the suitable distance between homes that was so important in establishing independence and stable family interactions. In 1861, minister John Ware offered a New England perspective on the issue, which reappeared in other regions as well. Ware described the ideal home situation as a geographic place, decidedly rural in aspect, but "not far from some thrifty New England village." Also writing in a religious magazine, the *Christian Examiner*, H. W. Cleveland pointed out that railroads made proximity a relative term, and that a home "ten, twenty, or thirty miles' distance" from any city could offer residents the best of both urban and rural worlds.[44]

This geographic location led some nineteenth-century commentators to go a step further, and to picture the ideal home as a haven from the cares, concerns, and influences of the outside world. In 1860, for instance, the *Monthly Religious Magazine* suggested that the ideal home should be a school, a church, and a place of quiet industry, but most importantly a "strong-hold against the intrusions of a world ever bent upon usurpation and eager to occupy every spot on earth with its industries and its pedantries, with its excitements and its pageants." Nearly fifteen years later, *Hearth and Home* described home as a place to escape undesirable human influences like "broils in the street," or "fierce competition in the market," and all "turbulence at the polls and in the state." The home even protected families from natural disasters like drenching rains, comets, or volcanoes. "A fine, cozy place its is," *Hearth and Home* concluded, "shut in from all the rest of the universe."[45]

In many descriptions, ornamental plants enhanced this vision of home as haven. Often modified with descriptors that signified both strength and shelter, drooping branches, draping vines, and many other plant forms served as mediator between the vicissitudes of life and the family gathered safely in the home's interior. In the early 1840s, Andrew Jackson Downing voiced the concept in his architectural design book for "cottage residences" and their grounds. Downing urged home owners to cultivate "tall trees and wreathed vines" around their dwelling, so that as they grew and flourished, they might figuratively "shut out whatever of bitterness or strife may be found in the open highways of the world." As historian Jackson Lears has emphasized, mid-nineteenth-century embodiments of the domestic ideal as a vine-covered cottage used horticultural imagery to define "rootedness," and to establish a protected, still point in a rapidly gyrating world.[46]

The composite image of the ideal home that gradually emerged from the pens of reformers and social critics offered literal space for ornamental plant embellishments and articulated many benefits associated with trees and flowers. Effective and influential homes were properly ornamented, were surrounded by spacious homegrounds, and had a comfortable proximity to the natural world. Ornamental plants enhanced that link, bringing the moral messages and benign influences of nature closer to the doorstep or even inside the home, and filled the available homeground space with just the right touch of refinement and beauty. From the perspective of a diverse array of domestic commentators, it was impossible to define the ideal home without including at

least some level of ornamental plant embellishment in the descriptive picture.

Despite their repeated attempts, reformers' well-reasoned and carefully framed appeals sometimes fell on deaf ears. On many occasions, the most ardent reformers found their proposals discredited by individuals who rejected ornamentation on utilitarian grounds. The objection, some agricultural reformers suggested, was particularly prevalent among the rural population, who tended to resist change and who often viewed the world in very practical terms. This hesitation or outright resistance on the part of some Americans led resourceful agricultural advisers to integrate their suggestions for beautification into more practical concerns with farm management. Often the argument centered on the moral economy of beauty. Reformers claimed that the judicious investment in ornament was far more valuable to the future well-being of the home or family than stocks or money hoarding. Frequently, trees, shrubs, and flowers emerged from these commentaries as the most practical and economical embellishment to select. In his 1855 book, *Practical Landscape Gardening*, Ohio landscape gardener G. M. Kern typified this argument, suggesting that with very little effort, ornamental plant embellishments could transform the farm into a paradise that made family members "thinking, feeling, human beings," rather than mere producers of grain and pork.[47] Other advisers attempted to touch the farmer's pocketbook directly by adding up the costs of embellishment, and arguing that it was, in fact, a bargain. Walter Elder, an experienced horticulturist and frequent contributor to the well-respected *Gardeners' Monthly*, described the trees, shrubs, and vines on a handsome property and then calculated their value at $100. If the farmer spent his money on buildings, Mr. Elder argued, he would have emerged with little to show for his investment. The presence of $100 worth of ornamental plants, in contrast, transformed the appearance of the farm, giving the farmer and his family handsome returns on their investment.[48]

Other writers looked to long-term economic gain, and suggested that the homeowner would profit from ornamental plant embellishment when they came to sell their homes. In 1856, the *American Agriculturist* stressed that argument as it compared two fictional farms, both with fertile soil, solid buildings, and ready access to markets and social institutions. One home had a "fine yard or grassy lawn with a moderate supply of well selected shade trees and flowery shrubs, and a garden stocked with fruit trees of various kinds." In contrast, the other home had an "onion bed, a

Figure 2.1. As this real estate advertisement indicates, Midwesterners recognized the economic value of ornamental plantings. From *Adrian Times and Expositor* (29 May 1873).

potatoe [*sic*] patch, a few hills of corn" in the front yard, but no other ornamentation. Clearly, the *Agriculturist* argued, prospective buyers would be drawn to the ornamented home, and willing to pay for its improvements.[49]

Real estate advertisements confirmed that reformers did, in fact, have a point, and suggested that many Midwesterners recognized the economic value of ornamental plants in their dooryards. Mid-nineteenth-century property advertisements often emphasized the presence of ornamental plants right along with descriptions of location, condition of the dwelling house, and types of outbuildings—all presumably critical information to enhance the sale. An advertisement for an Adrian home, for example, stated that the house was "commodious," in "good repair," had all the "necessary surroundings," and in addition, there was a "fine variety of

fruits and shrubbery, on the lot." A similar advertisement for a farm home provided an even more detailed accounting of yard ornamentation. The former owner had "fitted up" the place "regardless of expense," and the grounds boasted an orchard with apples, pears, peaches and other trees, a grapery, various kinds of small fruits, and a "handsome variety of shade and ornamental trees."[50] (See Figure 2.1.)

As reformers honed their many arguments for the importance of horticultural embellishments, another kind of meaning for trees and flowers emerged in the advice literature. Reformers had routinely insisted that ornamental plants added shade, comfort, and beauty to domestic landscapes and, by their presence, developed desirable moral virtues among the families who planted and tended them. Eventually, however, some reformers suggested that ornamental plants not only fostered those vir-tues, but symbolized them. According to these advocates, trees and shrubs or a well-tended flower garden represented important social values like love of beauty and order, or even hard work, intelligence, and a moral outlook. The transferal of meaning often went one step further, leading horticultural proponents to insist that the presence or absence of ornamental plants in the dooryard expressed the refinement, moral character, progressive attitudes, or respectability of household members.[51] In the hands of horticultural advocates, homegrounds adorned with trees and flowers functioned as "sociotechnic tools," objects that in their familiarity and pervasiveness supported social order and communicated cultural values.[52]

The process of imbuing ornamental plants with cultural meaning was neither straightforward or necessarily explicit, but added a tremendous potency to reform arguments and their influence. On the one hand, as ornamental plants came to signify refinement, industriousness, and other virtues, they took on their meaning as symbols in the landscape. In that capacity, they brought both a literal order to human surroundings and a less tangible, but no less significant, cultural design that was ubiquitous, readily visible, and easily understood. This transmutation of cultural ideas into the explicit reality of trees and flowers was particularly effective, for as several scholars have pointed out, their very physicality, "naturalness," and position in the landscape made the values they represented seem both permanent and inevitable, communicating a kind of "temporal depth." As landscape components, trees and flowers, and the cultural norms they symbolized, seemed a part of long-standing tradition, but with the added advantage of maintaining a very solid presence in the day-to-day world.[53]

A Reason for Planting   43

As It Is   AN UNPLEASANT HOME
BEFORE PATRONIZING THE NURSERYMEN

As It Will Be   A PLEASANT HOME
AFTER PATRONIZING THE TREE DEALERS

Figure 2.2. Horticultural advocates often made their case with graphic comparisons of adorned and unrefined homes. From *Cabinet Fruit and Flower Plates* (Rochester: Stecher Lithograph Co.). Courtesy of the Department of Rare Books and Special Collections, University of Rochester Library, Rochester, New York.

Horticultural advocates also played on the power of social approbation and enhanced community relationships to solidify the potency and persuasiveness of landscape symbols. Planting trees or tending flowers enabled citizens to enhance their status or gain social approval in a relatively simple, yet very public way. From the opposite perspective, families that

chose no horticultural ornamentation, or allowed their garden to run to weeds, bore the burden of widely articulated disapproval, and the suspicion that they lacked character and respectability. By establishing the presence of ornamental plants as a way to measure character, moral virtue, or social attitudes, reformers offered a new incentive to conform to their standards. The choices were obvious, clearly defined, and, for those who worried about social standing and public opinion, hard to resist. (See Figure 2.2.)

Reformers often portrayed this transferal of cultural meaning through illustrations, but many also received the message through fictional comparisons of proper and improper homes. These short accounts, brimming with moral lessons for all, might appear in general interest magazines, women's journals, or even local newspapers, but were an especially popular technique among agricultural reformers who published their ideas in journals like the *Michigan Farmer*. In an 1859 example, Mrs. L. B. Adams, the editor of the *Michigan Farmer's* household advice column, typified the practice by comparing the work habits and the homes of two neighboring women. The fictional Mrs. Willis lived in a comfortable home, accomplished a tremendous amount of work, encouraged her children to assist in household chores, and usually wore a "fresh, clean gingham dress." Her cool and inviting home was surrounded by a profusion of roses, honeysuckle, and hop vines. Her neighbor Mrs. Tompkins, in turn, wore a fashionable dress that looked as if it had "seen at least a week's service in the kitchen," allowed her children to idle the day away, and lived in a "sultry, untidy" home. Needless to say, Mrs. Tompkins's yard boasted no thriving vines, flowers, or shade trees.[54]

Other domestic commentators and advisers articulated the connection between good character and plant culture quite precisely, and went on to consider the implications of their pronouncements. In his widely circulated floral guide, New England horticultural adviser and seedsman Joseph Breck typified a common observation when he told his readers that the intelligence and "pure-mindedness" of a family were revealed by the floral ornamentation of their home. "The house of the more intelligent was surrounded with flowers," Breck wrote, "the windows displayed them—vines were twined with care and taste over the dwelling." Breck made his judgment on the basis of a flower garden, potted plants, and sheltering vines, but other commentators suggested that plants communicated character in much simpler ways. According to the *Prairie Farmer*, it was "not so difficult to judge of the character and habits of the family by the single vine trained at the doorway, or the single plant, transplanted to a wooden box on the door stone."⁵⁵

While both Breck and the *Prairie Farmer* emphasized ornamental plants as signs of good character, moral virtue, or intelligence, other agricultural advisers added their own twist to this transferal of meaning, and argued that ornamental plants were sure signs of adherence to progressive farm practices and the improving spirit. An essay on farmers' gardens published in the *Report of the Commissioner of Agriculture for 1863* clearly linked the presence of ornamental plants with the range of progressive values. "Cultivated flowers," the *Report* maintained, "are a sort of floral thermometer, indicating in some degree the intelligence and refinement of the people." Flowers, the *Report* continued, went along with "more order about the buildings, an air of tidiness, thrift and comfort, and better farming generally."⁵⁶

The lack of horticultural embellishments, an inappropriate display, or badly tended homegrounds also revealed personal information, and social critics warned that passing observers could easily draw negative rather than positive conclusions about a family's moral stature. The *Michigan Farmer* made the point with a graphic comparison of a well-tended and neglected front yard. On the one hand, a neatly ornamented yard impressed observers with "marks of order, neatness and taste," and enabled them to "place a high estimate upon the moral and social qualities of the favored occupants." In the contrasting yard, however, "marks of negligence, filth, confusion and disorder" were everywhere, and the *Farmer* asked rhetorically, "what sort of sentiment do you form in advance of the moral and social condition of that family?" In case readers failed to get the message, the *Farmer* concluded that "the appearance of things around the family residence, is a pretty fair index ... of the condition of the occupant's affairs generally."⁵⁷ Unfortunately, the appearance of the home had implications for the children's future, and placed an onus on their character long before they reached maturity. "You know, as soon as you approach some houses," reformer John Ware pointed out in an 1864 issue of the *Monthly Religious Magazine*, "that the children within them are growing up in an atmosphere of coarseness and vulgarity."⁵⁸

While they willingly promoted ornamental plants as public signs of moral virtue, respectability, or the improving spirit, some commentators sensed they were treading on dangerous ground, and worried that only a thin line separated signs of true character from a veneer of respectability that was display and nothing more. Repeatedly, commentators warned

against pretentious or false display, and insisted that people should make "gardens and pleasure-grounds first and foremost, for their own comfort and gratification." According to the *Horticulturist*, the man who adorned his home solely for fashionable purposes was left with only the "glitter of a toy," not the "shining light" of real improvement.[59]

As they worried over the image conveyed by home and its surroundings, commentators often discussed the idea of "appropriateness," and suggested that it was key to establishing a true and meaningful portrayal of personal realities. Most advisers viewed appropriateness in one of two ways. Many simply expected the home to present a truthful picture in which its outward appearance matched and communicated its interior function. An appropriate home, in other words, must look homelike. Other reformers suggested that the home and its surroundings should reflect the owner's economic status or occupation, and were concerned that the home communicate a true, unpretentious image of the family's circumstances. "A house to be a true home must be strictly adapted to the owner's position in society, his calling and means," John Ware wrote, and continued, "the house of the laborer, the mechanic, the merchant, the professional man, must differ as their callings do."[60] Along with the home itself, horticultural embellishments must present an accurate picture of personal realities, and commentators routinely insisted that elaborately planted homegrounds surrounding a modest dwelling were no more appropriate than a stately mansion bereft of trees and flowers.

This concept of "appropriateness" added a curious and seemingly paradoxical twist to reform arguments, for advocates had to resolve how the same embellishments could communicate both differences in social standing and shared cultural values.[61] Many critics solved the problem by insisting that in terms of judging a family's character, differences in scale of embellishment remained simply irrelevant. The amount or specific type of horticultural display mattered very little if the homeowner had made a sincere attempt to create a well-ordered, refined, and beautiful home environment. Commentators like horticulturist Edward Sprague Rand, writing in his 1869 *Flowers for the Parlor and Garden*, repeatedly argued that simple plants in "the poor man's window" carried just as much cultural weight as the most elaborately planted homegrounds. In some situations, the simplest of embellishments might actually send the strongest message. Ornamental plants were unusually eloquent when they graced the homes of the "poor, the careworn, and those struggling for life and its comforts in new places," the *Prairie Farmer* insisted.[62] With the

help of shade trees or a pot of flowers, the poor or "inmates of a humble dwelling" could gain the respect and approval of their more affluent neighbors.[63]

Along with the poor, farmers and other rural residents were also subject to these comments, for agricultural reformers often promoted the appearance of farm homes as a way to enhance the image of the profession and its practitioners. In 1880, horticultural commentator Jason Woodman provided a graphic assessment of the issue for participants in a Michigan Farmers' Institute. If farmers persisted in planting potatoes in the front yard, Mr. Woodman warned, the farmer and his home would be "the laughing stock of more refined and elevated people." Mr. Woodman insisted that vocation was not the source of the farmer's dismal social image; instead the failure to express ability and intelligence through appropriate public symbols made farmers vulnerable to social criticism and public derision.[64]

Eventually, advisers tightened the link between ornamental plant embellishments and character even further by insisting that individual family members had a moral obligation to plant trees and flowers. Their willingness to do so, according to advocates, was a direct measure of how well they fulfilled their prescribed role in the family or community. Frequently, commentators urged their readers to look beyond their typically utilitarian impulses and responsibilities, and to furnish adornments that added to the home's beauty and refinement. "It is not enough that we provide food, drink, clothing and shelter for those dependent upon us," Michigan horticulturist Charles Garfield told members of the Michigan State Pomological Society in 1872, "we are to look after not only the necessities of life, but also those attributes that add so materially to the happiness and love of the dearest spot on earth to us—our home." Garfield also warned his readers that there was a real danger in ignoring those objects that enhanced home associations. "Do not alienate yourself from the confidence, respect, and love of your family," he continued, "by a heedless indifference to the objects which really add so much to the loveliness and happiness of a home."[65]

As advisers described the obligation to provide a refined and ornamented home, the question of who was actually responsible for home embellishment often surfaced in the prescriptive literature. The issue was both complex and controversial, involving gender roles, the actual meaning of home beautification, and a redefinition of domestic space. A number of historians have explored the typical gendered divisions of re-

sponsibility in nineteenth-century households, and have suggested that men often controlled the allocation and use of the exterior home space while women were responsible for the interior domestic sphere. In theory, men also dominated productive aspects of the household economy, while women provided nurture and tended to the household's physical and spiritual needs.[66]

The issue of home adornment, particularly with ornamental plants, jumbled those accepted roles, for the planting of trees, shrubs, and flowers around the family home involved a new use of space, and moved an area typically construed as a part of household production into the realm of nurture and refinement.[67] As early as 1837, the *New England Farmer* suggested that front yards were often neglected because they were outside the traditional spheres of both the farmer and the farm wife. Some people thought the front yard was "part of the farm and under the supervision and control of the husband," the *Farmer* pointed out, while others considered it "'part and parcel' of the house, and being such, [was] within the jurisdictional limits of the wife." The *New England Farmer* solved the dilemma by concluding that the "husband owns the rights of soil such as building and repairing fences, planting and pruning shrubbery," but both husband and wife had "a right of entry and possession."[68] In addition to a new use of space, home embellishment often required expenditures and the allocation of household resources for something besides the family's direct economic well-being. While male heads-of-household were accustomed to providing food and clothing, ornamentation was a new kind of necessity and their obligation to provide it less clearly defined.[69]

Despite the ambiguity, many commentators thought home embellishment so important that it was "plainly the duty" of all "good men" to ornament their homes. Others, like newspaper editor and rural adviser Jane Swisshelm, pointed out that men did not always fulfill their responsibility, and that women were justified in moving out of their traditional sphere to get the job done. "True the outside decorations should chiefly belong to man's labor," Swisshelm told her readers, but then warned, "very many men will not plant a tree or flower." The results of home beautification were too important to ignore, and Swisshelm urged women to busy themselves planting "half a dozen trees, a grape vine, and some flowers every spring." While Swisshelm thought women should take action, a commentator in the *Ladies' Floral Cabinet* solved the issue by suggesting that both parents were responsible for home adornment, since it

so clearly enhanced the lives of their children, and advertised the character and social standing of the whole family.[70]

The gendered aspect of ornamental plant embellishments took another twist for some reformers and commentators. Many argued that men had a particular responsibility to their wives to provide for ornamental plants and flower gardens around the home. They based their comments on the routinely articulated idea that women were particularly fond of beauty and flowers, and home adornment was especially important in enriching their lives. Michigan horticultural enthusiast Mrs. M. P. A. Crozier echoed a widely held concept when she noted that many women would live a week on "corn bread and cold water for the gift of a climbing rose at her doorway, or go without butter for a fortnight for a bed of glorious verbenas before her window."[71]

By the time reformers had finished assigning responsibility for ornamental plant culture, trees and shrubs had absorbed another public meaning. The presence of a well-tended garden or a nicely shaded lawn told the neighbors that family members understood their roles as good fathers and mothers, husbands and wives. To meet the exigencies of the changing nation, reformers insisted, those roles had grown more complex. If they were to carry out their family obligations adequately, men and women needed to consider the emotional, intellectual, and spiritual needs of other family members right along with the evening meal and a safe place to sleep.

These concerns did not end at home, however, for as some observers noted, interest in horticulture and its positive influences quickly spread from neighbor to neighbor. Andrew Jackson Downing enthusiastically pointed out that landscape gardening and the taste it engendered would gradually creep beyond the estate or farmstead and reappear in the "pot of flowers in the window ... of the humblest cottage by the wayside." Promoting a more activist stance, others suggested that men of moderate means, along with the wealthy, could exert a moral influence over their own neighborhoods or the general aspect of the country by the judicious promotion of horticulture. In 1868, the *Horticulturist* emphasized that individuals with already embellished homes needed to encourage interest throughout the neighborhood, and suggested several ways to do it. "No true horticulturist," the journal claimed, "should ever permit himself to ride or walk daily past an unsightly place or building" without offering to plant a few trees, share flowering shrubs, or bring over a few extra rose bushes.[72] Homeowners, reformers insisted, not only had a moral responsi-

bility to extend horticultural interest throughout the neighborhood or village, but if they failed to do so could find their own good reputation and character threatened. In the hands of horticultural advocates, a well-tended, neatly embellished community reflected the moral stature of those who lived and worked within its boundaries, and helped to define a new and very visible collective identity.

All these arguments for ornamental plants and their role in influencing and reflecting character expressed a common theme. Shade trees and flower beds had come to symbolize the good character of a family or individual, and their adherence to a set of shared cultural values. As they discussed the spreading influence of embellishment, reformers pointed out that eventually those values would be so clear, so public, and so widespread that they would create a homogeneous, moral, progressive community, and could even define who moved into an area. In his award-winning essay on the cultivation of flowers, Illinois agriculturist O. Ordway suggested that "persons of intelligence are often guided in their choice of residence" by noting the presence of ornamental plant embellishments throughout an area, and by considering its implications for the refinement and intelligence of the local landowners. Michigan horticulturist T. T. Lyon made a similar observation for readers of the *Michigan Farmer* when he noted that embellished neighborhoods attracted "a better and more refined class of settlers." A community or neighborhood composed of like-minded individuals, sharing common values and expressing them in the landscape, provided a powerful incentive to conform or go elsewhere.[73]

With their personal image on the line, and the well-being of the nation at stake, how could the citizens of Lenawee, Jackson and Washtenaw Counties resist? Sugar maples along the roadway or the lilac in the dooryard offered more than cooling shade or delightful scent. From a myriad of sources and in many ways, Midwesterners learned about the importance of an embellished home and its moral influence. They read and heard of pleasant, long-lasting associations, those mysterious amalgamations of sense and memory that linked individuals to their home and its moral and spiritual principles. And they learned that proper home associations required beautiful objects and tasteful arrangements to most effectively establish meaningful memories.

The desirability and practicality of ornamental plant use sounded clearly through the chorus of advice on proper home embellishment. According to advocates, plants were beautiful, unpretentious, readily avail-

able, exemplified the moral order of the natural world, and at least in the case of trees and shrubs, were very public in their expression. Citizens had a moral and civic responsibility to tidy up the dooryard, mow the lawn, and plant lilacs. Reformers underscored that responsibility by linking the properly embellished home to larger social issues, suggesting that the ornamented home encouraged social stability, developed good character, and instilled appropriate moral values.

A recurring and consistent theme ran through all of those explanations and exhortations. American society was changing. With growing cities, new work patterns, increasing materialism, and constant movement, traditional social structures and guidelines, like the church and the patriarchal family, were falling away, and Americans struggled to create new sources of stability and moral guidance. The ideology of home accompanied by ideas of association and moral materialism filled a gap in the nation's psyche, offering reformers a viable way to encourage a modern individualism that was properly tempered with self-restraint and morality. Although they often promoted an image of home as a stable moral center, reformers also recognized that the concept was essentially portable, and that home influence could extend both geographically and temporally to meet changing individual and social needs. In addition, horticultural advocates understood that trees and flowers, as landscape symbols, were especially potent tools for communicating a particular social view. As they promoted ornamental plant culture, reformers hoped to imprint their vision of a progressive society on the landscape of the future.

For the residents of Lenawee, Jackson, and Washtenaw Counties and other Midwesterners, the ideology of home supplied compelling reasons for planting trees and shrubs or growing flowers, and offered social approbation for those who heeded the advice. In as much as they shared in the cultural climate of their day, whether they read general interest magazines, subscribed to agricultural papers, or simply talked with their progressive neighbors, these Midwesterners certainly knew what advocates thought ornamental plant embellishments could do, and had the assurance that both their families and their public image would benefit from shade trees and flower beds.

Despite its pervasiveness and influence, however, the ideology of home was just one factor influencing individual decisions and behavior. Midwesterners heard many other voices promoting ornamental plant culture, and sometimes encountered contrasting meanings for homeground embellishment. In addition, most Midwesterners considered the avail-

ability of plants, their cost, the realities of their daily lives, and their personal interests and inclinations before they took to planting trees or tending flowers. For a real understanding of expanding horticultural interest, we must blend these varied perspectives with the widely articulated concepts of home and its importance. Ornamental plants could indeed represent moral virtue and good character, but as complex cultural symbols they also communicated many other meanings to, and about, those who promoted, planted, and nurtured them.

Three

Hands-on Horticulture

Horticultural Advisers and Information at Mid-century

The widespread promotion of home as a stable moral center for American life left few stones unturned. By the mid-nineteenth century, residents of Lenawee, Jackson, and Washtenaw Counties, and other Midwesterners, could learn about home influences and the importance of horticultural embellishment from a diverse array of sources and from many perspectives. This idealistic portrayal stirred interest, and in 1858, the *North American Review* observed that the country abounded with individuals ready to improve their homegrounds, their heads full of the "prose and poetry of velvet lawns, leafy groves and thickets, groups and masses, statues and vases." The *Review* feared, however, that "prose and poetry" were not always an effective basis for cultivating plants, and pointed out that while people might know why they should embellish their homegrounds, they did not always know how to proceed.[1]

In some measure horticultural advisers had contributed to the "poetry" surrounding ornamental plant culture, but they also stood ready to provide a practical counterbalance to the rampant idealism. By the mid-

 nineteenth century, these knowledgeable and enthusiastic horticulturists offered their readers "how-to" advice on garden design, homeground layout, plant selection, and many other topics through a growing number of widely available horticultural and agricultural journals, advice columns, and specialized books. Their practical comments created an image of modest homeground embellishment, offered encouragement for those with little horticultural experience, and supplied the more knowledgeable with the latest horticultural information. This chapter explores these varied sources of horticultural information, evaluates the advice routinely offered, and considers how horticultural professionals helped to spread ornamental plant interest even more widely.

As it had for promoters of domestic ideology, the increasing availability of horticultural books, journals, and reports, spurred on by the mid-nineteenth-century publishing bonanza, enhanced the influence of professional horticulturists and made practical horticultural advice easy to come by. This burst of publishing enthusiasm, however, emerged from very sparse beginnings. Prior to the first decades of the nineteenth century, American horticulturists had made few attempts to advise their countrymen or women on horticultural or agricultural matters. For many aspiring horticulturists in the new nation, widely circulated farmers' almanacs provided the primary source of both horticultural and agricultural information, but typically offered only a limited repetition of traditional lore along with reminders of seasonally appropriate activities. In the last half of the eighteenth century, a few individuals turned to newly formed learned societies, like the American Philosophical Society for Promoting Useful Knowledge, founded in Philadelphia in 1769, and found in their reports some detailed and current horticultural information. The rather elite membership of these organizations, however, did little to extend horticultural interest beyond its traditional boundaries defined by wealth, education, or special interest.[2]

Some eighteenth-century American horticulturists and wealthy landowners visited notable European gardens or consulted European gardening books, and many found English landscape and garden traditions particularly appealing. Americans knew about the formal, geometric garden designs of the seventeenth and early eighteenth centuries, and followed the evolving English concern with naturalism and the picturesque landscape ideals. The idea of "ornamented farms," rural land holdings that combined the beauties of horticulture with the practicalities of agriculture, appealed to some Americans, and like Thomas Jefferson, they called

upon these concepts as they developed their own substantial properties. The ideas of renowned English landscape gardener Lancelot "Capability" Brown, promoting undulating expanses of lawn highlighted with simple clusters of shade or ornamental trees, had their American interpretations in the last decades of the eighteenth century. Wealthy Americans also embraced Humphrey Repton's late-eighteenth-century ideas of "embellished nature," and designed flower gardens or ornamented pleasure grounds to act as a transition between the home and the natural world. In the early decades of the nineteenth century, knowledgeable Americans read and discussed the works of English horticulturist John Claudius Loudon. Loudon built on Repton's ideas, and in a style he termed "gardenesque," emphasized the use of individual specimen trees and exotic flowers displayed in separate beds as a way to blend natural and cultivated landscapes.[3]

Although English gardening advice books and journals, like John Claudius Loudon's the *Gardener's Magazine*, were available to American horticulturists in the early nineteenth century, some practitioners found them wanting. Loudon attempted to popularize horticulture and to provide information suitable for small homegrounds as well as grand estates, but despite his efforts, readers still complained that most English or European gardening books were simply not suitable for American environmental conditions and social circumstances.[4] Recognizing the problems that these distance sources posed for American gardeners, several horticulturists attempted to fill the void in American gardening literature. Nurserymen John Gardiner and David Hepburn, for example, published *The American Gardener* in 1804, and another prominent nurseryman, Bernard M'Mahon, followed with *The American Gardener's Calendar* in 1806. M'Mahon's work was particularly influential, and remained a standard authority in American gardening literature for the next fifty years. Both books offered much more comprehensive gardening information than any of their predecessors and emphasized American growing conditions, but still drew heavily on English sources for specific horticultural and garden design practices.[5]

As the nineteenth century progressed, American horticulturists increasingly found their voice, and by the middle decades contributed horticultural advice to a variety of publications. In contrast to earlier authors, these mid-nineteenth-century horticultural writers focused on American resources and environmental circumstances, and tended to emphasize the importance of ornamental plants for people of all ranks and occupa-

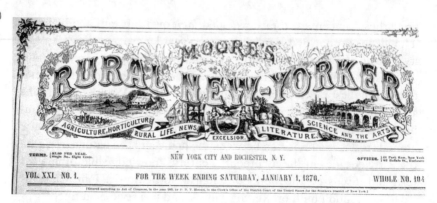

Figure 3.1. *Moore's Rural New Yorker,* a popular regional agricultural paper, often included horticultural information. From *Moore's Rural New Yorker* (1 January 1870).

tions. The newly formed and rapidly expanding agricultural press was one of the first and most important purveyors of horticultural information. Beginning with the publication of the Baltimore-based *American Farmer* in 1819 and continuing until the Civil War, agricultural advisers and reformers established hundreds of new journals that stressed agricultural issues and rural improvements. Some, like the *American Agriculturist,* cultivated a national readership. Others, such as the *Michigan Farmer* or the *Prairie Farmer* of Illinois, reflected a more regional orientation. Still others like *Moore's Rural New Yorker,* had a regional base, but a wide circulation in the Midwest as well as in New York State. These regional agricultural papers became a medium for the exchange of localized agricultural information, but also offered readers a national perspective on important agricultural and horticultural issues.[6] (See Figure 3.1.)

As part of their broad interest in rural reform and improvement, agricultural journals routinely discussed stock raising, proper field crop cultivation practices, improved crop varieties, and other distinctly agricultural subjects, but many also included information on topics like orchard management, floriculture, and homeground layout or garden design that typically fell within the realm of horticulture. Among the first of the agricultural papers to devote extensive space to horticultural issues, the *New England Farmer* and its editor, Thomas G. Fessenden, actively promoted horticultural interests.[7] Gradually, many other agricultural journals took up the cause, some simply publishing occasional articles on ornamental plants or garden practices, and others including regular horticultural columns or departments. Occasionally, agricultural magazines even added

a horticultural editor to the journal's staff. Thomas Meehan, editor of the well-respected *Gardener's Monthly*, noted the trend, and in 1861, published a list of twenty-two reputable agricultural journals that routinely featured significant amounts of horticultural information. The list included a number of Midwestern journals like the *Wisconsin Farmer*, the *Ohio Farmer*, the *Indiana Farmer*, and the *Minnesota Farmer and Gardener*.[8]

Two widely read Midwestern agricultural journals typified this interest in horticulture. In its second year of publication, *Michigan Farmer* editor D. D. T. Moore changed the paper's official title from the *Michigan Farmer and Western Agriculturist* to the *Michigan Farmer and Western Horticulturist*, emphasizing that "more attention was paid to the culture of gardens and orchards."[9] Beginning in the late 1840s, the *Michigan Farmer* featured a regular section devoted to horticultural topics, and for at least a portion of the time, enjoyed the services of an experienced horticultural editor. For a number of years, two prominent and knowledgeable Michigan horticulturists, S. B. Noble, an Ann Arbor nurseryman and fruit grower, and J. C. Holmes, a Detroit nurseryman and eventually secretary of the Michigan State Agricultural Society, alternated as the *Farmer*'s horticultural editor. R. F. Johnstone, the general editor from the mid-1850s until 1880, had early experience as a typesetter with landscape gardener A. J. Downing's well-known journal the *Horticulturist*, and prominent Michigan horticulturists considered him well-versed in horticultural issues.[10]

Like the *Michigan Farmer*, the *Prairie Farmer* discussed horticultural matters as well as agricultural concerns, and in the 1850s, John Kennicott, a well-known Illinois nurseryman and practicing physician, filled the position of horticultural editor for the *Farmer*. In a retrospective of its progress in the late 1860s, the *Prairie Farmer* reaffirmed its commitment to horticultural interests, suggesting to its readers that the "next step in the field of progress is to give aid and comfort to horticulture." The *Farmer* justified its increasing interest by asserting that horticulture was the "handmaid" of agriculture, and would eventually make agricultural pursuits "more attractive, healthful, and profitable." Eventually, the practice of including horticultural issues in agricultural journals became so widespread that more specialized horticultural magazines worried about the competition. In 1872, *Horticulturist* editor Henry T. Williams complained that "the agricultural journals of the present day have stepped over into the field of horticulture and by engaging horticultural editors, writers, etc., draw away a great many from the patronage of the horticultural magazines."[11]

In addition, local newspapers encouraged the widespread interest in horticulture by regularly publishing horticultural and agricultural advice columns. In 1861, the *Gardener's Monthly* proclaimed that "almost every country paper has now its agricultural column," and several years later noted that the need for agricultural and horticultural advice was so great in city papers, that a number of major newspapers were "casting about" for qualified agricultural or horticultural editors.[12] Residents of southeastern Michigan were among the many who could read about horticulture in their local newspapers. The Lenawee County *Adrian Times and Expositor* ran a regular column entitled "Farm and Garden" for many years. Others, like the Washtenaw County *Ypsilanti Commercial* or the Lenawee County *Tecumseh Herald*, frequently published articles on horticultural topics.[13]

Although Midwesterners could obtain horticultural information in a variety of sources, the most detailed accounts and greatest emphasis on practical advice appeared in specialized horticultural journals or in books on particular horticultural topics. Like the agricultural press, horticultural journals blossomed in the middle decades of the nineteenth century. Seven horticultural journals established in the 1830s and 1840s had ceased publication by 1850, but between 1850 and 1880, readers could select from at least twenty-five periodicals devoted exclusively to horticultural issues. Some of these journals were long-lived, had a nationwide distribution, and were very influential, while others appeared for a year or two and then died away.[14]

Charles M. Hovey, a well-known nurseryman and strawberry breeder, founded the first and one of the most successful journals in 1835. The *Magazine of Horticulture, Botany, and All Useful Discoveries and Improvements in Rural Affairs*, published in Boston, continued under Hovey's editorship until 1868 when it merged with the *American Journal of Horticulture* to form *Tilton's Journal of Horticulture and Floral Magazine*. During its long life, the *Magazine of Horticulture* regularly included a retrospective look at the progress of horticulture, detailed appropriate "horticultural operations" for each month, provided in-depth articles by leading horticulturists and landscape gardeners, and offered information on the latest horticultural advances. Of the long-running horticultural journals, the *Magazine of Horticulture* remained the most consistent in its efforts to provide high-quality, comprehensive horticultural information, causing some readers to complain that the magazine was "too scientific" and "too far in advance of the popular taste." In 1859, Charles Hovey admitted that the magazine was not necessarily aimed at amateurs, but instead tried to

Figure 3.2. One of the premier horticultural journals of the mid-nineteenth century, the *Horticulturist* was dedicated to improving rural landscapes and taste. From *Horticulturist*, 24 (June 1869).

"record new and superior modes of culture, new facts, new observations and new experiments."[15]

Hovey's *Magazine of Horticulture* continued as the primary conduit of detailed horticultural information until 1846, when the immensely influential nurseryman and landscape gardener Andrew Jackson Downing established the *Horticulturist and Journal of Rural Art and Rural Taste* in Albany, New York. In contrast to the *Magazine of Horticulture*, the *Horticulturist* changed editors frequently, but attracted a particularly influential and competent group to the position. Downing continued as editor until his death in 1852, and was followed by Patrick Barry, a well-known Rochester, New York, nurseryman and European-trained horticulturist. Horticulturists J. J. Smith, Peter Mead, G. E. Woodward, and Henry T. Williams all took their turn at editing the journal until its merger with the *Gardener's Monthly* in 1875. The place of publication also varied, moving from Albany to Rochester, New York, then to Philadelphia, and eventually to New York City. The *Horticulturist*, like the *Magazine of Horticulture*, emphasized detailed horticultural information and the "scientific pursuit of horticulture," but had an additional interest in improving the "arts of taste." The journal prided itself on catering to a rather elite clientele, and frequently told potential advertisers, "We reach everywhere those who have fine places to embellish and beautify, and who have money to spend."[16] (See Figure 3.2.)

Figure 3.3. Specialized horticultural publications flourished in the mid-nineteenth century and the *Ladies' Floral Cabinet* was one of the most popular. From *Ladies' Floral Cabinet* (December 1871).

In contrast to the *Horticulturist*, the *Gardener's Monthly*, a third long-running and widely circulated horticultural magazine, emphasized its low cost, availability, and usefulness to all horticulturists. Thomas Meehan, a Philadelphia nurseryman, started the journal in 1859, and served as editor until it merged with the *Horticulturist* in 1875. As he evaluated the scope and direction of the journal, Meehan often observed that he aimed to "furnish knowledge, at a cheap rate, to those whose limited means prevented them from subscribing to more expensive journals." Like its more scientific and tasteful counterparts, the *Gardener's Monthly* attempted to present and analyze the "latest improvements in the art" of gardening, as it stimulated a widespread interest in horticultural advancement.[17]

While magazines like the *Gardener's Monthly* or the *Horticulturist* covered a wide spectrum of horticultural issues for a broad readership, a few national publications cultivated a more specialized audience. In the early 1870s, for example, *Horticulturist* editor Henry T. Williams unveiled a new periodical called the *Ladies' Floral Cabinet and Pictorial Home Companion*. Williams noted that the journal was richly illustrated, published articles on flower gardens, flower arranging, and indoor plants, and filled the demand "for a popular ladies' floral and home journal." Because Williams encouraged readers to draw on their own experiences and write for the *Floral Cabinet*, many women found their horticultural voice and penned articles for the magazine. Like a number of other specialized mid-nineteenth-century horticultural journals, the *Ladies' Floral Cabinet*

had a relatively short run, continuing publication into the 1880s, and eventually merging with the *American Garden*.[18] (See Figure 3.3.)

A number of regional horticultural magazines also sprang up in the middle decades of the nineteenth century. Several focused on Midwestern cultural conditions, and attempted to provide specific information for gardeners or orchardists in the region. The *Western Farmer and Gardener*, published in Cincinnati from 1839 to 1845, was among the longest lived while others, like the *Western Horticultural Review*, also published in Cincinnati, lasted only a few years. Some regional journals specialized in certain aspects of horticulture. Published in Des Moines, Iowa, and Leavenworth, Kansas, the *Western Pomologist*, for example, emphasized the concerns of Midwestern fruit growers.[19]

Often available for a few dollars a year, a subscription to an agricultural or horticultural journal, or even a local newspaper, opened access to horticultural information available in many other sources.[20] Sometimes newspapers or journals republished entire horticultural articles, and at other times summarized content or directly quoted large passages of important information drawn from other books or journals. In May 1873, for example, the Washtenaw County *Dexter Leader* published an article describing the importance of tree planting that first appeared in the *California Farmer*. In a similar instance, the *Adrian Times and Expositor* summarized an article on the progress of horticulture first published in the *Gardener's Monthly*. Newspaper and journal editors also critiqued other periodicals, or reviewed important new horticultural books. In a typical instance, the *Tecumseh Herald* reported on the *Michigan Farmer*, telling readers, "We have received the September number of this valuable monthly. Its list of contents presents an attracting [*sic*] variety of useful and agreeable matter suitable for evening family reading as nights are becoming longer and cooler."[21]

Along with the expanding number of publications, the actual circulation of agricultural and horticultural journals offers another way to assess the dissemination of horticultural information at mid-century. In 1870, more than 770,000 Americans subscribed to a range of horticultural or agricultural journals. While smaller than that of literary or political magazines, the circulation figure for horticultural and agricultural magazines exceeded that of advertising, commercial, sporting, and several other categories of periodical literature. Nationally, widely advertised journals like the *American Agriculturist* dominated circulation numbers. In 1866, the *Agriculturist* claimed to have between 150,000 and 200,000 subscribers. At

a local level, regional agricultural papers often remained the most influential and widely read of any periodical, and the *Michigan Farmer* was no exception. By 1870, the census reported the *Michigan Farmer*'s circulation at five thousand, a number quite typical for many other agricultural journals.[22]

Particularly in their early years, the more specialized horticultural journals struggled with a much smaller subscriber base than agricultural magazines. In 1855, nurseryman and *Horticulturist* editor Patrick Barry worried that horticultural magazines had "an aggregate circulation that [did] not exceed ten or fifteen thousand." Circulation of these journals expanded as the century wore on, and by 1865, the *Horticulturist* editor described the journal's circulation as "large." In 1874, editor Henry Williams announced that the current issue of the *Horticulturist* would reach over five thousand new readers, and that the *Ladies' Floral Cabinet* had a circulation of nearly thirty thousand.[23]

Actual readership for all of these newspapers and magazines may have been far larger than these circulation figures indicated. In 1854, for example, the *Michigan Farmer* stated that the newspaper had forty thousand readers, and although the editors did not reveal the source of this estimate, circulation of the journal among family members, neighbors, and friends certainly extended its influence far beyond paid subscribers. In the 1860s, the *Horticulturist* noted the practice of sharing newspapers and magazines and the concomitant extension of influence, and also pointed out that the journal was freely distributed to "every public library and every horticultural society that furnishes its address to us."[24]

If Midwesterners needed additional horticultural guidance, an ever-growing number of general guide books and specialized monographs were available by mid-century to supply their wants. Agricultural historian and horticulturist Liberty Hyde Bailey has estimated that over two hundred books on horticulture and gardening topics were published in America between 1850 and 1880.[25] Some of these works went through numerous editions, extending their influence on horticultural practices or garden design until the end of the nineteenth century. Andrew Jackson Downing's *A Treatise on the Theory and Practice of Landscape Gardening,* first published in 1841, appeared in multiple reprints and revisions through the next four decades, and was a primary example of just such a pivotal work. Downing articulated ideas about landscape design and the importance of homeground embellishment that, although not completely original, struck a new and very resonant chord in American society. His contemporaries continually remarked on the importance of the work, and its effect on American domestic landscapes, often noting the book was "to be found everywhere, and nobody, whether he be rich or poor, builds a house or lays out a garden without consulting Downing's works."[26]

While Downing's influence persisted through much of the nineteenth century, some critics claimed his ideas were actually "too refined and ornate for the mass of improvers," and other books on landscape design gained in importance. In 1855, for example, G. M. Kern, a Cincinnati landscape gardener, published *Practical Landscape Gardening*, and several reviewers pointed out it provided information far more suitable "for the mass of men who desire to fit up their places in a not expensive yet agreeable way." In 1870, Frank Jessup Scott, a young Toledo real estate agent who had worked with Andrew Jackson Downing and trained as an architect, published *The Art of Beautifying Suburban Home Grounds of Small Extent*. His influential book detailed practical solutions to landscape design, and moved the emphasis to smaller dooryards and city or suburban lots. Citing its usefulness for people of modest means, the *Horticulturist* recommended Scott's book as the "best book on ornamental trees, landscape gardening or homeground layout" that was currently available.[27]

Several noteworthy works also provided general information on plant culture. Joseph Breck, prolific writer and owner of the New England Agricultural Seed Store in Boston, published *The Flower-Garden; or Breck's Book of Flowers* in 1851, and the book went through a number of printings in the next several decades. Philadelphia florist and seedsman Robert Buist wrote the *American Flower-Garden Directory* in 1831, and that work also provided basic horticultural information for a number of years, with many reprints and new editions published through the 1850s. In the late 1860s, Peter Henderson, a successful New York florist and seedsman, contributed several widely read and frequently reprinted horticultural guides including *Gardening for Profit* and *Practical Floriculture*.[28]

Along with general horticultural works, a number of specialized monographs appeared between 1850 and 1880, and a few examples suggest the range of topics horticultural writers considered. Edward Sprague Rand, a lawyer and well-known amateur horticulturist, published popular works on bulbs, garden flowers, orchids, and indoor gardening. *Horticulturist* editor Henry T. Williams assembled a treatise on window gardening that went through fourteen editions in twelve years. Trees and shrubs also received their share of attention. In 1858, John A. Warder, editor of the *Western Horticultural Review*, published an account of evergreen and

hedge culture, advocating their use and providing detailed cultural instructions.[29]

Although these sources of professional horticultural advice were widely dispersed, they did not reach everyone. Hoping to fill the void and wanting to extend the reach of sound information even further, both national and state governments provided freely distributed reports that often contained horticultural information. The federal government entered the fray when, in 1839, Congress allocated a small sum to the United States Patent Office for the collection of agricultural information and the dissemination of seeds. Encouraged by that show of support, Henry L. Ellsworth, Commissioner of Patents, issued a short report detailing Patent Office activities and containing some agricultural information. With continued Congressional backing, the agricultural section of the Patent Office report gradually increased in size and often included information about weather conditions, crop yields, cultural practices, and horticultural issues along with important innovations reported to the agency by local farmers or regional "agents." The Commissioner justified the interest in horticulture by noting that the Patent Office received frequent requests for information on vegetable, fruit, and flower culture, and pointed out that its reports reached "thousands of people in retired or remote parts of the Union" who had little access to horticultural books and journals.[30]

With the formation of the Department of Agriculture in 1862, interest in disseminating reputedly current scientific information on both agricultural and horticultural topics continued.[31] By 1871, *Tilton's Journal of Horticulture* speculated that the report was probably "the most widely circulated agricultural work in the country," and hoped that its content warranted such widespread exposure. Those comments reflected years of criticism from members of the horticultural and agricultural press, who, perhaps disliking the competition, frequently pointed out when the government published outdated, irrelevant, or inaccurate information. Critics also insisted that whatever useful information appeared in the reports was easily available in other sources.[32] Some Lenawee, Jackson, and Washtenaw County farmers apparently disagreed with the critics, for over the years, they responded to Patent Office circulars seeking information on regional conditions and agricultural practices, and several residents wrote to the Commissioner requesting a copy of the report.[33]

By the mid-nineteenth century, individual states also began to take an active role in disseminating agricultural and horticultural information through newly organized statewide agricultural societies. These organizations published annual reports that included organizational news and addresses given before the society, but in addition, frequently contained fair reports, news, and essays submitted by county or regional associations. Although emphasizing agricultural topics, agricultural society reports also contained material on horticultural issues and concerns. Like many other Midwestern societies, the Michigan State Agricultural Society, founded in 1849, published its transactions with financial assistance from the state of Michigan.³⁴

A number of states also boasted statewide horticultural organizations dedicated to the dissemination of horticultural information. Between 1850 and 1880, nineteen state horticultural societies, including eleven from the Midwest and adjoining states, published reports that detailed horticultural procedures and practices. As they did with agricultural society reports, state governments sometimes provided funds to publish these costly works, and to encourage their free distribution. The Illinois legislature, for example, allocated $2,000 per year to the state horticultural society to pay for premiums, for procuring "scientific investigations relating to horticulture," and for publishing the transactions of the society. The Illinois State Horticultural Society published its first report in 1856, and for a number of years, included regional Illinois horticultural society reports in the state transactions. The Indiana State Horticultural Society issued its first annual report in 1866, and the Wisconsin Horticultural Society began publication in 1871. The Michigan State Pomological Society, founded in 1870, issued its first report in 1871, and like other Midwestern states depended on the state to publish the transactions.³⁵ The number of Michigan State Pomological Society Reports published and distributed yearly gradually increased until in 1879, the state printed 5,747 copies, and the Pomological Society distributed 4,076 volumes to members, libraries, schools, other horticultural organizations, and to interested individuals.³⁶

Agricultural and horticultural societies routinely exchanged copies of their organization's reports, giving their members access to the latest information from across the region or the nation. In 1880, the Michigan Pomological Society library illustrated that practice, with its inventory of reports from the American Pomological Society, fourteen state horticultural organizations, and two Canadian societies. In addition, the Society's library housed agricultural society reports from fifteen states and from the United States Department of Agriculture. Most reports began with the

founding of the Michigan Pomological Society in 1870, but others extended back to the 1850s.[37]

Local residents appeared to value all of these resources and the information they supplied. Agricultural and horticultural journals along with specialized books on horticultural topics often appeared in town and village libraries or in agricultural and horticultural society collections. In 1844, for example, the Cincinnati Horticultural Society listed Mrs. Loudon's *Ladies' Companion to the Flower Garden*, Downing's *Cottage Residences*, and Hovey's *Magazine of Horticulture*, the *Western Farmer and Gardener*, the *American Agriculturist* and a number of other horticultural books and journals among its holdings. The Michigan State Agricultural Society boasted a similar collection, including Downing's *Rural Architecture*, the *Ladies' Companion to the Flower Garden*, the *Michigan Farmer*, and many others. Even the Adrian Y. M. C. A. offered the *American Agriculturist* among the selections of journals available to citizens who visited its reading room in the 1870s.[38]

State and local horticultural and agricultural societies also offered horticultural journals and books as premiums for prize-winning entries in their fairs and exhibitions. In 1873, the Michigan State Pomological Society went so far as to adopt a policy regarding the use of books and journals as prizes, and justified their decision by proclaiming "books on agriculture and horticulture, and papers devoted to the discussion of those subjects, are eminently calculated to extend the influence of the society, and promote the objects for which it was instituted." The Society then listed the journals, including *Moore's Rural New Yorker*, the *Gardener's Monthly*, the *Horticulturist*, the *Michigan Farmer*, the *Prairie Farmer*, and a number of others it thought suitable for prizes.[39] Well aware of the practice, some agricultural book publishers even placed advertisements reminding society officials to use books and journals as premiums.[40]

The growing enthusiasm for horticulture certainly owed a debt to the ready availability of these sources, but more importantly, horticultural advisers attempted to stimulate ornamental plant interest by emphasizing certain themes and horticultural practices. Although commenting on a wide range of topics, many mid-nineteenth-century advisers focused their attention on the horticultural problems and concerns of the "masses," people who were either "poor or only moderately wealthy," or people of moderate means who had a decided interest in ornamenting their homegrounds, and who yearned for practicality as well as refinement.[41] Even

journals with a somewhat elitist cast like the *Horticulturist* prided themselves on supplying practical information. In contrast to their European counterparts, the *Horticulturist* asserted in 1852, Americans "want the ornamental and useful together,—we require facts as well as theories,—we build houses to live in as well as for effect—we cultivate gardens for profit, as well as beauty."[42]

Horticultural commentators often began the discussion of homeground improvement at a most basic level. They urged their readers to set aside space for ornament, and then discussed or illustrated practical and appropriate layouts for the home, outbuildings, and surrounding domestic landscapes. Because homeground embellishment involved an essentially a new use of space, advisers presented a variety of plans to help their readers make appropriate choices. An 1855 *Michigan Farmer* plan entitled a "Design for a Farm House, with Grounds and Farm Buildings" typified recommended layouts. The *Farmer*'s plan included the arrangement of rooms in the house and barn, the position of outbuildings, orchards, and the kitchen garden, and suggested patterns for placing garden beds and walkways in front of the home. An 1854 *American Agriculturist* plan for a half acre "mechanics garden" designated the house, wood yard, privy, walkways, and small fruit plots, but then indicated the position of more than eighty-seven different plant species. The *Agriculturist* admitted the plan offered detailed instructions and extremely varied plantings, but claimed, "we have given the names of the varieties not so much to recommend them, as for a guide or index to *new* hands at cultivating small plots."[43]

Some landscape reformers described the initial layout of homegrounds verbally, and their discussions tended to place a stronger emphasis on the ideas or principles underlying a particular design strategy. With the publication of his seminal work, *A Treatise on the Theory and Practice of Landscape Gardening Adapted to North America* in 1841, A. J. Downing set the tone for a discussion that continued throughout the century. An understanding of landscape gardening must guide the layout of homegrounds, Downing insisted, and that knowledge, in turn, was based upon "a highly graceful or picturesque epitome of natural beauty." For Downing, landscape design aimed to create beauty, sometimes through a simple imitation of natural patterns and compositions, but more frequently by a harmonious abstraction that expressed the spirit and essence of nature. The emphasis on beauty and the imitation of natural forms had a number of practical im-

plications, and led Downing and many others who shared his concepts to advocate layouts with curving lines, clustered plantings, and a diversity of plant forms and sizes.⁴⁴

Although Downing's concepts provided a springboard for much of mid-nineteenth-century landscape design, not all horticultural advisers agreed with either the basic purpose or the efficacy of the natural style and its suitability for most American homegrounds. Frequently, American horticultural commentators tempered Downing's philosophical pronouncements with practical considerations, suggesting, for example, that homeowners should be interested in not only beauty, but the level of maintenance and the cost of implementing a large-scale landscape plan. Many argued for simplicity in design, claiming that it led to a well-kept appearance, neatness, and order, the most important attributes of any landscape plan. This interest in simplicity and the desire to meet the design needs of a wide range of families encouraged some advisers to advocate a "truly American system of gardening." In 1852, a commentator in the *Horticulturist* called for a new concept of "suburban gardening," a distinctly American variation of the more complex landscape gardening, and one designed to promote "simple good taste, and the habits of a republican people."⁴⁵

Along with an articulation of basic principles, commentators often described just how much space should be included in homeground ornamentation. In the tradition of large European estates, Andrew Jackson Downing initially recommended an area ranging from five to one hundred acres as an appropriate size for American landscape gardens, and suggested that the true expression of landscape gardening could not be applied with "equal success to residences of every class and size in the country." Many horticultural advisers quickly pointed out that such grand plans depended upon large estates, wealthy patrons, and cheap labor, and were simply not practical for most Americans. Although they occasionally complained about farmers who fenced off "perhaps ten rods of ground" in a wide open country, most critics urged their readers to embellish a few acres or a small city or suburban lot and developed design plans suitable for those conditions.⁴⁶

Some commentators even turned the idea of a smaller ornamented space into an advantage, telling readers it was easier to maintain in "perfect order," and that it did not contribute to prideful display or unseemly pretensions. Beginning in the 1860s, several horticultural advisers suggested an innovative way to increase the sense of space around crowded

city lots or in suburban settings. "Throw the front grounds of several buildings into one enclosure, and have them embellished with trees and flowers," the *American Agriculturist* urged in 1862. Under such a plan, a number of families could enjoy a "considerable park or pleasure ground." In 1870, Frank J. Scott presented the concept of adjoining open front lawns, unencumbered by fences or hedging as a democratic and neighborly approach to landscape design and as an important part of his new ideas for suburban homegrounds of "small extent."[47]

After stressing the importance of a good basic layout, horticultural advisers turned their attention to particular elements of homeground adornment. Building on Andrew Jackson Downing's concern with grass and trees, many mid-nineteenth-century horticulturists suggested that a well-kept lawn was the single most important element in creating a pleasingly embellished home.[48] Because few Americans of middling or lower rank had lawns surrounding their homes, many authors prefaced their practical remarks with reasons why they thought an expanse of well-tended grass so important. Often, their claims simply emphasized that a "broad sweep of the smooth green turf" was itself a lovely sight, and that it lent "a charm to all surrounding objects" by accentuating the beauty of trees, shrubs, and flowers. Nurseryman Patrick Barry suggested that a lawn also had a positive influence on a family's "gardening taste," and gradually led to additional improvement around the home.[49] Many horticultural advisers also emphasized that a well-tended lawn fell within reach of "farmers, mechanics, professional men, indeed all who have homes to adorn, with a few feet of spare ground around their dwellings." Lawns, advocates claimed, were the most democratic of embellishments, requiring neither excessive space or labor, and offering almost everyone the pleasure of "cheap luxuries."[50]

Despite these "democratic" claims, discussions of lawn care revealed that the process of establishing and maintaining a lawn was neither as simple or labor-free as advocates suggested. To begin with, horticultural commentators insisted that farmers and townspeople would enjoy the benefits of a lawn only if they developed and cared for it properly. To help their readers understand the issues, advisers supplied detailed information on establishing the lawn, including proper soil preparation and the process of sodding or seeding, and offered additional advice on manuring and weed control. Once the lawn was in place, advisers adamantly insisted that adequate maintenance required regular mowing, "sweeping," and rolling at weekly or, at most, two-week intervals. Lawn mowing

created a major stumbling block for many rural and urban residents. Some farm families solved the problem by allowing sheep or cattle to graze near the house, but in the process endangered any other ornamental plantings in the same area. Most horticultural advisers urged their readers to routinely scythe the lawn, but operating a scythe with the precision needed to produce a smooth surface required great skill. Farmers sometimes ran horse-drawn mowing machines, widely available in the 1860s, over the ornamental grounds, but as their critics pointed out, the machines were designed to cut hay, not create velvety green expanses. In addition, mowing machines did nothing to solve the lawn maintenance problems of urban dwellers or those with small homeground spaces.[51]

Beginning in the 1850s, the horticultural and agricultural press noted advances in English hand lawn mowers, and urged Americans to consider the machines as an answer to both their mowing and lawn rolling problems. In 1859, the *Horticulturist* proclaimed the virtues of the new hand mower, and speculated that its widespread use would "revolutionize lawns all through the country." The *Gardener's Monthly* fervently hoped that as hand mowers improved, Americans could look forward to the "abolition of the Scythe institution." American manufacturers finally turned their inventive genius to small hand lawn mowers, and by 1870, American machines, known for their durability and easy operation, successfully competed with English models. At first, these hand mowers ranged in price from $15 to $200, but with competition and technological advances, the price dropped, and in 1872, the *Michigan Farmer* remarked that hand mowers could be had "cheap and easily at every implement store." Once homeowners could readily obtain inexpensive hand lawn mowers, the old arguments about ease of lawn maintenance had a more realistic ring. Even long-time lawn advocate Rochester seedsman James Vick admitted that the availability of lawn mowers made a tremendous difference. "These Lawn Mowers are a real blessing," Vick told his catalogue recipients in 1873, "for not one in ten thousand can cut a lawn properly with a scythe, and therefore our lawns, before the introduction of these Mowers, always looked wretched."[52] (See Figure 3.4.)

Along with a well-tended, verdant lawn, most horticultural advisers considered shade and ornamental trees an essential part of embellished homegrounds, and constantly reminded readers of their importance. A. J. Downing again set the stage in 1841 by proclaiming trees the most highly ornamental, most indispensable, and most easily managed of all embellishments. The "judicious employment of trees," Downing insisted, pro-

THE CHARTER OAK LAWN MOWER

Took the First Premium at the Michigan State Agricultural Fair, September, 1875, in competition with the "Excelsior" and "Philadelphia."

---o---

BUY THE BEST.
---o---
THE NEW
"Charter Oak" Lawn Mower.
---o---
Light Draft, Easy Running, Durable, and does Splendid Work.
---o---

Its chief features are simplicity of construction, perfectness in its manufacture, ease of operation, easy way of sharpening the knives when dull, and particularly its adapting itself to slopes, undulating lawns, ridges and valleys.

We wish to call the attention of the public to the great and important improvements in this Mower, the results of which are its great durability, the power with which it moves through heavy grass, requiring less power to operate it than any other machine made; its noiseless, positive rachet; fine steel cutting edges, and various changes in its construction, which places it FAR IN ADVANCE OF ANY MOWER IN THE MARKET. The Mower *par excellence* of all others—combining all the improvements that years of experience in the manufacture could suggest. It is supplied with a three-blade, solid, Revolving Cutter, with steel knives, noiseless rachet, close covered gears, which will not clog; reversible driving rollers, allowing the Machine to turn to the right or left without injury to the sod. Patent journals and boxes to the Revolving Cutter, preventing grass winding around the cutter (a valuable and important improvement); and an iron handle, adjustable to the height convenient to the operator. Simple in construction, and not liable to get out of order. *In a word, it is the most beautiful and perfect Lawn Mower ever offered to the public.*

WE WARRANT THEM IN EVERY PARTICULAR.

☞ *Sizes and prices as follows:*

10-inch cut		$15 00
13 "		18 00
15 "	Standard Size	22 00
18 "	for Large Lawns	30 00
28 "	Pony Mower	100 00
32 "	Horse Mower	125 00

The means of sharpening a Lawn Mower when it becomes dull, is of the greatest importance to every one owning or buying a Machine. This valuable feature in the "Charter Oak" Lawn Mower, by which it can be sharpened without taking out the knives, cannot be over estimated, and can be easily and quickly accomplished by placing oil and emery on the cutters, and reversing the motion (see Directions for Use), thereby giving the knives a smooth and keen cutting edge, and keeping them perfectly true, without which no Lawn Mower will cut grass clean and do good work, experiencing none of the trouble and vexation caused by following the advice of manufacturers of other machines, of taking out the knives and grinding them on a grindstone, resulting in spoiling the knives, and having to send to the agent to buy a new one—as any practical person will at once see the impossibility of grinding on a stone by hand and keeping the edge true. A Lawn Mower that has no way of sharpening the knives, except by taking them off and grinding them on a grindstone, would be a source of expense and annoyance to its owner and the operator of it.

DIRECTIONS FOR USE.

Before using the Mower, see that the ground is free from stones or other obstructions. The Machines are sent out properly adjusted and ready for immediate use. The knives should just touch lightly when passing. If too close, the Mower will not work freely. When the Machine requires to be readjusted, the under cutter is raised toward the revolving cutter, by slightly turning the set screws at each end of the cutter bar, inside the frame, one each side, with the wrench which is supplied with the machine. Regulate the height of cut required, by the wheels in front, on either side. This is done by sliding the wheels up or down. Adjust the handle to the height of the person using the Machine. To sharpen the cutters, place oil and fine emery (No. 90) on the edges of the knives; raise the back end of the Machine and use the handle-pin as a crank, and taking off the outside cover to the gears, and placing it in the hole in the rim of outside driving wheel, and turn forward. Always clean the Machine after cutting wet grass. ALWAYS OIL THE MACHINE, *especially the rachet driving-gear, before using, with good sperm oil,* and it will last for many years. When the season for use is past, clean and oil the Machine, place it in the case, and put it away in a dry place, ready for the next season. To make a fine, thick, velvety lawn, the grass should be cut once a week, or oftener, according to season. Each part of the Mower is numbered or lettered, and made to interchange in case of breakage, and can be supplied by us or our Agents. Directions for use are furnished with each Machine.

☞ *We supply any extra parts of these Mowers.*

Figure 3.4. Lawn mower advertisements, increasingly common after the 1870s, usually emphasized ease of operation and quality of the mower. From *D. M. Ferry & Co.'s Illustrated, Descriptive and Priced Catalogue of Garden, Flower and Agricultural Seeds* (1877), 131. Original in Clarke Historical Library, Central Michigan University, Mount Pleasant, Michigan.

THE KILMARNOCK WEEPING WILLOW.

Figure 3.5. Many advisers recommended weeping trees, noting the grace and beauty they added to the dooryard. From *Second Annual Report of the Secretary of the State Pomological Society of Michigan* (Lansing, MI: W. S. George and Co., 1873), 425.

vided the "greatest alterations and improvement within the scope of Landscape Gardening." In 1855, the *Horticulturist* seconded Downing's remarks, asserting that well-placed trees and a healthy lawn were "abundantly ample, without any artificial aid, to give the greatest enjoyment to every lover of rural art."[53] Like lawns, trees had a number of attributes that made them suitable for domestic landscapes of almost any size and style. Horticultural advisers eloquently praised the beauty of deciduous trees, and the shade they offered during the warm summer months. (See Figure 3.5.) Many also recommended conifers or other evergreen trees, noting that they added diversity and life to winter landscapes. Articles on evergreens and their virtues highlighted the range of trees available along with their desirable characteristics, but most advisers, citing beauty, hardiness, rapid growth, and ease in transplanting, admitted that nothing

Figure 3.6. Some horticulturists advocated flowering shrubs, pointing out that they were easy to care for and, like this *Prunus*, supplied floral beauty year after year. From *Second Annual Report of the Secretary of the State Pomological Society of Michigan* (Lansing, MI: W. S. George and Co., 1873), 421.

PRUNUS TRILOBA.

could surpass the Norway spruce as the evergreen for the millions.[54] In addition to their beauty, tree enthusiasts suggested that both coniferous and deciduous trees were an inexpensive, manageable adornment, and that once established, provided years of pleasure with very little work.[55]

Some horticultural advisers included flowering or evergreen shrubs in their general recommendations about trees, and although admitting they were less frequently planted, argued that they shared the same attributes of easy cultivation and persistent beauty. In addition, many shrubs required little space, adding charm to small homegrounds that lacked room for larger trees. For residents unable to spend time or money on flower beds, hardy flowering shrubs offered a very practical and beautiful alternative.[56] (See Figure 3.6.) Vining plants, many horticultural advisers insisted, also made an important contribution to homeground embellish-

ment. Like trees and shrubs, they added a welcome note of graceful beauty and required very little labor to maintain. Some advisers urged homeowners to grow their favorite vines on cedar poles or trellises in the yard. Others suggested planting vines to shade an arbor, to add privacy to a porch or verandah, or as a beautiful frame for a window. In 1869, the *Horticulturist* labeled climbing plants a "help to home adornment," noted their use in hiding an "unsightly fence or wall," and their importance in softening the contours of a newly built or unornamented home with the "genial touch and presence of nature." Over the years the horticultural press recommended a variety of vines as the most desirable and interesting to grow, suggested a number of locations where they could most effectively ornament the yard, and promoted a range of trellises and arbors to support them. Most agreed, however, that vines provided a huge ornamental boost for very "little time or money."[57]

As they searched for appealing and practical ways to stimulate horticultural interest, some experts promoted flower gardens, but often tempered their enthusiasm with warnings to proceed with care. In contrast to trees and shrubs, commentators usually assumed that once the family established flower beds, women of the household were responsible for their planting and upkeep, and to some extent the cautionary tone stemmed from that understanding. In his 1833 advice manual, the *New American Gardener*, Thomas Fessenden voiced a theme that reappeared many times in the next fifty years. Farmers, he observed, did not have a taste for flowers, but "the fair daughters of America" took a very active interest in flowers and flower gardens. Since women needed to fit flower gardening into a multitude of both indoor and outdoor household tasks, many commentators cautioned their readers to begin their floricultural undertakings slowly. In their enthusiasm for beautiful embellishments, advisers claimed, women were very likely to attempt too much, and instead of well-tended flower beds, could end up with "illy-planned, badly planted" gardens. Misplaced or poorly tended garden beds often "marred" an otherwise beautiful lawn, and could far too easily detract from, rather than enhance, an embellished homegrounds.[58]

Despite the reservations and warnings, a significant portion of horticultural advice literature described garden types, suggested planting patterns, and advised readers on the best flowers to plant. Paradoxically, many articles focused on flower garden styles, which between 1850 and 1880 underwent several major changes in design and layout that increased both cost and labor. According to some flower garden proponents,

the old style of gardening, combining flowers with vegetable and fruits in a square or rectangular "large-bed" or kitchen garden, while practical, no longer served ornamental needs, and they urged their readers, if they hoped for "any degree of distinction," to establish separate flower beds. As early as 1840, the *Magazine of Horticulture* advocated flower cultivation in beds cut out of the turf or lawn, but observed that very few Americans had attempted these "cut-in" gardens. In the next several decades, advisers continually urged their readers to plant vegetables behind the house, and to highlight embellishment in the front yard with cut-in beds, often described as "little fancy figures and shapes of earth" in the lawn, planted with "precious [flowers] seeds and roots."[59]

Once homeowners decided to establish a separate flower bed, they had a number of other choices to make. In 1860, the *American Agriculturist* described two "leading systems" of garden design. The first emphasized a variety of perennials, biennials, and annuals planted in mixed beds and arranged for diversity in form and flowering time. Many considered long, narrow border gardens lining front walkways or enhancing pathways to outbuildings as the most appropriate setting for those mixed plantings. A second, far more fashionable garden design used a single plant variety and often a single color in each bed, arranging adjacent beds for a colorful, yet harmonious contrast. By mid-century, floriculture experts recommended these "massed flower gardens" much more commonly than mixed borders, and touted their "striking and brilliant effect."[60] Attractive mass displays required plants with uniform growth forms and showy flowers or colorful leaves that persisted through much of the growing season. An expanding array of annuals with long blooming spans, tender perennial "bedding plants" like verbenas or geraniums, and large showy plants like cannas, castor bean, Japan lilies or gladioli filled the bill, while long-favored perennials and biennials took a back seat. A few commentators even suggested that perennial gardens should be partially concealed from view so that their "beauties may wax and wane without disturbing anybody's fastidious taste."[61]

Over the years, horticultural advisers touted a number of variations in mass planting designs. Most commonly, residents planted their flowers in simple cut-in, geometrically shaped beds placed in various portions of the yard. Worried that such random placement of garden beds showed little unity or harmony, some writers advocated parterres in which a number of garden beds, linked by gravel or grass pathways, formed an often complex geometric design. Gardeners usually planted a single type of flower

Plan for Flower Garden.

1. Verbena (blue).
2. Verbena (white).
3. Pansies, of the fine showy sorts.
4. Portulaca (white).
5. Tom Thumb Geranium.
6. Verbena (striped).
7. Portulaca (golden).
8. Campanula Capartica, with Tree Rose in the centre.
9. The same.
10. Tom Thumb Geranium.
11. Portulaca (white).
12. Verbena (striped).
13. Portulaca (golden).
14. Pansies, of the fine showy sorts.
15. Verbena (white).
16. Verbena (blue).
17. Ageratum.
18. Heliotrope.
19. Tom Thumb Geranium.
20. Verbena, Sunset (rose).
21. Portulaca (golden).
22. Portulaca (scarlet).
23. Same as No. 8.
24. Geranium, Lucia Rosea (pink).
25. Tom Thumb Geranium.
26. Tom Thumb Geranium.
27. Geranium, Lucia Rosea (pink).
28. Portulaca (scarlet).
29. Tom Thumb Geranium.
30. Heliotrope.
31. Verbena, Sunset.
32. Portulaca (golden).
33. Ageratum.
34. Same as No. 8.
35. Vase, or Statue. If a vase, to be filled with Verbenas, Petunias, etc. If a statue, to be surrounded with a circle of Oxalis Floribunda.

Figure 3.7. Typical of many ornate mid-nineteenth-century garden designs, this "Plan for a Flower Garden" emphasized geometric patterns and designated appropriate plants. From *Horticulturist,* vol. 26, no. 295 (January 1871): 21.

within each section of the parterre, but might repeat species in various beds adding to the effectiveness of the total plan. The horticultural and agricultural press occasionally offered parterre designs along with lists of suggested plants to fill each space, but most commentators carefully noted that the successful implementation required a great deal of taste, knowledge, and time. (See Figure 3.7.) The *Michigan Farmer* advised its readers to start with small beds of verbenas, petunias, pansies or phlox, and to aim at perfecting their culture before undertaking more complex designs. The *American Agriculturist*, in turn, pointed out that every curve and pathway required labor, and unless the gardener was inclined to do a lot of weeding and trimming, simple garden beds were far more desirable.[62]

In the early 1860s, horticultural advisers announced a new kind of mass planting design called ribbon gardening, and claimed it would be a handsome addition to "even the humblest yard." Often rectangular with a curved or wavy perimeter, ribbon beds boasted "ribbons" or bands of flowers in contrasting colors. Again, bedding plants of uniform height and

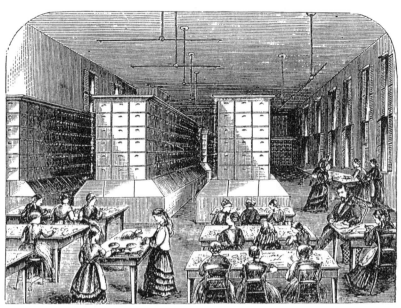

PACKING ROOM.

Figure 3.8. Carpet bedding designs emphasized varying foliage colors. From *Seventh Annual Report of the Secretary of the State Pomological Society of Michigan* (Lansing, MI: W. S. George and Co., 1878), 392.

constant bloom provided the most effective display, and commentators recommended varieties like petunias, verbenas, ageratums, and German asters. Although ribbon beds created a pleasing effect in a small space, the necessity of purchasing these plants and replacing them each year called advocates' claims that they were appropriate for households of all ranks into serious question.[63]

The same reservations applied to another widely touted bedding-out system that gained in popularity in the late 1860s. To answer complaints that massed plantings lacked diversity or presented a garish display of color, some horticultural writers advanced a planting system called carpet bedding. According to its promoters, carpet bedding used a large number of more somber or "neutral-tinted" plants to cover the surface of the garden, and then interspersed brightly colored bedding plants at intervals among the carpet plant background. Sometimes carpet beds emphasized varying foliage colors rather than flowers to create designs, and brought plants like the gray-leafed santolina or a yellow-leafed pyrethrum into prominence. The new plan offered both variety and brilliance, but kept

flower gardens from degenerating into a "mere chromatoscope." Like ribbon beds, however, carpet bedding required the purchase of large numbers of plants and a good deal of planning and labor to keep the beds in top form.[64] (See Figure 3.8.)

Through all the hoopla about bedding plants and ribbon gardens, some horticultural advisers openly expressed concerns about the functionality of the new systems, and a few continued to advocate the practical aspects of "old-fashioned" perennials and biennials. Occasionally, critics noted the costs incurred through the repeated purchase of bedding plants. In 1878, seedsman James Vick calculated that a typical carpet bed ranged in price from $15 and $30 for the plants alone. Admittedly, Mr. Vick made his living selling seeds, and he did not hesitate to point out that gardeners could save money by planting annuals from seed, but others echoed the issue he raised.[65] The amount of labor these beds required was also a concern, and in 1867, with carpet bedding at its peak, the *Horticulturist* pointed out that perennials offered the best choice for the rural flower garden where labor was likely to be at a premium. According to their promoters, perennial plants required far less effort to maintain, and yet provided a beautiful variety of flowers for people of "all classes and ages."[66]

Horticultural experts occasionally offered their readers information on specialized garden types, but again framed their comments with warnings about practicality and good taste. Descriptions of rockwork gardens or rockeries, for example, appeared sporadically in mid-nineteenth-century horticultural literature and often included designs, suitable plants, and cautions. Initially designed for the types of plants that "succeed[ed] best when planted among rocks," rockeries required the skillful and tasteful arrangement of boulders to create a naturalistic effect. Most authors advised placing a rockery in a sheltered spot where the massing of boulders could blend into the surrounding landscape, and most effectively imitate natural stone outcroppings.[67] The much more common practice of arranging a pile of rocks on an open expanse of lawn, filling the cracks with soil, and planting a variety of plants elicited unfavorable comments from a range of advisers. S. O. Knapp, a well-known Jackson, Michigan, amateur horticulturist, reflected a frequently expressed opinion when he told the Michigan Pomological Society, "Rockwork as it is generally employed is detrimental to the appearance of a place. There is no particular beauty in a rounded boulder upon a lawn, or a regular pile of them put up with mechanical exactness." Mr. Knapp contended, "it is hard to find one who

can handle rockwork to any good purpose, and it should be used with great caution."[68]

In their efforts to explore all avenues of adornment, horticultural writers sometimes offered advice on horticultural accessories like outdoor vases, garden furniture, architectural structures, or other decorative items for the homegrounds. These "unnatural" ornaments posed a problem for some advisers, because they had so consistently derided ostentatious display and promoted simple, practical homeground designs. Journals like the *American Agriculturist* eased their own concerns and tempered their readers' enthusiasm for unsuitable ornaments by offering justifications for horticultural accessories along with advice on their use. In 1861, for example, the *American Agriculturist* published an article on yard vases, and after emphasizing their long-standing proclivity for practical advice, proclaimed the "propriety" of using vases and other ornamental figures to ornament the lawn and garden. These items, the *Agriculturist* declared, added beauty and a special "air of refinement and finish" to the domestic landscape. More importantly, however, these obviously manufactured objects carried the "spirit of the [home] architecture" into the surrounding grounds, and formed a visual and thematic link between the home and its surroundings. Horticultural ornaments, in other words, helped to subdue the "violent contrast" between the precise architecture of the house and the irregular forms of the surrounding trees and shrubs.[69]

Horticultural columnists discussed a number of possible ornaments including sundials, fountains, and occasionally statuary, but many found ornamental vases the most acceptable form of adornment. They urged residents to fill the vases with well-chosen plants, to tend them carefully, and suggested their placement between the house and roadway as a pleasing accent, in the center of a flower bed, or in the densely shaded, dry areas under trees where little else could grow. Traditionally, ornamental vases were sculpted from marble in classical or very formal designs, and were far too expensive for most homeowners, but by the mid-nineteenth century, new materials increased availability and made them more interesting to homeowners of modest means. Some advisers pointed out that newly available cast-iron vases were beautiful, durable, and much less expensive than their marble counterparts. By the early 1870s, vases molded from an artificial stone offered an even thriftier alternative to cast iron. The Michigan Ornamental Stone Company, located in Detroit, described its "sand-stone" vases as far superior to cast iron in their ability to grow "any kind of plants, vines, flowers or shrubs," and insisted they be-

RUSTIC ADORNMENTS.

Figure 3.9. To avoid unseemly display, many advisers recommended "rustic" adornments like these plant stands. From *Gardener's Monthly*, vol 3, no. 5 (May 1861): 128.

longed in every front yard. At least some Lenawee County residents agreed, for in 1870, the *Adrian Times and Expositor* reported that "our citizens are decorating their grounds with the handsome vases."[70]

Despite the increasing availability of cast-iron and stonework vases in classical designs, many landscape critics had reservations about their appropriateness for most American homes. By the early 1850s, some horticultural advisers turned their attention to "rustic work" as a much more flexible and unpretentious alternative. They used this term to describe any kind of ornamentation that had a surface or shape "left in the natural condition," and eventually advocated the use of unpeeled logs, old stumps or roots, "crooked and knotty sticks," grapevines and cedar branches, and many other materials to construct a variety of rustic-work ornaments. Because homeowners, in theory, could make rustic-work ornaments themselves from natural materials, and use them in the "most unpretending grounds with good effect," rustic work fit the ideal of practical ornamentation available to all. Ironically, designers eventually created molded and cast-iron ornaments that imitated actual rustic-work pieces. A number of advisers suggested that rustic-work vases were a par-

Hands-on Horticulture 81

ticularly appropriate embellishment for the farm home, where their natural forms readily blended into rural landscapes.[71] (See Figure 3.9.)

Although their comments appeared infrequently between 1850 and 1880, a few horticultural advisers maintained that lawn furniture provided beautiful and appropriate additions to the homegrounds, especially if constructed using rustic-work techniques. They urged their readers to consider seats encircling trees, or chairs and benches made from unpeeled branches, tree stumps or other natural materials, pointing out they supplied needed resting spots, and encouraged "a happy enjoyment of the quietness and contentment of rural life." Some horticultural advisers thought that structures like arbors and summer houses offered an even more obvious retreat and a stronger incentive to enjoy the natural world; they urged residents to consider the garden "a country parlor" with appropriate amenities for entertaining friends and visitors.[72] While seemingly at odds with the usual emphasis on utility and practicality among farm families, these recommendations in the agricultural press grew out of the same reform spirit. Rural advocates hoped that the farmer and his family might recognize the beauty of their rural surroundings, and outdoor architectural structures or simple lawn furniture presented a tempting invitation to sit down and enjoy the view. For urban dwellers, those same amenities promoted at least a fleeting contact with the natural world, a most desirable counterpoise to the frenzied pace of mid-nineteenth-century urban life.

While advisers attempted to provide practical outdoor gardening advice with wide appeal, there was an inherent limit to the "democratic" nature of their recommendations. Gardening or landscape ornamentation demanded space and presumed either home ownership or access to homegrounds in a rented property. No matter how widely they promoted inexpensive, easily maintained embellishments, these requirements left some people out of the picture. Growing plants indoors, however, really could include urban dwellers or those too poor to own a home, and horticultural advocates grasped every opportunity to encourage indoor plant culture. Frequently, plant experts framed their practical comments with glowing accounts of the benefits of houseplants, arguing that they offered easy access to nature's moral influence, and in addition, encouraged a love of beauty, a desire for order and neatness, and a pleasant, cheerful disposition. The cultivation of window plants, Edwin A. Johnson suggested in his *Winter Greeneries At Home*, promoted refinement and really enabled gardeners to "cultivate [them]selves." On a more practical level, advisers

insisted that indoor plants were the most effective means of interior decoration, readily available at little cost, and an appropriate adornment for both luxurious and very unpretentious surroundings.[73]

For all their assurances that house plants or "window pets" would make a beautiful contribution to any home, horticultural advisers had several problems to address. Providing adequate light, suitable and reasonably regulated temperatures, good ventilation, and proper amounts of water were challenging at best, and as the century wore on, innovations in home heating and lighting made the problems even more serious. Commentators in journals like the *Horticulturist* or the *Gardener's Monthly* claimed that in contrast to fireplaces and wood stoves, "hot air" furnaces created a dry, intense heat that reduced humidity levels so severely that houseplants gradually "pined away and died." Ironically, while many advisers were noting the decline of indoor plant culture among prosperous families, others pointed out that the "poorer classes," lacking new furnaces, still grew "healthy luxurious plants, with innumerable flowers." In 1871, *Michigan Farmer* offered a very telling commentary on the difficulties of indoor plant culture. "We like to see flowers in the parlor when we can admire them," the *Farmer* observed, "but not when the only emotion excited is that of compassion for the sickly stilted plants."[74]

In contrast to concerns with overheated rooms during the daylight hours, horticultural advisers often warned their readers about the danger of frost as indoor temperatures dropped during the night. Some advised moving plants away from windows and into the center of the room or even taking them to the cellar on particularly cold evenings. Others recommended covering the plants themselves or lining windows with newspapers. The *Jackson Daily Citizen* even suggested that a bucket of water, placed on the plant stand at night, would warm the air enough to keep plants from freezing. Despite the precautions, prized indoor plants did freeze in mid-nineteenth-century homes, and the horticultural press offered many tips on resuscitating frozen specimens. The *Adrian Times and Expositor* typified the comments when it suggested dipping frozen plants in cold water, and then removing them to a cool dark spot until they regained their vigor.[75]

Despite these challenges, horticultural advisers were undaunted, and vowed to supply practical, simple plans that would enable their readers to grow beautiful, healthy plants indoors. As they had with outdoor ornamentals, most commentators advised a modest beginning and urged their readers to aim for a few well-chosen specimens. They carefully outlined

the basic necessities of successful indoor plant culture, advising their readers on soil mixes, the use of porous clay pots, insect control, and the importance of keeping plant leaves free from dust and grime. Advisers also offered tips on appropriate amounts of water, suggested when and how it should be applied, and encouraged readers to supply fresh air for the plants whenever possible.[76]

Providing adequate light presented the most serious problem for most indoor gardeners, and commentators frequently urged them to place plants in a south, southeast, or west window where light conditions were most favorable. Bay or "bow" windows were particularly desirable, because they exposed plants to light from several directions. Some writers pointed out that readers could transform bay windows into a miniature greenhouse or conservatory by adding a glass door or curtain between the window and the room's interior. Houseplants could then enjoy both the benefits of adequate light and a more humid atmosphere. During the summer months, many horticultural commentators argued, robust houseplants needed an airing and would benefit from outdoor light and rain. Most recommended leaving houseplants in their pots, but suggested plunging pots directly in garden beds. Others urged their readers to display favorite houseplants on porches, in pots under trees, or in another location with at least some protection from sun and wind.[77]

While cultural conditions brought on many indoor plant failures, horticultural advisers also warned homeowners to choose their houseplants judiciously. Certain plants were simply unsuited for window culture, and required the more stable and refined conditions of a greenhouse if they were to thrive. As the century wore on, commentators urged their readers to select from a number of tried and true varieties including calla lilies, geraniums, Chinese primroses, cyclamen, fuchsias, Jerusalem cherries, or bouvardias, rather than more difficult specimens like camellias or many types of roses. Advisers also pointed out that despite the interest in flowers, foliage plants were far easier to cultivate, and new introductions from Japan, California, and Australia broadened the choice significantly. In 1856, the *American Agriculturist* recommended that beginners try dracaenas, crotons, abutilons, begonias, and several other varieties that were likely to thrive under indoor conditions. In the early 1870s, *Horticulturist* editor Henry T. Williams noted a shift away from flowers to decorative foliage plants and encouraged cultivation of palms or the India rubber trees whose "thick, coriaceous" leaves could withstand the rigors of window culture year around. Because they flourished under the arid indoor condi-

Figure 3.10. Plant stands (left) minimized problems with watering, while enclosed bay window arrangements (right) helped to regulate temperature and humidity for indoor plant culture. From "Illustration of the Month," *Ladies' Floral Cabinet and Pictorial Home Companion* (November 1874): 168; and Charles W. Garfield, "Beautiful Homes. Their Influence and How to Make Them," *Second Annual Report of the Secretary of the State Pomological Society of Michigan* (Lansing, MI: W. S. George and Co., 1873), 434.

tions, cacti and other succulents were particular favorites among many advisers.[78]

To make indoor culture as easy as possible, plant experts offered a variety of suggestions on how to best house or display indoor plants. Some experienced gardeners used benches to line bay windows or shelving at various levels across windows as convenient spaces for potted plants. For even greater simplicity and to prevent water damage to floors, carpeting, or windowsills, several advocates suggested window boxes or flower tables, lined with zinc and filled with soil, so that a number of plants could be grown together in a single container. The *American Agriculturist* pointed out that such an arrangement was not only pretty, but made it much easier to water plants and to move them if necessary.[79] (See Figure 3.10.)

Others noted the increasing popularity of hanging baskets, and promoted their use as an easy yet beautiful way to grow indoor plants. In 1869, Peter Henderson pointed out that "the taste for hanging baskets" extended to "every town and hamlet throughout the land," and that baskets made from wire, earthenware, or rustic work were readily available. Their popularity, Henderson speculated, was probably due to their display at the influential 1851 London Crystal Palace Exhibition, and had quickly spread from England to the United States. According to most proponents, their popularity was completely justified. Hanging baskets provided a suitable ornament for both elegant and extremely modest homes, and individuals could easily construct them using rustic-work techniques. Most importantly, plants commonly grown in hanging baskets like ivies, hoyas, or Madeira vine usually flourished despite difficult indoor growing conditions.⁸⁰

Among the many devices for encouraging indoor plant culture, Wardian cases or ferneries received the most frequent and fervent promotion from horticultural advisers. Like hanging baskets, they came to widespread public attention during the 1851 Crystal Palace Exhibition and almost immediately caught the fancy of indoor gardeners in both England and the United States. Essentially glass-enclosed plant containers, Wardian cases provided a humid, well-regulated atmosphere for indoor plant culture. In an operating Wardian case, water vapor, lost from plant leaves in the process of transpiration, condensed against the glass of the container and eventually fell back on the soil to be reabsorbed by the plants. Accidentally discovered in 1829 by N. B. Ward, an English "amateur cultivator," Wardian cases offered a number of advantages. They protected plants from the "baneful influences of the impaired atmosphere" caused by burning coal or gas lighting systems, and once established and operating properly, the cases required very little attention. Wardian cases also expanded the types of plants available for indoor culture. Ferns, native woodland species, and other plants that thrived at low light levels and in moist, sheltered spots were particularly appropriate, and their popularity burgeoned as people established Wardian cases.⁸¹

Many horticultural advisers, however, found Wardian cases most significant as democratizers of indoor plant culture. Although ornate cases were available, indoor gardeners could purchase simple but adequate Wardian cases at very reasonable prices. According to some advisers, inventive indoor gardeners could even construct a homemade case from readily available materials. Horticultural writer Edward Sprague Rand suggested that "Florence flasks" used for salad oil might be easily and beautifully converted into a simple Wardian case. Best of all, a number of horticultural advisers agreed that Wardian cases "introduced living specimens of some of the most curious and beautiful forms of vegetation into the parlors and windows of the crowded dwellings of cities," where urban conditions like "dust and smoke and darkness" often prevented indoor plant culture.⁸² For those who staunchly promoted contact with the nat-

ural world as a panacea for the nation's ills, indoor plant culture and especially Wardian cases crossed one last and very difficult boundary. That tiny spot of green, sheltered in a glass enclosure, offered poor urban dwellers a chance at both appropriate moral guidance and respectability.

These comments on indoor plant culture, descriptions of appropriate flower beds, and detailed appraisals of lawn maintenance techniques reflect a very important reality critical to our understanding of mid-nineteenth-century horticultural advocacy. The residents of Lenawee, Jackson, and Washtenaw Counties, and their Midwestern counterparts, could turn to a myriad of widely circulated journals or more specialized resources for horticultural information. In those sources, practical advice designed to meet the needs of a diverse population was the order of the day. Many horticultural experts shared with domestic reformers a belief in the social importance of an attractive, orderly home environment. Many also adhered to a progressive, "scientific" approach to home and farm management, and continually evaluated their own "how-to" horticultural advice on the basis of efficiency and economy as well as beauty. With these two strains of thought shaping their actions, horticultural experts dedicated themselves to supplying reasonable information and setting attainable standards so that Americans of all ranks and occupations could enjoy horticultural refinements. Although from our vantage point, their claims of creating a practical horticultural transformation did not always reflect reality, their advocacy and their advice did indeed promote an expanding enthusiasm for ornamental plant culture.

For all their importance in supplying practical information, horticultural advisers required help to advance their cause. By the mid-nineteenth century, the American horticultural industry had also come into its own and could offer citizens a vast array of plants along with additional incentives and encouragement to take up plant culture. In some cases, widely read horticultural advocates were also nurserymen or seed dealers, and while they may have fervently believed in the benefits of horticultural pursuits, promoting a love of trees and flowers enhanced their own economic prospects. To further understand the blossoming of horticultural interest among Midwesterners, the practical advice of professional horticulturists must be woven into the commercial realities of the mid-nineteenth-century ornamental plant trade.

Four

Commercial Realities

Supplying the Needs

Despite its pervasiveness, all this well-intentioned horticultural advice and encouragement lacked a critical element. Midwesterners needed a source of ornamental plants if they were to cultivate and display them. By the 1850s, the once negligible American nursery and seed trades were well on their way to prominence, and stood ready to supply the growing need while working to extend horticultural demand even further. Increasingly, Midwesterners turned to the horticultural industry for inspiration and an ever-expanding array of ornamental plant materials. As skilled horticulturists and aggressive businessmen, many nurserymen and seed dealers promoted their wares with increasingly sophisticated advertisements, beautifully illustrated catalogues, and through the first-hand sales efforts of itinerant plant peddlers. While developing their markets, commercial horticulturists, like domestic reformers and horticultural advisers, imbued ornamental plants with additional layers of public meaning. Trees and flowers, their commercial promoters insisted, were available to all, easily obtained and cultivated, economically priced, and most importantly, widely recognized as signs of good taste and the latest fashion. This chapter explores traditional sources of

ornamental plants, documents the rise of America's horticultural trade and its westward expansion, and considers the challenges of reaching distant markets with horticultural merchandise. Chapter 5, in turn, investigates the promotional strategies of the mid-nineteenth-century horticultural industry.

When America's nurserymen and seed dealers attempted to drum up customers and promote interest in ornamental plants, they confronted a long tradition of informal exchange and non-commercial horticultural transactions. Many families filled their horticultural needs by visiting nearby forests or fields, and routinely transplanted native trees or shrubs and flowers to line the roadsides or to shade and embellish their front yards. In 1862, William Green, an agent of Illinois nurseryman John Kennicott, noted that the practice was widespread in newly settled regions of Wisconsin, and wrote to Kennicott warning that certain trees and shrubs would not sell there because the "woods abound with the greatest variety, including mountain ash and evergreens." Residents, Green claimed, were likely to dig them for ornamental purposes.[1]

Diaries, letters, and published reminiscences also suggest that families packed "flowering roots," cuttings, and seeds as a part of their household goods, and brought them along when establishing new homes in the Midwest. In 1874, Hortense Share, a *Ladies' Floral Cabinet* correspondent, described the practice at her family's new Minnesota homestead, noting that as soon as the farmyard was in order, the family planted "all we brought with us, shrubs, roots and bulbs."[2] Once settled, friends and family also shared cuttings, exchanged seeds, or divided up their favorite herbaceous perennials. The practice of "begging of neighbors" was so widespread that some nurserymen and their agents feared there would be little market for commonly cultivated plants, even in newly settled regions. Local newspapers like the *Tecumseh Herald* of Lenawee County encouraged plant exchanges, and even urged homeowners to grow enough trees, shrubs, and flowers to supply their less progressive neighbors. On occasion, the horticultural press advised their readers to cultivate "one bush of each kind" just to provide plant slips for friends who called.[3]

By mid-century, several unique sources of ornamental plants helped to bridge the distance between free exchange among friends and neighbors, and the purchase of trees and shrubs from commercial and sometimes distant nurserymen or seed dealers. In the early 1850s, many agricultural and horticultural journals developed seed "giveaway" programs, and used the pages of their journals to let readers know of the opportunity. These

Commercial Realities: Supplying the Needs 89

promotions certainly stimulated interest in the journal and may have boosted circulation, but the process also encouraged readers to obtain ornamental plant materials in a new way. Some newspapers, like the *Michigan Farmer* and *Moore's Rural New Yorker*, promoted seeds as "premiums" to agents and individuals who sold new subscriptions or formed subscription "clubs" among their friends or neighbors. Others offered free seeds as a reward for current subscribers.

While grander in scale than most, the *American Agriculturist* operated an extensive seed distribution program that typified the practice. Beginning in 1857, with the "undreamed of" success of a free sugar cane seed offer to subscribers, the *Agriculturist* resolved to expand the selection and to make free seed distribution a yearly event for their readers. The next year, as promised, the *Agriculturist* published a greatly extended list of seeds including Norway spruce and arborvitae seeds, over twenty-five flower varieties, and a large selection of vegetable and field crop seeds. Along with a plant list, the journal included descriptive information, plant illustrations, and simple cultural directions to aid any inexperienced gardeners who might participate in the program. Subscribers could select five types of flower seeds or three flower packets in combination with other choices from the list, and would receive enough seeds to plant a small garden. The *Agriculturist* urged readers to send in a stamped self-addressed envelope for their share of the free seeds.[4]

Over the years, these promotional offers were immensely popular. By 1859, the *Agriculturist* reported the distribution of a half million seed parcels, and routinely noted increasing demand until the general free distribution program ended in 1864.[5] Rochester seedsman James Vick had introduced a similar program while serving as editor of the *Genesee Farmer* in the late 1850s, and as he reminisced about the program's success years later, noted that it developed horticultural interest by "scattering" thousands of packages of "choice annual flower seeds" across the nation. The fact that the seeds were free boosted the popularity of the distribution programs, and *Moore's Rural New Yorker* suggested that the press actually performed a service for those who could not afford plants from commercial sources.[6]

Although the claim that they provided a public service for the poor stretched the point a bit, the practice of offering free seeds provided a new source of ornamental plants, and encouraged many individuals with little gardening expertise to develop at least some horticultural interest. The process also promoted habits that might eventually lead to commer-

cial exchanges, and familiarized individuals with obtaining plant materials from distant sources. The *American Agriculturist*'s annotated and illustrated list of free plants, for example, resembled a typical seedsman's catalogue, and offered readers a glimpse of resources available in even larger quantities from commercial seed houses. Like commercial catalogues, the *Agriculturist* presented plant selections in very enticing terms, calling attention to plants like the double balsam, a "choice annual, desirable in the smallest collection of flowers," morning glories, "covered with large flowers of white, blue, purple and varied color," or cypress vine, "admirably adapted for the conical trellis, or training upon strings arranged as fancy may dictate."[7]

The *American Agriculturist* also assumed that its readers, long accustomed to direct local transactions, needed help with ordering plants. The journal routinely featured an illustration of a properly addressed and stamped envelope, and accompanied it with additional detailed written instructions. Hoping to facilitate order processing, editors begged their readers to forego detailed correspondence in favor of simply marking the number of the seed types they wanted on the outside of the envelope.[8] For the uninitiated, the *Agriculturist*'s careful instructions simplified a system that might seem both foreign and complex. Once the pattern of ordering was established, obtaining plants from a distant source was no more formidable than trading with neighbors, and had the added allure of a free gift arriving by mail or express.

In addition to agricultural and horticultural press promotions, the federal government provided a source of free ornamental plants and encouraged individuals to request them. Beginning in 1819, with a Treasury Department circular letter urging American Consuls to send useful plants back to the United States for propagation, testing, and distribution, the federal government had taken an active, if somewhat erratic, interest in horticulture. In 1827, President John Quincy Adams issued a second directive through the Treasury Department, again requesting that U.S. Consuls and the captains of Navy vessels identify and send home seeds, roots, or cuttings of plants that might be successfully cultivated in the United States. While these requests resulted in a few new trials and introductions, government horticultural interest found its most consistent expression through the agricultural division of the Patent Office.[9]

In 1839, Congress made its first appropriation to the Patent Office, designating $1,000 for the collection of agricultural statistics and for the introduction and distribution of seeds and plants. Oliver Ellsworth, a prominent Indiana landowner, headed the Patent Office at the time, and carried out the initial horticultural mandate with enthusiasm. Interest continued through the tenure of several commissioners, the transferal of the Patent Office from the Department of State to the Department of the Interior in 1849, and even after the formation of the Department of Agriculture in 1862.[10] Field crops and vegetables often dominated the distribution program, but in addition, the government sent out huge numbers of flower and other ornamental plant seeds. The program, Commissioners persistently maintained, was designed to awaken interest in agricultural and horticultural improvement, collect horticultural information, and introduce improved plant varieties across the nation.[11] Annual reports regularly documented the extent of the program. In 1861, for example, the Department distributed nearly two and a half million seed packets including 154 vegetable seed varieties and 230 types of flower seeds. In 1863, the number of "parcels" dropped to 950,000, and continued to fluctuate over the next decade. In 1871, the Commissioner reported the distribution of 183,259 packages of flower seeds including 54 varieties along with 1,571 packets of tree seeds. In 1874, Agricultural Commissioner Frederick Watts noted the steady increase in the numbers of seeds distributed during his tenure, and described the efforts of the Department to conduct the program efficiently and economically.[12]

The government also ventured into ornamental plant distribution through its experimental gardens program. As early as 1842, the Patent Office constructed a government greenhouse for preserving botanical collections sent to the Department, and by 1857 also operated a "propagating garden" for cultivating introduced species. In 1862, Agricultural Commissioner Isaac Newton appointed William Saunders, a well-respected horticulturist, to head the government's "propagating and experimental gardens," and under his able leadership, the Department was soon distributing thousands of "vines, bulbs, cuttings, and plants" to both members of Congress, and agricultural and horticultural societies.[13] Although the experimental gardens and plant distribution program added a new source of ornamental plants, it never reached the scale of the Department's seed initiative.

The Department distributed seeds in a variety of ways. In order to compensate for lack of funds, Ellsworth had initially taken advantage of the Congressional postal franking privilege to distribute seeds free of postage. Gradually the seed distribution program evolved into a method for rewarding Congressional supporters regardless of their horticultural or

agricultural interests. The Department also sent seeds to agricultural or horticultural societies for distribution among their members, to "statistical correspondents," and to individuals who wrote requesting seeds. The 1864 distribution breakdown illustrated the general pattern of seed dispersal, with approximately five hundred thousand packets dispersed by members of Congress, two hundred thousand to agricultural and horticultural societies, thirty-five thousand to statistical correspondents, and about three hundred thousand to individuals who "applied directly for them."[14]

Government officials expected seed recipients to plant the seeds, harvest the crop, save seeds for next year's planting, give some to neighbors, and report the results of their harvest to authorities in Washington. Recipients, Commissioner of Patents Charles Mason declared in 1856, should consider themselves "agents ... for the public benefit," and conduct careful experiments with the seeds they received. While the Department documented some legitimate reports from those who tried the seeds, they often expressed disappointment in the experimental aspects of their program. Occasionally, the Commissioner suggested that reports on the quality of seeds and their success in a particular locale lacked validity simply because of "inexperience, or neglect or want of skill in their culture" on the part of those who received them. Many seemingly responsible recipients often failed to report their findings, or even worse, used up all the seeds, leaving none for additional plantings in future years or for distribution among friends and neighbors.[15]

In addition to negligent recipients, the seed distribution program suffered from wavering government support. The Patent Office Commissioner or the Commissioner of Agriculture often had to stretch very limited resources over a wide range of tasks, and experienced frequent appropriation cutbacks and funding uncertainties. Even after the formation of the Department of Agriculture, agricultural officials lacked adequate resources, and working conditions for the Department were difficult at best. During the Civil War, Mrs. L. B. Adams, the former domestic department editor for the *Michigan Farmer* served as a clerk in the Agricultural Department, and in 1864, reported on a Congressional visit to the Department and its seed packing facilities for the *Detroit Advertiser and Tribune*. Adams found the Department in an "old, low building situated in the immediate vicinity of stables and filthy back alleys." Inside, male workers sorted and packed seeds in a crowded room functioning as both work and storage space. In an adjoining narrow room, forty women,

Commercial Realities: Supplying the Needs   93

crowded together, "bent over the work of filling and stitching up the piles of little seed bags" that would soon be distributed to "country homes far away." Adams pointed out that although workers labored under adverse conditions, the packages of free seeds meant a great deal to the thousands of homeowners in the "distant and sparsely settled West" where the government's beneficence enriched newly established yards and gardens.[16]

Comments from recipients suggest that they did, in fact, appreciate the government largesse. In 1851, George W. Bowlsby, a farmer from Lenawee County, wrote to the Patent Office reporting on local agricultural conditions in the region but ended his letter with the comment that he would receive any seeds the Office had for distribution with "extreme pleasure." Several years later, O. H. Kelley, the Corresponding Secretary of the Northwood Farmers' Club in Wright County, Minnesota, described how the need to distribute Patent Office seeds actually encouraged local farmers to form an efficient, active educational and social organization. In 1870, B. G. Buell, a Cass County, Michigan, farmer corroborated Kelley's point when he noted that the local Volinia Township Farmers' Club received a shipment of government seeds, and that an unusually large turnout of members assembled to divide up the shipment. Buell's observations suggested that, at least in southwest Michigan, local farmers were curious about what the government had sent out, and willing to take time from their busy day to collect their seed parcels.[17]

Despite support of recipients and its long-term potential for promoting a national horticultural marketplace, the agricultural and horticultural press had little sympathy for the Department's seed distribution program. Almost from its inception, critics launched a persistent campaign to close down the government "seed store" and use agricultural resources in a more productive way. They accused the program of a variety of wrongs. Some disliked the types of seeds distributed, arguing that they were readily available from most seedsmen or had been widely dispersed long before the government made them available. Others claimed that the quality of the seed was poor, that few germinated, and some were even infested with "chinch bugs" or other pests. Of all the complaints, however, one of the most persistent revolved around how the government procured its seeds. In 1858, the *American Agriculturist* suggested that seed purchase represented a "jobbing affair, for the benefit of a few favorites," and decried not only the poor quality of the seed but the seeming favoritism in government purchases. The press was also outraged at the tendency to procure seeds from European seedsmen when, they pointed

out, the varieties were readily available at home.[18]

Rhetoric aside, much of this criticism of the "government seed store" had its origins in the increasingly competitive horticultural marketplace. As the number of American nurserymen and seed dealers grew, they struggled to carve out or maintain their share of the horticultural trade, and the free distribution of government seeds seemed threatening to an already slender profit margin.[19] Concerns about competition were well-founded, for by the mid-nineteenth century, the few long-established American commercial nurseries and seed houses felt pressure from many sources, especially as new firms recognized the business potential and entered the ornamental plant market. The numbers of individuals calling themselves nurserymen, seedsmen, or florists provides one indication of just how rapidly the trade expanded. In the 1850 population census, 8,479 Americans labeled themselves gardeners, florists, or nurserymen. By 1860, the number had jumped to 21,788, and in 1870, reached 32,520. In 1880, 56,032 Americans listed their occupation as florists, gardeners, nurserymen or vinegrowers.[20] Nineteenth-century statistics on the actual number of ornamental plant nurseries, florists selling greenhouse plants, and seed firms were notoriously vague, but several reports provide a general idea of their numbers and their rapid increase at mid-century. In 1859, the Illinois State Agricultural Society published an admittedly incomplete list of the state's nurseries and seedsmen. Of the forty-six firms that reported selling ornamental plants, all were established after 1846, and most began operation in the early 1850s. In a second list of "nurserymen, fruit-growers, agents, dealers and publishers," assembled in 1867, the Illinois State Horticultural Society recorded 331 enterprises selling ornamental plants or seeds nationwide, and 61 firms in Illinois.[21]

In the late 1860s and 1870s, agricultural book publisher Orange Judd of New York brought out the *American Horticultural Annual*, a self-proclaimed "year-book of horticultural progress." The *Annual* attempted to keep up with commercial enterprises, but admitted that "all attempts to make complete lists of dealers in horticultural stock have been unsatisfactory," and finally resorted to listing those firms who advertised in horticultural journals or who sent in a catalogue. Even with that limited sample, the numbers were impressive. In 1871, the *Annual* recorded 855 American nurseries, seedsmen, or florists who sold ornamental plants, and also listed 37 European and one Japanese nursery who advertised widely in the United States.[22]

Embedded within those numbers and contributing to their ambiguity

Commercial Realities: Supplying the Needs 95

was an even more complex story of industry growth on a national, regional, and local level. Through much of the eighteenth century the horticultural trade centered along the Atlantic coast or around large eastern cities, with only a few nurseries available to supply public wants. Early nurserymen usually selected sites with favorable soil and climactic conditions to enhance plant culture, but in addition, needed large population centers to support their business. Before transportation advances like canals and railroads eased shipping problems, access to natural waterways also contributed to site selection.[23] Established in Flushing, Long Island, in 1737, the Prince Nursery was one of the first and most prominent of these early firms, and almost from its founding, sold a varied selection of fruit trees and ornamental plants. Beginning in the 1770s, the Prince family published nursery catalogues, and eventually turned to newspaper advertising, hoping to expand horticultural interest and sales beyond their limited local market. Although it gradually lost its competitive edge, the Prince Nursery remained in business until the mid-1860s. Further south, French and German immigrants established several well-known nurseries and seed houses in the last decades of the eighteenth century. Peter Bellet, an itinerant seed and bulb peddler, operated a nursery that supplied grafted fruit trees, ornamental plants, bulbs, and seeds at Williamsburg, Virginia, while Maximillian Heuisler, a German nurseryman, opened a similar establishment in Baltimore beginning in the early 1790s. The Richmond Nurseries founded by Franklin Davis, and Berckmans' Nursery in Augusta, Georgia, also flourished at the turn of the century.[24]

In the early decades of the nineteenth century, several more nurserymen supplied ornamental plants to a gradually increasing clientele. Near Boston, the Winship Nursery began operations in 1816. Thomas J. Hogg and Sons opened a nursery in New York City in 1822, and by the 1860s had gained renown as an importer of Asiatic plants. In Newburgh, New York, the Downing family opened a small nursery in 1801, and vastly expanded its offerings over the next several decades. Parsons and Company, established on Long Island in 1838, operated a successful nursery for much of the nineteenth century. The firm specialized in native trees and shrubs and in the importation and distribution of Japanese plants. In the early 1830s, Robert Buist entered the nursery and florist trade in Philadelphia, and his Rosedale Nurseries soon attained international renown for the cultivation of roses, camellias, and greenhouse plants. Buist was especially well-known for the culture of verbenas and the commercial introduction of the poinsettia. A little further west, David Thomas began a

fruit tree nursery near Cayuga Lake, New York in the early 1830s.[25]

Many of the prominent early seed firms also concentrated around large eastern cities. In 1784, for example, David Landreth established a seed importation business in Philadelphia, and under the guidance of his son, the firm remained a major player in the seed industry for much of the nineteenth century. At mid-century, the Landreths became the first American firm to grow their own seeds, and over the years developed a sound reputation for quality and reliability. Another early Philadelphia seed house operated by Bernard M'Mahon specialized in native American tree, shrub, and flower seeds. Henry Dreer, also of Philadelphia, opened a seed business in 1838, and eventually emphasized the mail-order distribution of flower seeds. Grant Thorburn, a Scottish immigrant, operated a seed business in New York City in the early nineteenth century, and although the firm struggled for many years, it eventually became a major seed supplier with a national following. Also widely known for his dahlias, Thorburn brought their culture to its "highest perfection." Horticulturist and author Joseph Breck operated a seed store in Boston. The firm opened in 1836, and continued in business throughout the nineteenth century. Another long-established seed firm, B. K. Bliss of Springfield, Massachusetts, and New York City, an early promoter of the mail-order seed business, produced the first major seed catalogue with colored illustrations in 1853.[26]

The Shaker seed industry was an important exception to the concentration of early seed companies around eastern cities. Beginning in the late 1700s, Shaker settlements at Watervliet and New Lebanon, New York, grew seeds commercially and distributed them through a complex network of local merchants. The Shakers pioneered a number of marketing techniques. Their representatives delivered quantities of individual seed packets to village and city stores, sold them on commission, and then returned to pick up any unsold merchandise at the end of the growing season. The process assured customers that Shaker seeds were always fresh and viable, and they developed a reputation for quality. Although widely distributed, Shaker seeds, at mid-century, included few ornamental plants among their many vegetable and herb offerings.[27]

While many eastern nurseries remained prominent through much of the nineteenth century, the center of the nursery trade gradually moved westward and by the late 1850s, was well established in Rochester, New York. The region offered an ideal location both in terms of plant culture, and for distribution into the newly settled Midwest. The climate, tempered by Lake Ontario, provided a long and relatively mild growing season that complemented the area's fertile soils and plentiful rainfall. The Erie Canal, access to Great Lakes steamers through Lake Ontario ports, and, by the 1840s, a well-developed rail network enabled nurserymen to quickly obtain new plant material and to move their trees and shrubs into western markets much more rapidly than their eastern competitors. These transportation advances, the increased use of advertising, and the proliferation of well-illustrated nursery catalogues meant that nurserymen were no longer tied to large cities to insure a customer base.[28]

Rochester establishments like the Monroe Garden and Nursery began to cultivate fruit and ornamental trees, shrubs, and flowers in the early 1830s, and the proprietor, Asa Rowe, even published a catalogue of his offerings in an 1833 *Genesee Farmer*. George Ellwanger and Patrick Barry established their Mount Hope Nursery in the early 1840s, and by 1860 had turned their organization into the largest nursery in the world. The company aimed at serving the horticultural needs of the West, and as the proprietors pursued their goals, they increased the size of the nursery from forty-three acres in the 1840s to 650 acres at its peak in 1871. Ellwanger and Barry prided themselves on producing more trees and shrubs than any other two American nurseries combined. In addition to Ellwanger and Barry, at least eighteen major commercial nurseries cultivated thousands of acres of fruit and ornamental trees, shrubs, and a variety of flowers in and around Rochester by the mid-1860s. Firms like A. Frost's Genesee Valley Nurseries, the Rochester Wholesale Nurseries, and the 350-acre Old Rochester Nursery soon enjoyed a nationwide reputation.[29]

By mid-century, Rochester was also well on its way to becoming the center of the nation's seed industry. In the late 1830s, Rochester resident C. F. Crosman began experimenting with seed production, and by 1880 Crosman Seeds was one of the largest seed companies in the world. The Briggs family opened a Rochester seed house in 1849, and over the years worked hard to provide reliable and economical flower and vegetable seeds. In 1855, horticultural editor, publisher, and writer James Vick turned his formidable talents to seed production and soon transformed a small seed garden on Rochester's Union Street into one of the nation's best-known and respected seed companies. The growth of the Vick Seed Company was phenomenal. By 1867, the firm filled nearly a thousand orders each day, and in 1870, the Rochester postmaster noted that letters for James Vick comprised about one-fourth of those received at the Rochester post office. The firm advertised in over three thousand news-

papers nationwide, and by the early 1870s, sent out over two hundred thousand catalogues yearly. Even with Vick's success, the volume of trade continued to entice others into the seed business. In 1879, Joseph Harris, the proprietor of another very large and famous seed house, began his business in Rochester. Several other firms, including William H. Reid Company and the Mandeville-King Seed House, also flourished in Rochester beginning in the 1870s.[30]

Many of the well-established eastern firms and the new Rochester nurserymen and seed dealers cultivated a national clientele, but regional nurseries serving a more localized market also emerged in the mid-nineteenth century. Like their competitors to the east, these firms selected locations to take advantage of favorable growing conditions, transportation networks, and a newly expanded market as settlers moved west and established orchards and ornamented homegrounds. Regional nurseries often advertised in local newspapers and agricultural journals, sometimes specialized in plants especially suited to the region, and usually enjoyed sales in a somewhat more restricted area than their national competitors. By the mid-nineteenth century, Midwesterners could choose from a number of prominent regional nurseries, with especially high concentrations in central Illinois and around several major Ohio cities. John Kennicott, a physician and horticultural editor of the *Prairie Farmer*, started one of the first commercial nurseries in Illinois in the early 1840s. Located northwest of Chicago, Kennicott called his home and nursery "The Grove," and from the beginning sold a variety of ornamental trees, shrubs, flowers, and bulbous plants. Kennicott served customers throughout Illinois and in newly settled regions further west, with many fewer customers in Michigan or other adjoining states to the east.[31]

By the early 1870s, Bloomington, in central Illinois, had taken over as an important regional center for the nursery trade, boasting over sixteen ornamental plant establishments. One of the most successful nurserymen, F. K. Phoenix opened his Bloomington Nursery in 1852, after gaining experience in the nursery business in Delavan, Wisconsin, in the 1840s. By 1860, Phoenix had 120 acres under cultivation, and offered a large selection of fruit trees, greenhouse plants, evergreens, and other ornamentals. The firm bragged that it would take advantage of "unexcelled" rail connections for the prompt delivery of plants, and promoted their collection as "Western Trees for Western Planters." Phoenix regularly documented the nursery's growth in his advertisements, boasting of four hundred acres and ten greenhouses in 1868, and by 1872, six hun-

Commercial Realities: Supplying the Needs 99

dred acres and twelve greenhouses.³² Other Illinois nurserymen made their mark in the trade by specializing in particular varieties. Robert Douglas of Waukegan, Illinois, for example, refined the propagation of coniferous trees from seed to the point that they could be efficiently and economically produced. Douglas's nursery was soon known as one of the major and most reliable sources of conifers in the nation.³³

Ohio also supported a number of well-known regional establishments. M. B. Bateham, a horticulturist, editor, and seed store proprietor from Rochester, New York, opened a widely advertised nursery in Columbus, Ohio, in the mid-1850s and continued in operation until 1864. The Columbus Nursery's 1859 wholesale catalogue noted that the nursery covered nearly one hundred acres and that the stock of trees "was the most extensive and complete ever grown in the West." Bateham sold evergreens, ornamental trees and shrubs, dahlias, roses, greenhouse and bedding plants, and fruit trees especially suited for culture in southern Ohio and northern Kentucky. Elsewhere in Ohio, Storrs, Harrison and Company operated a large and very successful nursery in Painesville, just northeast of Cleveland. The Toledo area also supported a number of prominent nurserymen. Among them, the 170-acre Fahnstock Nursery supplied ornamental trees adapted to western conditions, but was particularly renowned for the culture of roses. In the southern part of the state, a number of large nurseries flourished around Cincinnati.³⁴

By mid-century, several Michigan nurseries also operated on a regional scale with most concentrated in and around Detroit. One of the first and most extensive, Hubbard and Davis began operation in Oakland County in 1836, but in 1851, consolidated their business in Detroit. The firm published a catalogue for many years, and in the 1846 edition listed over sixty ornamental trees and shrubs, a variety of vines and creepers, roses, peonies, iris, and seventy-three other herbaceous perennial flowering plants. The 1853 catalogue pointed out that with the completion of the railroad, the firm enjoyed a direct rail link with Chicago, and could supply the horticultural needs of western families much more readily than eastern nurseries. In 1860, Hubbard and Davis published an abridged descriptive catalogue in the *Michigan Farmer*, and the extensive listing of ornamental plants indicated how the firm had grown. The nursery advertised eighty-one rose varieties, over a dozen Chinese chrysanthemums, and over thirty-five varieties of dahlias. Particularly well-known for verbena and fuchsia culture, Hubbard and Davis also offered their customers many other ornamental trees, shrubs, and flowers.³⁵ (See

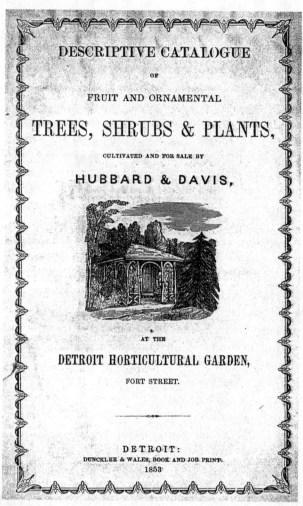

Figure 4.1. Hubbard and Davis Catalogue (1853). Original in Clarke Historical Library, Central Michigan University, Mount Pleasant, Michigan.

Figure 4.1.)

William Adair, a second Detroit nurseryman with a regional clientele, began operation in 1844, and for many years specialized in greenhouse plants. In 1849, the *Michigan Farmer* found his facilities to be the "most spacious and well stocked of any in the region," noted that he grew over twenty-five hundred potted plants, and marveled at the two hundred varieties of roses and fifty different geraniums in Mr. Adair's collection. In

Commercial Realities: Supplying the Needs 101

Figure 4.2. William Adair Catalogue (1872). Original in Clarke Historical Library, Central Michigan University, Mount Pleasant, Michigan.

1860, the *Farmer* observed the construction of new greenhouses and remarked that Adair's "business is growing." Adair regularly advertised fruit and ornamental trees, flowering shrubs, roses, dahlias, and many other plants in the *Michigan Farmer*, and urged readers to write for a free catalogue. In 1871, citing the "earnest solicitation of many friends," and perhaps feeling the competition from the expanding seed trade, Adair moved into the seed business and began publication of his *Descriptive*

Catalogue of Choice and Select Flower, Vegetable, and Agricultural Seeds. The 1872 catalogue included both ornamental plants and seeds and, sustaining the long-standing emphasis on greenhouse plants, listed nearly two hundred varieties for indoor culture.[36] (See Figure 4.2.) A third Detroit company, the Elmwood Garden and Nursery, offered William Adair some local competition, for according to the 1853 *Michigan Farmer*, Elmwood cultivated the most "numerous collections of green-house and ornamental plants in the state." The proprietor, E. G. Mixer, openly expressed his desire to out-compete eastern firms, and over the years advertised a wide range of trees, shrubs, and flowers adapted to Midwestern climatic conditions.[37]

In Monroe, just south of Detroit, I. E. Ilgenfritz operated the well-known Monroe Nursery. Founded in 1847, the nursery increased both its acreage and ornamental plant stock until, in 1857, the firm advertised over one million trees for sale, and invited orders from other nurserymen and wholesale dealers. In 1863, the *Michigan Farmer* reported the nursery was "as large as any within the state," and hoped that soon Monroe might successfully compete with Rochester as a regional nursery center. The area never reached that scale, but when members of the Michigan State Pomological Society visited the Monroe Nursery in 1874, they reported that the establishment extended over three hundred acres, included extensive facilities for grafting and "heeling in" trees in preparation for shipping, a shipping box manufacturing area, and over nine acres for the display of ornamental trees and shrubs. Impressed with the operation, committee members urged Michigan tree planters to "give these gentlemen a trial."[38]

Detroit was also home to a prominent and rapidly growing seed company. In 1854, a Rochester, New York, seedsman named M. T. Gardner moved to Detroit, and sensing a business opportunity, took over a small local seed store. Several years later, Gardner, along with Detroit businessman E. F. Church, and another Rochester native, Dexter Mason Ferry, established M. T. Gardner and Company, and went about building an innovative seed distribution system. From the beginning, the company produced attractively packaged seeds, but also supplied a "commission box" that provided storekeepers with a simple, convenient, and eye-catching way to display the seed packets. Representatives of the company distributed commission boxes to stores and businesses in the spring and returned in the fall to pick up the box and any unsold seeds. The idea caught on and from about five hundred boxes distributed in 1856,

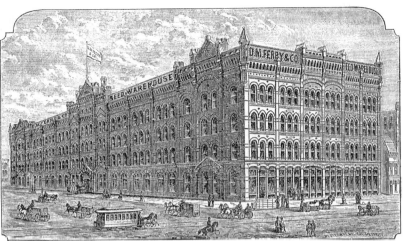

D. M. FERRY & CO'S NEW MAMMOTH SEED STORE AND WAREHOUSE.

Figure 4.3. D. M.. Ferry built this five-acre seed store and warehouse in downtown Detroit in 1879. From *D. M. Ferry & Co. Seed Annual* (1881), 2. Original in Clarke Historical Library, Central Michigan University, Mount Pleasant, Michigan.

the company sent out over two thousand by 1862. After several reorganizations, the firm emerged as D. M. Ferry and Company in 1867, and entered a long period of growth and prosperity. The Ferry Company continued to base its business on commission box sales, but also published beautifully illustrated catalogues and advertised widely in local, regional, and national publications.[39] (See Figure 4.3.)

The expansion of commercial horticulture on both a nationwide and regional scale offered Midwesterners many opportunities for purchasing ornamental plants, but Jackson, Lenawee, and Washtenaw County residents could also find trees, shrubs, and flowers closer to home. Throughout the Midwest, small local nurseries sprang up almost as soon as settlers moved into a region. Some enterprises developed through the efforts of ardent horticulturists who, as they moved west, brought along fruit or ornamental plants to supply their own needs. Once established, they then helped their neighbors plant an orchard or embellish their dooryards, but soon went out of business as initial demand died away. Other local nurserymen intended to make a living at the nursery trade, and sometimes agreed to act as agents for national or regional firms or very actively developed and disseminated their own nursery stock. To increase their customer base, these horticulturists sometimes offered to deliver trees or

actually peddled ornamental plants and fruit trees throughout the neighboring countryside by wagon.[40] Some individuals saw the nursery trade as a lucrative opportunity, and although they had very little horticultural knowledge or experience, hoped to cash in on the bustling ornamental plant trade. In a typical example of this entrepreneurial spirit, R. O. Thompson of Nebraska Territory, wrote to well-known Illinois nurseryman John Kennicott, offering to trade Kennicott his supply of "boots and shoes" for nursery stock. Another correspondent, Andrew L. Siler of Utah Territory, told Kennicott that he had made his living as a cabinet maker, but had "become heartily tired of jack planes and rip saws" and thought he would "find the labor of fruit and flower growing certainly more healthy and pleasant and in all probability equally as remunerative." Siler lacked not only horticultural experience, but capital, and asked Kennicott to send seeds and cuttings so that he might begin his nursery business with little expense.[41]

Like their counterparts throughout the Midwest, the local nurserymen of Lenawee, Jackson, and Washtenaw Counties started their establishments under a variety of circumstances. E. D. Lay of Washtenaw County was perhaps the first to establish a commercial nursery in the region, and in the spring of 1833, visited Michigan Territory with that purpose in mind. After selecting the "plains" near Ypsilanti in Washtenaw County, Lay and his brother Z. K. Lay brought a load of fruit and ornamental trees from Asa Rowe's Monroe Gardens in Rochester, New York, as the basis for their nursery. According to Lay, the stock included nearly twenty-five thousand plants, among them a variety of fruit trees, quinces, strawberries, grapes, and a "large assortment of ornamental shrubs, evergreens, roses, peonies, [and] herbaceous perennial flowering plants." In the fall of 1834, the Lay brothers built a small greenhouse, probably the first in Michigan, and just two years later, expanded the facility, quickly filling it with a "choice collection of tropical plants." By 1847, the Lays cultivated a fourteen-acre nursery and continued to supply an array of fruit trees, ornamental trees, shrubs, vines and creepers, bulbous flowers, perennials, and many greenhouse plants for indoor culture. The firm also produced a catalogue and offered to ship plants to distant customers. In addition, other merchants in the region served as outlets for trees and flowers from the Ypsilanti Nursery. In 1844, for example, the residents of adjoining Jackson County could obtain a good selection of Lay's trees, shrubs, and flowers "for gardens and Door yards ... by calling at their chandler's shop"

in the city of Jackson.[42]

E. D. Lay and his brother established their nursery at a time when settlers in the region were still clearing land, developing farms, and constructing their first homes, and before railroads linked the more populous eastern counties with other parts of southwestern Michigan. They offered not only the practical fruit trees, but a large selection of ornamentals and a greenhouse full of tropical plants. The strategy paid off, and even though the region was just emerging as a settled community, they found a substantial market for horticultural embellishment. In 1852, the *Michigan Farmer* reported, "We are glad to see this veteran pioneer nurseryman of our State prospering. His grounds are now adorned with a fine house, and we are told that he shows other unequivocal signs of having reaped the reward of his toil." Several years later, E. D. Lay retired from the nursery business, but remained an active promoter of horticultural and agricultural advancement in the area. He joined the Detroit Horticultural Society as a founding member in 1846, actively supported the State Agricultural Society from its founding in 1849, and over the years sometimes entered his prized plants in the annual Washtenaw County Fair.[43]

Not all early nurserymen in the three-county area enjoyed the success of the Ypsilanti Nursery. Horticulturist S. B. Noble established the Ann Arbor Garden and Nursery in the late 1830s, and Ann Arbor's *Michigan State Journal* kept a watchful eye on his progress. After a visit to the nursery in 1838, the *Journal* noted that Noble offered a variety of greenhouse plants for sale, and reported that their "editorial olfactories" came away "delightfully refreshed with the mingled fragrance of roses, lilies etc. in full bloom." The editor urged local residents, particularly the ladies, to patronize Mr. Noble. Several years later, the *Journal* published yet another description of Mr. Noble's greenhouses, noting his large selection of dahlias, geraniums, China roses, and lilies, and again urged readers to support this "desirable, but always precarious enterprise." Although Noble's Ann Arbor Nursery may have remained economically marginal, it did offer local residents a very diverse collection of ornamental plants. Like E. D. Lay, Noble actively promoted horticultural causes, and for many years served as horticultural editor of the *Michigan Farmer*.[44]

In Lenawee County, Benjamin W. Steere operated one of the region's earliest ornamental plant nurseries, continuing a family tradition of horticultural interest. Benjamin's father, David Steere, cultivated a variety of fruit trees at his home in Ohio, and when the family moved to Michigan

Figure 4.4. H. F. Thomas promoted his newly established nursery with advertisements emphasizing his honesty and personal service. From *Combination Atlas Map of Jackson County, Michigan* (Chicago: Everts and Stewart, 1874), 135; and Jackson Weekly Citizen (13 February 1872); and (18 February 1873).

in 1833, quickly set about establishing one of the first orchards in Lenawee County. Benjamin Steere took an interest in his father's work, helped him graft fruit trees, and after a stint at school teaching in Ohio, returned to Michigan to establish a nursery business. After several moves, the Steeres opened their nursery near the town of Adrian in 1850. To broaden his offerings, Steere obtained nursery stock from the Lays' Ypsilanti Nursery, but eventually purchased a variety of ornamental shrubs, trees, evergreens, and flowers from Ellwanger and Barry's huge commercial nursery in Rochester, New York. Benjamin Steere acknowledged that he went into the nursery business to make a living, but his zeal for horticulture and his interest in furnishing his customers the "very best of everything" encouraged him to experiment with a rich array of ornamentals. Like both E. D. Lay and Noble, Benjamin Steere took an active interest in Michigan's horticultural advancement. He participated in the Adrian Horticultural Society, frequently entered plants and flowers in the Lenawee County Fair, and often judged flower entries at the State Agricultural Fair.[45]

Between 1850 and 1880, a number of new local nurseries, seed dealers, and florist's establishments opened in the three-county area. In Jackson County, for example, Harwood and Dunning established the Jackson Nursery in the early 1850s, and cultivated fruit trees as well as ornamental trees and shrubs. They specialized in Norway spruce, Scotch pine, and American arborvitae, and, as the *Michigan Farmer* reported in the late 1850s, were busily building the foundation of an outstanding nursery. In nearby Napoleon, Mr. D. D. Tooker operated a local vegetable and flower seed business, and offered to fill orders by mail. In the early 1870s, H. F. Thomas opened his Oak Grove Nursery near the city of Jackson, selling plants from other nurseries at reduced prices until his newly planted trees and shrubs attained an adequate size. The year Thomas began operations, the *Jackson City Directory* listed just one other nursery in the city. Several years later, three firms had joined Mr. Thomas in the trade, and by 1883, Thomas remained in business with six competitors.[46] (See Figure 4.4.)

By the 1870s, several local firms competed for the ornamental plant trade in Lenawee County. One of the most prominent, Loud and Trask of Adrian, advertised widely in local newspapers, actively participated in the county fair or in the Adrian Horticultural Society, and over the years, occasionally established branch agencies to promote the ornamental plant trade in outlying Lenawee County towns and villages. D. M. Edmiston also operated a nursery and fruit farm near Adrian in the early 1870s. Edmiston specialized in "Adrian grown fruit trees," but in addition, cultivated some evergreens and other ornamentals.[47] In Washtenaw County, James Toms offered foliage and bedding plants, a variety of an-

Figure 4.5. Loud and Trask advertisement. *Adrian Times and Expositor* (26 March 1870).

nuals, verbenas, dahlias, and many other varieties at his Sion Greenhouse. Toms opened his business in 1858, and by the 1870s used prominent and very focused advertising to insure its continued success. In local newspapers such as the *Ypsilanti Commercial,* Toms prefaced his advertisements with enticing statements like "It Concerns Ladies Most." He then encouraged women customers by suggesting they deserved fresh plants after a hard winter bereft of beauty. Along with his rather unusual focus on female horticulturists, Toms's advertisements often included descriptions of special plant offers or combinations that he would sell at reduced prices.[48]

In the competitive ornamental plant market, local nurserymen like

Commercial Realities: Supplying the Needs   109

Toms, B. W. Steere, and Loud and Trask had several advantages. They operated within a well-known framework of local exchange where customers knew merchants and could examine the goods they were about to purchase. Local nurserymen also knew their neighbors, and could extend their circle of influence by participating in local organizations, and regional horticultural and agricultural societies. If reputable, nurserymen could depend on word-of-mouth or favorable endorsements to expand their clientele. Many nurserymen cultivated good relationships with prominent, influential customers, who would, in turn, recommend their services to friends and neighbors. Sometimes these local businesses turned that support to real advantage, and even used it in promotional advertising. In 1870, Adrian's Loud and Trask urged readers to consult with widely recognized local horticultural authorities like the Rev. E. P. Powell or Dr. Woodland Owen, a dentist and active member of the Adrian Horticultural Society, about the validity of their advertising claims. (See Figure 4.5.)

In addition to favorable personal relationships, local nurserymen could supply special services like plant delivery or door-to-door sales that posed a greater problem for their distant competitors. Hendrick's Garden in Ypsilanti, for example, advised customers to "Notice the mule team and hail the driver" if they needed any hot-house plants or vegetables. For Hendrick, the mule team not only served a practical function, but became a very obvious advertising icon for the business.[49]

Frequently, local nurserymen used their regional location to support claims that their plants were superior to those grown by eastern firms. An 1864 advertisement for Loud and Trask's Adrian nursery typified the argument that local nurserymen persistently voiced. "It is almost needless to remind the purchaser of the advantage to be gained in obtaining plants from our own vicinity, instead of Eastern nurseries," the advertisement stated. "The distance of shipping being so much less, trees can be delivered in better season, and in better condition," Loud and Trask contended and then added, "besides, many kinds flourish better when grown in the climate where they are to be permanently planted." Well aware that their distant competitors could offer customers a greater selection, local nurserymen often practiced a delicate balancing act, attempting to emphasize their local advantages without sinking into parochialism. In 1870, a Loud and Trask advertisement again typified the constant intermingling of the two worlds, as the Adrian nurserymen reported that they had abandoned the "old and worthless varieties," and that their stock of "new, novel, and desirable plants" could compete with that of any eastern

firm.⁵⁰

Competition between local nurserymen and distant firms remained a very persistent theme in the development of mid-nineteenth-century commercial horticulture, and in some measure, probably stimulated innovative practices from both camps that further encouraged horticultural interest. National and regional nurseries and seed houses usually offered a much larger and more varied selection of plants, and had better access to fashionable new varieties than their local competitors, but they did have to cope with a number of problems. The cost and vagaries of transporting live plant materials over great distances continued to plague the industry, and until the mid-nineteenth century, nurserymen depended on a rather haphazard system for delivering plants to distant customers. Poor or non-existent roads meant that transport to some areas was very slow or simply impossible. Water transport offered the most reliable and often the fastest alternative, but did not solve the problems of moving plants from wharves and harbors to homes and gardens. In her study of the eighteenth-century exportation of native American plants, Sarah P. Stetson noted that nurserymen, lacking a reasonable alternative, often bestowed packages of cuttings, bulbs, seeds, or scions on passengers in a variety of conveyances, and begged them to deliver the plants safely to waiting customers. The wealthy clients of eastern nurseries sometimes sent their own wagons to transport the plants they had ordered. Expensive, time consuming, and risky at best, these methods severely limited access to ornamental plants for those living away from nursery centers.⁵¹

The nineteenth-century expansion of transportation networks assisted nurserymen in their distribution problems, and enhanced ornamental plant availability. Inland canals, railroads, and the gradual establishment and improvement of roads opened some previously inaccessible areas, and both steamships and locomotives generally speeded the movement of vulnerable plant materials. Even with improved transportation networks, however, the problems persisted, for nurserymen needed a system to assure delivery once their packages left the nursery. Both the development of express companies, and improvement and changes in the postal service gradually eased the situation.

In the early days of railroading, rail companies were concerned about shipping large cargoes and transporting passengers, but made little accommodation for the small shipper and individual packages. As it had with the stagecoach driver before him, the responsibility for these goods

Commercial Realities: Supplying the Needs 111

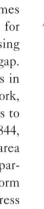

often rested with the conductor, and consequently packages sometimes literally fell through the cracks. Despite the problems, the demand for shipping small packages increased, and eventually individuals, sensing both a need and a lucrative business opportunity, moved in to fill the gap. One of the first, William F. Harnden, began advertising his services in 1839. Harnden offered to carry packages between Boston and New York, and finding business brisk, quickly established agents in other cities to extend the reach of his services. Others had similar ideas, and by 1844, over forty small express companies were operating in the Boston area alone. The nationwide need for similar services quickly became apparent, and small businesses gradually consolidated their resources to form large companies with agents across the country. The American Express Company began shipping merchandise from the east via western New York and the Great Lakes, and in 1854, the United States Express Company began a similar system using the New York and Erie Railroad and other western rail systems. The Adams Express Company grew out of the consolidation of a number of eastern and southern lines, while Wells, Fargo, and Company operated an express service along the West Coast, and soon connected eastern suppliers with western markets by both overland and water routes.[52]

Express companies offered nurserymen a new alternative for shipping plants to distant customers, but even their services sometimes ran into trouble. Many customers complained about the expense of shipping by express, and despite the slowness and greater risk, requested that their plants be sent as railroad freight. Express companies also made mistakes, sometimes shipping plants to the wrong agent or the wrong address, and as a result, increased the chance that perishable plant materials would die in transit. The comments of H. A. Terry of Crescent City, Iowa, a frequent correspondent of Illinois nurseryman John Kennicott, typified the concerns and problems that many experienced. In October 1861, Terry ordered a box of tulips from Kennicott, but several weeks later wrote to report that American Express had sent them to an agency far from his home. "So it seems that I am likely to lose another lot of tulips through the carelessness of those infernal blockheads," Terry complained. "They ought to be made to pay dearly for them." Terry was especially exasperated because he had experienced a similar problem with American Express the previous fall, and after both misadventures concluded that "they seem to have a set of fools for agents." Terry urged Kennicott to ship through U. S. Express, and vowed never to pay the "Am. Ex. Co. an-

other dollar if I can avoid it."[53]

Changes in the postal service also facilitated the shipment of plant materials across the nation. Improved and expanded railroads had speeded mail delivery to many regions by mid-century, but in addition, postal authorities adapted a series of measures designed to meet the competitive challenge of the express companies and to enhance the profitability of the Postal Department. Many of these reforms, instituted between 1845 and 1863, regularized and reduced postal rates and eventually broadened the definition of "mailable matter."[54] In an 1860 report, Postmaster General Joseph Holt gave the horticultural industry a boost when he proposed the inclusion of seeds and plant cuttings among many new items that could be mailed at very low postage rates. As he argued for the reforms, Holt noted that many of those articles were "now entirely withheld from the mails, owing to their being subject to letter postage." Reducing the postal rate, he insisted, would provide a "much desired" service to the public and by boosting demand, increase Department revenues as well. Joseph Holt was a lawyer from Breckenridge County, Kentucky, and came to the Post Office after several years as Commissioner of Patents. He may well have understood the importance of these issues for nurserymen and seed dealers. Evidence of his interest in what his constituents had to say surfaced in 1859, when as Commissioner of Patents, Holt invited a number of leading agriculturists to Washington to discuss agricultural and horticultural issues. Calling itself the "Advisory Board of Agriculture of the Patent Office," the group met for eight days, and came forth with a variety of suggestions to both improve the department and advance agricultural interests. Whether drawing on the comments of the advisory committee or his own experience, Holt put forth suggestions with far-reaching effects on commercial horticulture.[55]

Following the Postmaster General's recommendations, Congress passed postal regulations defining seeds or cuttings in packages less than eight ounces as mailable matter in 1861. The regulation established postage at a rate of $.01 an ounce for destinations less than fifteen hundred miles and $.02 an ounce for greater distances. In 1863, a new postal law initiated even more important changes, mandating a uniform letter rate based on weight and regardless of distance, and for the first time designating three classes of mail. The act defined letters as first-class mail, and newspapers and magazines as second class with a lower postage rate. A number of other printed items as well as "seeds, cuttings, bulbs, roots and scions" were classified as third-class mail. Under the new system, the

Commercial Realities: Supplying the Needs 113

postage rate on seeds and cuttings was reduced to $.02 per four ounces, and plant packages weighing up to four pounds could be sent through the mail.[56] Advertisers, in turn, could mail their circulars at the third-class rate, and pay only $.02 with up to three circulars in the same envelope. Nurserymen and seed dealers profited from both aspects of third-class mail.[57] Fearing that Congress might tamper with postal regulations, horticultural proponents regularly emphasized their importance to the industry. "Horticulturists and agriculturists probably make use of the mail to a greater extent than any other class," the *Gardener's Monthly* pointed out in 1873, and argued that "wise post-office laws" were intimately connected to horticultural progress. Horticultural consumers benefited too, for with the help of the post office, people in sparsely settled or remote areas could enjoy the benefits of "the best seed stores in the country" and the "best nurseries" right along with their more urban counterparts.[58]

Despite the expanded access to ornamental plants provided by the mail service, nurserymen and seed dealers still contended with the vagaries and dangers of transporting live plant materials. To counteract the problems, they developed very efficient and effective methods of packaging plants for shipment. Professional horticulturists often advised carefully washing the soil from plant roots, laying dry sphagnum moss around the damp roots, and then tightly wrapping the entire bundle in several layers of brown paper, preferably with a layer of oiled paper on the exterior. In preparation for long-distance shipping, many nurserymen "puddled" tree roots, dipping them in a mud bath of "good tenacious soil" and water, and letting the mud act as a protective layer to prevent drying. When sent by express, nurserymen sometimes shipped plants directly in pots, placing them closely together in wooden boxes, and surrounding the pots with straw or sphagnum to prevent movement.[59] Concerned about success once plants arrived at their destination, nurserymen advised acclimatizing trees and shrubs slowly to sun and wind. For indoor plants, they suggested careful potting, minimal watering, and shady conditions until the plant showed signs of new growth.[60]

No matter how they received their plants, this mid-nineteenth-century expansion of the American horticultural industry was a boon for Midwestern ornamental plant enthusiasts. They could easily obtain their favorite trees and flowers from local businesses or widely known firms with a national clientele. Commercial horticulturists offered a huge selection, and increasingly customers could count on an efficient delivery system that brought plant materials to their door. For nurserymen and seed deal-

ers, however, the increasingly competitive horticultural marketplace left little time to glory in their new-found popularity. Alert commercial horticulturists surveyed the competitive scene, and understood that their own economic future depended on stimulating the market for trees and flowers even further. They directed their entrepreneurial skills and their passion for horticulture to the problem, and emerged with diverse and innovative strategies that sent their seed packets or nursery stock into the hands of an ever-expanding clientele.

Five

Commercial Realities

Promoting Demand

In 1862, the *Gardener's Monthly* pointed out a marketing reality that plagued nurserymen and seed dealers in their increasingly competitive industry. Commercial horticulturists could not depend on a fickle public for their prosperity, the *Gardener's Monthly* claimed, but needed to "make [their] own customers." Enterprising nurserymen and seed dealers were bound to take on the roles of both "tradesmen and teacher," educating consumers in a way that promoted horticultural interest and demand for new plants.[1] Many commercial horticulturists took up the call with enthusiasm, establishing personal links with distant customers, fine-tuning horticultural advertising, promoting plant culture through richly detailed and often lavishly illustrated catalogues, and sending out legions of plant peddlers to drum up new and perhaps reluctant customers. This chapter explores those promotional strategies, and argues that as they developed varied and innovative marketing techniques, nurserymen and seed dealers gave new meaning to trees and flowers, assured an insatiable demand for the new and fashionable, and contributed even further to the transformation of Midwestern domestic landscapes.

To "make [their] own customers," some mid-nineteenth-century nurserymen and seed dealers began by cultivating a personal presence at the local level, often at some distance from their home nurseries. George Ellwanger of the famed Rochester, New York, Mount Hope Nursery, for example, participated in activities of the Michigan State Pomological Society, and in 1873, as chairman of the Committee on Ornamental Trees and Shrubs, presented a detailed report on the most desirable woody plants for home culture. Ellwanger's efforts were not unusual. Other nurserymen and seedsmen routinely visited state or county fairs, and, as a letter from James Vick to the Michigan Pomological Society indicated, sometimes used the occasion to reaffirm personal contacts in the region. "I had such a 'good time' at your State Fair that it did not seem like attending a fair at all," Vick wrote. "Old and new friends seemed to have congregated together, and I think I could tell a good deal more about what and who I saw than of the flowers I exhibited."[2]

Commercial horticulturists also used state and local exhibitions to showcase their own plants, and acquaint customers with what they had to offer. Local nurseries and florists and small regional firms often used the technique at county fairs, while larger enterprises tended to concentrate on statewide competitions. Always at the forefront of commercial promotion, James Vick took first prizes at the 1867 Michigan State Fair for the "best and greatest variety of flowers skillfully [sic] grouped," including German asters, seedling phlox, and a collection of gladioli. The fair committee noted that Mr. Vick's "fine contribution" added "much to the exhibition." Due to increased interest on the part of commercial horticulturists and in an effort to eliminate whatever advantages they might have, agricultural societies sometimes created separate divisions for professional and amateur cultivators. Even with those categories, enterprising businessmen made their presence felt. At a Michigan State Pomological Society exhibition in 1872, James Vick took twenty-two of seventy-five prizes offered in the bedding plant, shrubs, annuals, and bulbous plant categories.[3]

There was a danger in appearing too successful and discouraging local participation, and most commercial nurserymen knew when to draw the line. In 1872, Rochester seedsmen Briggs and Brothers wrote to the Michigan State Agricultural Society, offering to donate money they had received as premiums at the State Fair to the Society. "If we are regarded by your Society as having, by our efforts, aided in any degree in making your very successful exhibition more interesting," the firm wrote, "we are

sufficiently gratified." The Agricultural Society was indeed pleased and resolved to tell Briggs and Brothers of the "high satisfaction and exquisite delight which their very splendid display of flowers gave" to the Committee and fair visitors. A year later, James Vick told his catalogue readers that he would no longer risk the disappointment of friends and customers by competing with them for premiums. "I am quite willing to exhibit my flowers at any and all State Fairs that I can reach," Vick wrote, "but hereafter I shall exhibit only, and not compete for premiums—certainly not against persons residing in the neighborhood."[4]

James Vick went a step further, and in the early 1870s, began offering floral premiums for amateur competitors in fairs and exhibitions nationwide. He explained his purpose in an 1872 letter to the Indiana Horticultural Society, noting, "I am very anxious to increase the culture of flowers among the people [and] I desire also to show my special regard for my numerous customers in your State." Vick offered premiums ranging from $20 for the "best and finest collection" of cut flowers to $10 for the best collection of phlox, asters, balsams, dianthus, pansies, stocks, or gladioli. To encourage youthful interest in horticulture, he provided a special group of slightly lower premiums for "flowers grown by persons under twenty years of age." Initially the offer was limited to Vick's customers, but he eventually dropped that prerequisite, and opened the competition to all amateur horticulturists entering any state or territorial agricultural society exhibition. For county agricultural societies, Vick offered a "floral chromo" or chromolithograph for the "best exhibition of cut flowers." To avoid any taint of commercial heavy-handedness, Vick simply provided the funds or the prizes, and the local or state agencies did the judging using their own guidelines. Vick's premiums were larger than customarily given for floricultural or horticultural entries, and consequently made a significant addition to both the excitement and rewards of fair participation.[5]

Some nursery and seedsmen also enhanced their personal relation with customers by extending an open invitation to visit them at home, and then developed their private homegrounds or business landscapes as examples of horticultural embellishment for the public to enjoy. The horticultural and agricultural press encouraged the practice by publishing detailed accounts of visits to nurseries or seed farms, and occasionally featuring biographical sketches of the nurserymen themselves. Often, these descriptions included a survey of several nurseries and seed houses in a particular region, accounts of operations and facilities, and a reasoned

comparison of nursery stock or the specialties of the firm. They then urged their readers to become "horticultural pilgrims," and to consider commercial gardens as travel destinations offering both education and inspiration.[6]

Letters to Illinois nurseryman John Kennicott suggest that at least some commercial nurserymen succeeded in their efforts to develop a personal relationship with distant customers. Customers wrote to Kennicott to order plants, complain about crop failures, describe horticultural successes, or sometimes simply for advice. The correspondents seem to presume not only that Kennicott would understand their needs, but that he would take the time to respond in a thoughtful and very personal way. In a typical inquiry, J. Griffen of Prairie City, Illinois, asked Kennicott for comments on how to plant a "little plot of ground" adjoining his home. Griffen admitted he had no experience in growing flowers, but wanted to plant roses, dahlias, verbenas, and hardy perennials, and asked Kennicott how they might be arranged to "give an appearance of taste." Mr. Griffen included a sketch of his homegrounds and asked that Kennicott not only select appropriate plants, but mark where they should be planted on the diagram. Other customers requested that Kennicott pick out the most fashionable plants to send them, preferably some that were "a little ahead of everybody else" in the area. Still others asked for information on establishing hedges, identifying shrubs, or planting trees. A Chicago customer was so confused about ornamental tree culture that he invited Kennicott to visit so that he might see the "little place" in person and offer appropriate advice.[7]

Beyond efforts to establish a personal presence in local communities, advertisements provided the most obvious link between distant nurseries and their widely dispersed customers, and nurserymen and seed dealers were among the first to recognize their importance and power. From somewhat sparse beginnings, horticultural advertising gradually increased in volume, and by the 1850s, Midwesterners could find nursery or seed company advertisements in many widely circulated horticultural or agricultural newspapers or journals, as well as their local newspapers. The nationally circulated *American Agriculturist* offered a prime, if somewhat expensive, spot for horticultural advertisements, and the increase in nursery and seed company notices in its pages illustrated the general trend. Horticultural advertising peaked in the late winter and early spring months, and in March 1858, the *American Agriculturist* featured twelve advertisements for ornamental plants or flower seeds. By March 1874, the number

Figure 5.1. By the 1870s, an increasing number of horticultural advertisements appeared in papers like the *American Agriculturist*. From *American Agriculturist*, vol. 33 (March 1874): 110.

of ornamental plant advertisements had grown to thirty-eight, with large western New York State firms or other eastern companies dominating the advertising pages. A few large Midwestern firms like F. K. Phoenix of Bloomington, Illinois, also routinely advertised in the *Agriculturist* while other regional nurseries purchased only an occasional advertisement.[8] (See Figure 5.1.)

Midwesterners were also likely to see horticultural advertisements in regional agricultural papers like the *Michigan Farmer*. From its beginnings in the early 1840s, the *Farmer* carried advertisements for the Prince Nursery in Flushing, Long Island, and the Buffalo Nursery and Horticultural Garden in western New York, E. D. Lay's Ypsilanti Nursery, and J. C. Holmes's Detroit Nursery. As the decades rolled on, the number of notices gradually increased, but the broad mix of advertisements from both large eastern nurseries and smaller regional establishments continued. Michigan firms like the Hubbard and Davis Nursery, E. G. Mixer's Elmwood Gardens, and the D. M. Ferry Seed Company routinely advertised in the *Michigan Farmer* along with a number of Illinois, Ohio, and Canadian nurseries. In the early 1850s, the Genesee Valley Nurseries in Rochester and the Syracuse Nurseries in Syracuse, New York, began a long and consistent run of advertisements in the *Michigan Farmer*, and by the mid-1850s, Ellwanger and Barry were regular advertisers, sometimes running a number of notices in a single issue. J. M. Thorburn of New York City began advertising flower seeds in the late 1850s, and in 1859, the well-known French nursery of André Leroy also placed advertisements in the *Farmer*. Seedsman James Vick joined the fray by the early 1860s, and by the early 1870s, Briggs and Brothers of Rochester frequently advertised in the *Farmer*.[9]

In addition, local newspapers, often with a very limited circulation, carried a surprising number of advertisement for distant, regional, or local horticultural firms. The Lenawee County *Adrian Times and Expositor*, for example, had a weekly circulation of about fifteen hundred subscribers in 1870, but over the years featured advertisements for Ellwanger and Barry, Briggs and Brothers, James Vick, and the Rochester Central Nurseries all of Rochester, New York. Peter Henderson from New York City, F. K. Phoenix and Robert Douglas from Illinois, and several Ohio firms like Storrs, Harrison and Company and the Columbus Nurseries also advertised in the *Times and Expositor*.[10] The *Jackson Weekly Citizen*, with a circulation in 1870 of 1,275, boasted a similar line-up of distant advertisers.[11] The *Dexter Leader*, a Washtenaw County weekly with a circulation of six hundred subscribers, also featured advertisements from F. K Phoenix, James Vick, Landreth's Seeds of Philadelphia, and the Hovey Seed Company of Chicago. James Vick even advertised in the tiny *Tecumseh Herald*, a Lenawee County weekly with a circulation of five hundred.[12]

In the mid-nineteenth century, product advertising, particularly for firms without personal links to the local community, was uncommon, and agricultural and horticultural journals played a role in enhancing consumer acceptance of the practice. The *Michigan Farmer* exemplified a typical strategy when its editor insisted that the journal accepted only advertisements for appropriate agricultural or horticultural products. This policy had advantages for both consumer and advertiser, the *Farmer* maintained, for it developed consumer trust and maximized the effectiveness of advertising dollars. Understanding the importance of first-hand knowledge, the *Michigan Farmer* also claimed that it would "not print a nursery advertisement in 'our paper' for any man whom we cannot recommend, as both 'honest and capable.'"[13]

Typically, newspaper or journal editors enhanced the credibility of advertisements even further by discussing them in one of the non-advertising sections of the paper. In 1858, for example, the *Michigan Farmer* "called attention" to the advertisements of William Adair, Hubbard and Davis, and Ellwanger and Barry, urged subscribers to read the ads, and then appended rather personalized notes of approval for the firms and their merchandise. Ellwanger and Barry, according to the *Farmer*, offered opportunities to "obtain choice selections of all the new and beautiful plants," while Mr. Adair of Detroit sold plants most suited to Michigan's climate. In a similar commentary on the Toledo Nurseries, the *Michigan Farmer* advised readers that the firm stood "ready to supply northern Ohio and southern Michigan with trees that will grow. They are responsible, and near at hand to correct all mistakes."[14]

Many newspaper and journal editors routinely expressed a very sincere interest in promoting horticultural improvement, and they understood the importance of commercial interests in that process, but the promotion of advertisements was not totally altruistic. Advertising dollars helped some of the struggling journals and newspapers stay afloat, and whether they recognized the implications or not, promoting commercial horticulture enhanced their own bottom line. Nursery and seedsmen certainly understood the powers of the press to act as a conduit for their cause, and often rewarded editors who "puffed" their products. In 1878, a customer wrote to James Vick complimenting him on his skill in sending plants, but then noted, "I am the wife of a California editor, one of the fortunate ones who are made happy every year by your generosity to your 'brethren in the press'." The editor's wife did not indicate whether she referred to the routine distribution of free catalogues to the press, or whether her family had actually received complimentary seeds from Vick.[15]

The practice of distributing free seeds and plants to newspaper editors

Figure 5.2. Storrs, Harrison and Company advertisement. From *American Agriculturist*, vol. 33 (March 1874): 111.

was certainly common, and as a letter to John Kennicott revealed, had an effect on the attitude and opinions of the lucky recipients. Kennicott advertised his nursery in the Chicago *Evening Journal*, and in 1862, the editor John L. Wilson made a specific request of the nurseryman. "If you can send me some hardy roses for my own front yard I would like them," Mr. Wilson wrote. "Send to 50 Dearborn Street, Journal Office." Assuring Kennicott that he had taken appropriate editorial action, Wilson pointed out, "I gave the [nursery] prospectus a ventilation and administered a kick to the tree pedlars [sic]."[16]

Between 1850 and 1880, the content and visual quality of horticultural advertising changed right along with the quantity of the ads. Like most

early advertisers, commercial horticulturists of the 1850s tended to use rather uniform advertisements dominated by text, and they often provided a vast amount of information in a small space. In a typical 1852 *Michigan Farmer* advertisement, J. C. Holmes, the proprietor of the Detroit Nursery, described its location, asserted it was a "very pleasant drive or walk from the city," detailed efforts to provide a diverse selection of healthy and correctly labeled plants, and described facilities for shipment or delivery of orders. Holmes also invited interested horticulturists to visit the grounds, and assured readers that their orders would receive immediate attention. Most commonly, advertisements included information on types of plants available, their prices and any special bargains, and collections that the company might offer. A few companies even published entire catalogues as an advertisement. In an 1860 *Michigan Farmer* advertisement, Detroit nurserymen Hubbard and Davis listed hundreds of trees, shrubs, and flowers, and provided instructions on how to order plants, and other commentary.[17]

As the years went by, many advertisers also attempted to enliven the format with innovative content or eye-catching presentations. Using a typical strategy, an 1864 James Vick advertisement began with the striking, "Lovers of Flowers, Attention!" in large, bold letters, and in smaller type added, "customers badly swindled." In 1874, Storrs, Harrison and Company used a similar technique, but called attention to their advertisement with the assertive, "Ladies Read This." Increasingly, more and more nurserymen and seed dealers moved away from densely worded notices to some sort of illustration or logo that gave their advertisement visual distinction.[18] (See Figure 5.2.)

Eventually, many horticultural advertisements promoted the company's catalogue rather than providing specific information about available goods. Both nurserymen and seed dealers encouraged readers to send for a new catalogue, and often used advertising text to describe what it offered. In a typical 1864 *Michigan Farmer* advertisement, Mr. Vick touted his catalogue as a "beautiful work of fifty large pages," including "*twenty-five* fine engravings and one splendid color plate." As his catalogues became more extensive and lavishly illustrated over the years, Mr. Vick carefully documented their evolution in his advertisements, and always encouraged readers to enjoy their beauties first hand. Ellwanger and Barry, and a number of other firms, produced multiple catalogues with specific offerings in each, and used their advertisements to direct customers to the appropriate catalogue number. Commercial horticulturists

frequently included directions for obtaining the catalogue as a part of the advertising text, and emphasized that horticultural enthusiasts could receive all those beautiful illustrations and important horticultural information for free or at very low cost.[19]

Nursery and seed catalogues were in fact closely allied to paid advertisements in their ability to link distant nursery or seedsmen with their customers, and to promote ornamental plant sales. The earliest catalogues, produced by a few nurseries in the last decades of the eighteenth century, were often a simple listing of plants and their prices, but as the horticultural trade grew in the mid-nineteenth century, the number of catalogues and their complexity increased substantially. The agricultural and horticultural press promoted nursery catalogues with even greater enthusiasm than they had horticultural advertisements. Commercial horticulturists routinely sent their catalogues to newspaper and journal editors, and they, in return, reviewed the catalogue as they would a newly received horticultural book or guide. Some journals featured a regular section that simply listed catalogues received and offered brief comments on the type of information they supplied or their overall quality. In other cases, editors printed a full descriptive paragraph about the catalogue along with very subjective comments on its merits. In a typical appraisal, the *Horticulturist* noted the unusual qualities of the 1865 Vick's catalogue, and then told readers, "This is one of the most complete catalogues that has come under our notice, and is evidence of the energy and business talent of Mr. Vick."[20]

With their strong local ties, catalogue commentary in widely read regional journals or local newspapers was particularly influential. Both the *Michigan Farmer* and the *Prairie Farmer* regularly reported on catalogues received, often detailing the attributes of local nursery catalogues and those of more distant firms like James Vick or the Syracuse Nursery. The *Michigan Farmer* even invited readers to visit their office to view catalogues in person. Local newspapers also regularly reviewed catalogues, and urged readers to send away for them. The *Ypsilanti Commercial* typified the practice in its detailed account of Peter Henderson's 1873 catalogue, noting that the 175-page catalogue was "beautifully illustrated," contained "a colored plate of the new 'Fountain Plant'; and also a handsome lithograph of a group of new fuchsias." These endorsements gave distant nursery and seedsmen a very important boost, established their credibility, extended their range of contacts, and, most importantly, linked their products with well-known local authorities.[21]

PLANTS FOR WINTER BLOOMING.

Pteris, argyea. 25 cents to $1.00.
" critica alba lineata. 25 cents.
" serrulata. 25 cents.
" " cristata. 50 cents.
" tremula. 25 cents to $1.00.
Pella, hastata. 25 cents.
☞ We have also a great variety of other Ferns and Selagrenellas, too numerous to name.
Ficus, Elastica. (India Tubber Tree.) From East Indies; foliage large, dark green and shining; fine house plant. $2.00 to $5.00.
Ficus, Parcellii. A splendid decorative plant, with beautiful, variegated foliage. The leaves, which are of large size, thin texture and coarsely serrated on the margin, are of a bright green color, profusely but irregularly blotched with creamy white and dark green. $1.00 to $5.00.
Fragaria, Indica. Bright red strawberry ; used for rock work or baskets. 20 cents.
Fuchsia. (Lady's Ear Drop.) In shaded situations, with plenty of air and light, they make elegant summer-blooming plants for massing or grouping in beds. We have a choice collection of both single and double, which have been carefully selected with a view to obtain the best varieties in habit, growth and distinct shade of color. 20 cents each, except where noted. Larger plants 25 to 50 cents.
Fuchsia, Black Prince. Bell-shaped ; very fine.
" Champion of the World. (New.) Dwarf, very free-blooming, double variety, with scarlet tube and sepals, violet corolla. Best fuchsia out. 30 cents.
" Day Dream. Double ; corolla maroon, sepals crimson.

Fuchsia, Fulgens. A fine old sort, very heavy foliage.
" Golden Fleece. Foliage golden yellow; free, early bloomer.
" Lustre. Scarlet corolla, blush sepals.
" May Queen. Blush sepals, scarlet corolla.
" Mrs. Marshall. Corolla carmine, sepals white ; winter bloomer.
" Pearl of England. Cherry-red corolla, rosy-white sepals.
" Prince Imperial. Dwarf; corolla purple, sepals recurved, crimson; one of the very best of the dwarf varieties.
" Speciosa. Corolla scarlet, sepals blush.
" Sunray. Beautiful, tri-colored variety ; great novelty. 30 cents.

GERANIUM, Zonale. Scarcely rivaled as bedders of the surest growth on all soils, succeeding to perfection during the heat

Geraniums, Zonale. Double.

and drought of summer; of rich, massive, and often gay-zoned foliage ; compact growth, and profuse, constant, brilliant bloom in scarlet, crimson, salmon, and white ; they are especially valuable and beautiful for winter-blooming in the house. 20 cents each; $2.00 per dozen; $10.00 per 100, by express.
Bicolor. White, deep salmon eye.
Bridesmaid. Rosy-salmon.
Bridal Beauty. Salmon, very free bloomer.
Col. Holden. Extra large truss; crimson-scarlet.
Dazzler. Intense scarlet, white eye.
Delight. Rose.
Duchess of Sutherland. Scarlet, white eye.
Dr. Denny. New; color rich purplish magenta.
Emily Vaucher. White.
General Grant. Scarlet; very large truss.
General Scott. Salmon.
Glorie de Corbenay. Salmon-pink.
Helen Dick. Pink and white.
Jean Sisley. Intense scarlet, white eye.
Kate Nicholson. Rosy-pink.
Louis. Crimson-purple; extra.
Mad. Muzzard. Magenta.
Marginetta. White, pink eye and border.
Master Christine. Rich pink.
Mrs. Lowe. Crimson, splashed with pink.
Mrs. Leaver. Fine pink.
Mary. New; white shaded salmon; very free.
New Life. By far the finest striped variety ; color deep scarlet, striped and blotched with white.
Queen of the West. Rosy-scarlet.
White Clipper. Pure white, pink stamens; one of the best white grown.
Wood Nymph. New; clear pink, free bloomer.

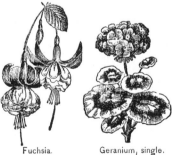

Fuchsia. Geranium, single.

Figure 5.3. Increasingly, catalogues supplied readers with a great deal of horticultural information. From *Catalogue of Dutch Bulbs and other Flowering Roots, also Seeds and Plants* offered by D. M. Ferry and Co. (1881), 18–19. Original in Clarke Historical Library, Central Michigan University, Mount Pleasant, Michigan.

As they had with advertisement endorsements, journal and newspaper editors certainly had an economic stake in promoting nursery catalogues, for as the horticultural industry thrived so did their advertising revenues. Beyond their own interests, however, many editors understood catalogues as a useful and readily available educational tool. They frequently pointed out the importance of supplying cultural directions, proper plant names, and good illustrations, and argued that all helped to educate the public in horticultural practices. Thomas Meehan of the *Gardener's Monthly* even went so far as to suggest that catalogues were "as valuable in many respects as the current horticultural literature of the day," and urged all readers to take advantage of the free or very low-priced opportunity to obtain horticultural information.[22]

By the mid-nineteenth century, many nursery catalogues warranted press approbation, for they increasingly supplied sound horticultural descriptions and cultural information. At the most basic level, catalogues listed available plants, offered both common and scientific names, and included their prices. Some, like the 1860 H. A. Dreer Catalogue, presented plants in list form, but added slightly more descriptive information such as height, time of flowering, color, and whether the plants were annual, biennial, or perennial. Eventually, some of the most complete catalogues included paragraph-long descriptions of each plant, noting special cultural requirements, giving details of size and growth habits, or emphasizing particularly desirable or useful characteristics.[23] (See Figure 5.3.)

Beginning in the mid-1860s, many horticulturists enhanced plant descriptions and the attractiveness of their catalogues with at least some lithographic illustrations. In 1868, the Washburn Company featured occasional illustrations throughout its catalogue, and directed the reader's attention to descriptive information about the plant on the appropriate page. Nearly ten years later, the D. M. Ferry seed catalogue provided an illustration along with each plant description. Although the quantity and arrangement of illustrations varied considerably, most commercial horticulturists understood their effectiveness in attracting customer attention, and prided themselves on the number that appeared in their catalogue. In 1872, D. M. Ferry claimed its catalogue was the "largest and most complete" catalogue produced in the United States, was printed on "beautifully tinted paper," had 162 pages, and, most importantly, "eighty engravings."[24] James Vick carried the art of catalogue descriptions and illustration to new heights. By the early 1870s, his catalogues provided a plethora of information including comments on plants in their native

Figure 5.4. By the mid-1870s *Vick's Floral Guide* was lavishly illustrated. From *Vick's Floral Guide for 1876.* Courtesy of Department of Rare Books and Special Collections, University of Rochester Library, Rochester, New York.

habitats, the origin of plant names, and very specific cultural directions like how to "take up" a particular plant for indoor growth during the winter months. In addition, Vick's catalogue included an illustration with each plant description. Eventually his *Floral Guide* became so large and detailed that Vick published a *Quarterly Guide* with information on individual plants appearing only in the first number each year. In 1875, Vick

noted that all four numbers of his catalogue totaled over two hundred pages of horticultural information.[25] (See Figure 5.4.)

Most catalogues also provided basic cultural information in the introductory pages, describing how to lay out a garden bed, sow seeds, or care for newly arrived trees or potted plants. Commentary on manuring, weeding, mulching, and watering appeared routinely, along with more specialized information like how to build a hot bed or cold frame or techniques for establishing a lawn. Several circumstances prompted commercial horticulturists to supply this broad-ranging information. A convenient source of horticultural advice helped to insure customer success, and that, in turn, encouraged customer loyalty and perhaps generated new business from admiring friends and neighbors. At an even more practical level, satisfied, educated customers blamed the nursery or seedsmen far less frequently for any horticultural disasters they might experience. By taking a genuine interest in horticultural education and customer success, commercial seedsmen and nurserymen also enhanced their image and trustworthiness in the eyes of potential customers. James Vick aptly summed up his motivation and that of many others, when he told his customers, "My great desire is to give such information as will make fair success possible to all, and especially to induce those who fail, to search earnestly for the *cause*."[26]

In the pages of their catalogues, nurserymen and seed dealers sometimes used special marketing strategies to draw more customers into the horticultural fold. Perhaps the most widespread technique encouraged customers to join with their neighbors and friends in forming ordering "clubs." B. K. Bliss of Massachusetts described the typical procedure in his 1860 catalogue. Clubs, formed from a number of individuals, pooled their orders, and if they sent in $5 could select $6 worth of seeds from the catalogue. Those sending in $30 selected merchandise valued at $40, and then distributed the bonuses among club members. James Vick made a very similar offer, but agreed to send orders separately to individual participants.[27] In one sense, these clubs eased the burden of processing and shipping small orders, but by encouraging customers to bring their neighbors into the national market, also provided a sure and subtle way of expanding the horticultural trade. For the uninitiated, the influence of neighbors and friends stimulated horticultural interest, and at the same time, offered a local link to distant and perhaps unfamiliar plant sources.

In addition to advantages for clubs, many catalogues touted special promotions to enhance customer interest. Some listed collections of flow-

Figure 5.5. James Vick and other commercial horticulturists sometimes included chromolithographs in their catalogues. From *Vick's Illustrated Monthly Magazine*, vol. 2, no. 2 (February 1879). Courtesy of Department of Rare Books and Special Collections, University of Rochester Library, Rochester, New York.

ers designed to meet particular needs, and offered them at reduced or very reasonable prices. Most collections were aimed at inexperienced gardeners, or those who had never placed an order, and assembled diverse, attractive varieties that were also easy to grow. B. K. Bliss, like many others, promoted a number of general selections for the "uninitiated," and suggested that the most inexperienced gardener could order them "with-

Figure 5.6. B. K. Bliss combined two favorite promotions in his 1872 catalogue. From *Benjamin K. Bliss and Sons' Illustrated Spring Catalogue and Amateurs' Guide to the Flower and Kitchen Garden for 1872*. Original in Agriculture Collection/Special Collections, Michigan State University Libraries, East Lansing, Michigan.

out fear of disappointment." Commercial horticulturists were also among the first to understand the popularity of chromolithographs, and to use them as a promotional tool. By the mid-1860s, some large establishments regularly included at least one elaborate color print in their catalogue.

These "chromos," often printed as a frontispiece in the catalogue, sometimes featured a single flower, a composite of several specimens, or elaborate floral bouquets and arrangements. Their brilliant colors and attractive designs made them so popular that many customers ordered catalogues simply to obtain the print. (See Figure 5.5.) In 1872, B. K. Bliss combined two favorite promotions, and included a large, brilliantly colored chromolithograph of a bouquet of flowers. Each flower was numbered, and a key drew the reader's attention to where they might find the flower in the catalogue. To make the process even easier, Bliss offered the numbered flowers as a special collection for home culture. The unspoken but clearly communicated message implied that with this selected seed collection, customers could reproduce the illustration's exceptional floral beauty in their home gardens.[28] (See Figure 5.6.)

Always on the alert for business opportunity, nurserymen and particularly seed dealers promoted new horticultural amenities as their catalogs gradually expanded. Often beginning with lists of recommended books, many firms advertised related horticultural items like whale oil insecticidal soap or specialized gardening tools. Over the years, the number and diversity of these items increased until customers could purchase a lawn mower or a set of "ladies'" horticultural tools from their favorite seedsman or nursery. The 1872 B. K. Bliss catalogue exemplified the trend with nearly forty pages devoted to non-plant items. A Bliss customer could chose from a portable lawn sprinkler, seven types of lawn mowers, nine rustic baskets, over twenty pruning knives, garden trellises, hanging baskets, cast-iron bird houses, flower stands, window boxes, benches, croquet sets, and hundreds of other items.[29]

Catalogues, far more than newspaper or journal advertisements, gave nurserymen and seed dealers an opportunity to personalize their relationship with distant customers, and over the years, they developed scores of techniques to make sure their readers felt at home. Often, introductory remarks established a friendly, conversational tone that usually permeated the whole catalogue. Most seedsmen addressed their customers as "friends," and like the Washburn Company, usually made "a few friendly and familiar remarks" as they set the stage for what the catalogue had to offer. James Vick indulged in what he termed "gossip with customers," and carried on a casual and very personal conversation throughout the catalogue.[30]

Many nurserymen and seed dealers encouraged prospective customers to see themselves as members of a special group, distinguished only by

their love of plants and flowers. The technique certainly bridged the distance between customer and merchant, but more importantly overlooked any disparities in occupation, economic rank, gender, or educational background that might exist among those willing to purchase plants. Often, these "friendly" comments were a very effective means of developing what historian Daniel Boorstin has termed a "consumption community," a group of perhaps very disparate individuals linked by their interest in and consumption of a particular commodity. Some enhanced their egalitarian outlook by insisting that new customers would "be placed on the same footing as old," and carefully emphasized that horticultural experience was unimportant. To encourage those with little gardening knowledge, many seedsmen offered special collections of flower seeds that were easy to cultivate and were certain to produce a pleasing result, but emphasized that the collection would satisfy anyone, no matter what their experience, resources, or the size of their homegrounds. In an 1872 catalogue, for example, seedsman B. K. Bliss described his special collections of flower seeds as a way for those at a distance to enjoy the same degree of floral beauty as people with nearby nurseries or seedsmen. While clearly cultivating a rural and perhaps unsophisticated market, Bliss emphasized that any lack in floral taste or interest derived from remoteness and a scarcity of suitable plant materials, not a deficiency on the part of the customer.[31]

Some commercial horticulturists personalized their catalogues even further by publishing portions of letters or testimonials they had received from satisfied customers. Most firms included a broad spectrum of correspondents in their testimonials, publishing letters from all parts of the country, and from both men and women. In his 1865 catalogue, for instance, James Vick included a particularly long testimonial section with comments from eighteen men and twenty-eight women. The correspondents wrote from several Midwestern states, Canada, the Northeast, Kentucky, New York, and Pennsylvania. The preponderance of women in the sample may well have indicated an awareness of their increasing importance as horticultural consumers. Hoping to impress their prospective customers, Detroit nurserymen Hubbard and Davis published testimonials from the State Geologist, an Agricultural College faculty member, and a nurserymen in their 1871 catalogue. Seedsman H. A. Dreer offered an explanation for the use of these very personal endorsements. "Self-praise is not recommendation, nor are newspaper puffs reliable," Dreer wrote. "We hold that the unsolicited testimonials of our customers are far more

Figure 5.7. Vick's seed warehouse. From *Vick's Floral Guide for 1876*, 21. Courtesy of Department of Rare Books and Special Collections, University of Rochester Library, Rochester, New York.

satisfactory." The technique certainly enhanced the credibility of the firm and, as Dreer noted, proved the "practicability of packing and forwarding plants to a distance with safety." In addition, these individual accounts set a normative pattern of personal interaction between horticulturist and customer, and gave all readers the idea that distant nurserymen listened to their comments.[32]

Commercial horticulturists sometimes used their catalogue to familiarize customers with the size and scope of the business. Following a common practice, Ellwanger and Barry supplied directions to their Mount Hope Nursery, and assured their customers that they, or a trusted assistant, were "at all times ready to conduct strangers over the grounds, and to impart all necessary information."[33] For those unable to visit, catalogues often provided descriptions or illustrations of company facilities.

PACKING ROOM.

Figure 5.8. Vick's Packing room. From *Vick's Floral Guide for 1876*, 23. Courtesy of Department of Rare Books and Special Collections, University of Rochester Library, Rochester, New York.

In 1873, for example, James Vick published six lithographs of his firm, including the seed house, store and its interior, the order and packing rooms, and the bindery. He wanted all visitors to see his business, Vick explained, but since many could not visit in person, he resolved to show them the complete operation. (See Figures 5.7 and 5.8.)

In the 1870s and early 1880s, D. M. Ferry followed a similar policy, publishing illustrations of its "New Mammoth Seed Store and Warehouse," but particularly emphasizing the seed farm with a number of pictures of women carefully weeding or harvesting seed.[34] In a seemingly paradoxical contrast to the personal, friendly tone of the catalogue, these descriptions or illustrations often emphasized the size of the operation, the volume of business they conducted, and the efficiency and order with which the firm carried out a multitude of tasks. Both Ferry and Vick appeared to use their large-scale operation and obvious success as a way to insure customers of their legitimacy. Because they invited their customers to see the firm and admire its progress and efficiency, they not

only indicated they had nothing to hide, but included often distant customers in their vastly successful enterprise.

The information contained in nursery and seed catalogues offered an unprecedented opportunity to educate customers and promote business, but perhaps their most important contribution to expanding horticultural interest centered on price. Most nurserymen and seed dealers offered their catalogues free of charge, and if they required a nominal fee for first-time customers, usually promised to subtract the amount from the price of any merchandise ordered. Families who were unable to subscribe to a horticultural journal or agricultural newspaper, or who could not afford a gardening guide, could send for a catalogue, and in many cases were supplied with all the horticultural information they needed for tending their homegrounds. Although not everyone who received a catalogue ordered trees and flowers, the colorful pages and enticing descriptions connected them to a national marketplace and empowered them as potential customers.[35]

While they were at the forefront of modern promotional strategies like advertisements and catalogues, many nurseries also depended on the personal contact of nursery agents or "tree peddlers" to extend horticultural interest. Plant peddling belonged to a long tradition of itinerant salesmanship that for decades helped to stimulate consumer interest and distribute a wide range of goods across the American landscape.[36] Historian Barbara W. Sarudy has documented the presence of itinerant plant merchants in the eighteenth-century Chesapeake region, and Prince Nursery used nursery agents as early as 1818, but plant peddling came into its own as a horticultural sales technique with the mid-nineteenth-century rise of commercial nurserymen.[37]

In their efforts to cultivate Midwestern markets, Rochester nurseryman were among the first to exploit this sales strategy. Shortly after the founding of their nursery in 1840, Rochester nurserymen George Ellwanger and Patrick Barry sent out a cadre of trained agents to market their ornamental plants and fruit trees. Like peddlers, Ellwanger and Barry agents traveled through the countryside, visited individual homes and families, and sold goods through personal persuasion. Rather than carrying cumbersome plants with them, however, nursery agents took orders for trees, shrubs, and flowers and then delivered the plants at a later date. Other developing nurseries noted the successful strategy, and by the late 1850s, plant peddlers from a number of Rochester nurseries scoured the Midwestern countryside.[38]

Most of these early agents were employees of their respective home nurseries. Many worked on commission, but were also paid expenses and a monthly salary. They sold only stock from their employer's inventory, and eventually carried nursery catalogues or certification to verify their legitimacy. As the *Moore's Rural New Yorker* noted in 1863, orders generated through their own traveling agents and those received through the mail received identical treatment from nurserymen. With the nursery's reputation at stake, customers could assume that they would receive healthy, accurately labeled plants no matter how they placed their orders.[39] Although they might not want to employ agents, other nurserymen found plant peddlers so successful and the competition so severe, that they had no choice but to send their own agents into the countryside. In an 1859 report, Illinois nurseryman F. K. Phoenix explained the commercial reality to his fellow nurserymen. Phoenix was personally opposed to what he called the "itinerant vagabond way of selling," but noted that it "appeared to please the dear people so well" that he was forced to employ men "to travel and take orders on his account." The importance of plant peddlers in promoting sales continued to grow, and in the early 1870s, the *Horticulturist* announced with seeming surprise that "nearly three-fourths of the nursery stock sold throughout the United States [was] sold by personal solicitation of agents or dealers."[40]

Competing nurseries were not alone in observing the plant peddlers' success, and the nursery agent system quickly spawned some unexpected and controversial results. Farmers and merchants interested in supplementing their income, individuals looking for a lucrative trade, and former nursery agents hoping to increase their profits were drawn to the apparent success of the peddling system, and some "began to travel and solicit orders on their own account." These independent plant peddlers made a variety of arrangements with nurserymen to supply their needs. Some acted as a kind of "branch agent" by actually obtaining trees and shrubs from nurserymen and selling them from their store, village home, or farm. Other aspiring agents simply took orders from friends and neighbors, and delivered plants when they arrived. Many agents traveled only within a local region, while others covered extensive areas seeking orders for a specific nurseryman. Finally, some plant peddlers took orders throughout the countryside and filled their requests from a variety of nurserymen, often sacrificing correctly identified, healthy plants for lower prices and their own profit.[41]

Depending on what they sold and their relationship with their supplier, plant peddlers worked in several different ways. A few peddlers specialized in flowering bulbs, and these salesmen carried their wares with them. They showed their customers beautiful illustrations of the plants in flower, and if they made a sale, could deliver bulbs on the spot. Bulb peddlers frequently worked during the autumn months, the appropriate time for bulb planting.[42] Although they might also sell bulbs, most nursery agents specialized in trees, shrubs, and herbaceous perennials, and traveled during the winter months to solicit orders. If those orders were small, many agents sent requests to the nursery, received the plants by rail or boat, and then distributed them to their customers, either by holding the plants at a central location and waiting for customers to call or by delivering the plants directly to the village home or farm. If they had many orders or served a large territory, traveling agents usually went directly to the nursery, supervised the filling of orders themselves, and then delivered plants to their customers. According to critics of the profession, good peddlers had a responsibility for their plants even after they made a final delivery. In a heated discussion about plant peddlers at the December 1877 meeting of the Michigan Pomological Society, one fruit grower labeled all tree peddlers as "swindlers," but made an exception for a new agent in northern Michigan who took "a continuous interest in every orchard of trees he sells, assisting in planting and coming around afterwards to make suggestions as to further management."[43]

As plant peddling flourished, a new industry also developed to assist independent agents in their sales techniques In 1858, Dellon Marcus Dewey, a Rochester book seller, began producing brightly colored illustrations of fruits, flowers, and ornamental trees that he called "plates." Taking his cue from nursery agents who used preserved fruit specimens, newspaper illustrations, or pictures clipped from horticultural magazines to illustrate their wares, Dewey reasoned that professional-looking illustrations would heighten customer interest and the peddlers' credibility. The idea quickly caught on, and Dewey's inventory grew from 275 plates in 1859 to over 2,300 illustrations twenty years later. Early plates included the picture and often the name and some attributes of the particular variety to help the agent with his sales pitch. Dewey sold the plates separately, combined and bound at the agent's request, or in prebound collections. By the 1870s, pocket-sized books filled with colorful plant illustrations were most popular with agents.[44] (See Figure 5.9.)

Figure 5.9. Ellwanger and Barry supplied their Mount Hope Nursery agents with these popular pocket-sized plate books. Courtesy of Department of Rare Books and Special Collections, University of Rochester Library, Rochester, New York.

Not content to simply supply the tools of the trade, D. M. Dewey thought nursery agents needed additional guidance, and in 1875 published *The Tree Agents' Private Guide*. The *Guide* offered a plethora of advice ranging from how to sell nursery stock and the practical care of plants to a pronunciation guide for horticultural terms. Dewey directed much of the guide towards improving the agent's image, and warned against foppish dress, slang phrases, smutty jokes, and drinking and gambling, characteristics often linked to peddlers. Successful agents, Dewey insisted, must fit into the community, and should emphasize their domestic virtues while downplaying their seemingly unfettered lifestyle. Dewey also instructed agents on how to best use the plate book, urging them to methodically present it as a very special work of art, and to write down orders as they displayed each page. Particularly astute agents, ac-

Figure 5.10. Plant peddlers could eventually get plate books, like this *Cabinet Fruit and Flower Plates,* from a number of companies. Courtesy of Department of Rare Books and Special Collections, University of Rochester Library, Rochester, New York.

cording to Dewey, noted ornamental plants in the customer's yard and pointed out additional varieties that could enhance the premises. Most importantly, Dewey warned agents to take care of the plate books, because a "dirty, greasy, soiled book [was] not fit to place before a gentleman, and much less a lady."[45]

As they used increasingly sophisticated sales techniques and as the number of plant peddlers grew, so did the mid-nineteenth-century controversy surrounding their sales practices and influence. Many local nurs-

erymen resented their competition and the intrusion of distant nurserymen into what they considered home territory. Even though large-scale nurserymen depended on their own agents to penetrate western markets, they feared damage to their reputation from unprincipled independents and disliked the competitive pressures these roving salesmen brought to the industry. Critics of plant peddlers and their practices voiced their concerns very widely, and the press, as well as agricultural and horticultural society reports, carried repeated complaints and warnings. As James Vick sarcastically suggested, "From the deep mutterings and the outspoken complaints and maledictions one hears, apparently the greatest evils the agriculturist has to deal with are insects and tree-dealers; these two and the greatest of these is the tree dealer."[46]

Most critics focused on two issues. They interpreted the sales techniques that Dewey so proudly promoted as a sophisticated deception, and claimed that plant peddlers were masters of deceit. Beware of "oily-tongued fellows with florid prints of impossible fruits faithfully depicted between richly bound lids," Bates Harrington warned in his 1879 critique of the traveling agent industry. Harrington was particularly concerned that the plate books gave the agent an aura of respectability when, in fact, it was an "easy matter for irresponsible persons to get hold of them." (See Figure 5.10.) H. Dale Adams, a Galesburg, Michigan, nurseryman, painted a similar portrait at a local Farmers' Institute when he warned his audience that the typical plant peddler was good looking, well dressed, and likely to promote sales by playing on the farmer's pride.[47] Other critics argued that these roaming salesmen not only used deceptive sales practices, but actually perpetrated fraud on unsuspecting customers. Plant peddlers, their critics warned, sold blue roses and yellow peonies, and a myriad of other plant novelties that did not exist. In addition, they mislabeled fruit trees and other plants, and bought "old and worthless nursery stock" and then promoted it as the finest available. Some critics said peddlers sold too cheaply, undermining legitimate nursery sales, and others argued that their prices were outrageously high, often double what local nurseries charged their customers.[48]

Occasionally, first-hand accounts graphically confirmed the plethora of critics' complaints. In 1860, Levi Thumb of Irving, Illinois, wrote to John Kennicott proposing to sell trees among his neighbors, and described the activities of Rochester agents operating in the area. "We have the most meanest agents here selling trees," Mr. Thumb wrote, "the rubbish of some of the New York nurseries [*sic*] and most of them die immediately."

Commercial Realities: Promoting Demand   141

The agent, Mr. Thumb claimed, sold thousands of "leavings" from Rochester nurseries while claiming they had been purchased in Bloomington, Illinois. Not only was the source misrepresented, but the trees had been "dried and froson" [*sic*] before they were delivered to customers.⁴⁹ According to E. P. Powell, an Adrian clergyman and respected horticulturist, the residents of Lenawee, Jackson, and Washtenaw Counties were also vulnerable to the depredations of plant peddlers. In January 1870, Powell wrote a letter to the *Adrian Times and Expositor* pointing out that every winter the town was "infested" with agents that sold preposterous plants at exorbitant prices. Powell urged citizens to purchase their plants from reputable companies like Ellwanger and Barry, I. E. Ilgenfritz of Monroe, Michigan, or from Adrian's own nurserymen and florists, Loud and Trask.⁵⁰

Despite the controversy that surrounded their activities, both legitimate nursery agents and independent plant peddlers provided a very real service to the horticultural industry and their Midwestern customers. Because they sold by personal contact and persuasion, plant peddlers came to know the customers' concerns and desires, and could relay that local market information to their often distant suppliers. Frequently, agents' letters to the home nursery offered information on what was likely to sell, competing sources for ornamental plants, and what customers perceived as fashionable. Nursery agents also communicated information on pricing plants, what the competition was charging, and what prices their customers would tolerate. Concern for fair pricing, competition and a regulated marketplace often surfaced in nurserymen's conventions at midcentury, and information filtered through agents' experience helped nurserymen evaluate their options. Usually agents worried that the prices were too high, and warned the home nursery to sell at the lowest possible level if they really wanted to promote business and additional horticultural interest.⁵¹

For many Midwestern customers, peddlers offered types of plants or new varieties not readily available from local sources. The more reputable peddlers had at least a smattering of horticultural knowledge, could provide information on the most suitable plants for Midwestern conditions, and were able to actually select quality plants for their customers as they filled orders at the nursery. Some peddlers also expedited the always precarious transport of plants by delivering to farm or village homes, or by tending the plants until customers arrived at a central location to collect their orders.⁵² Most importantly for both commercial horticulturists and

Midwestern customers, plant peddlers played a significant role in the extension and democratization of horticultural interest. Thanks to the peddler, more people were interested in plants, knew something about them, and were ready to buy. As they traveled the countryside, reaching village and farm homes, peddlers made the unfamiliar and perhaps mystifying process of ordering plants unnecessary. Quite simply, plant peddlers transcended both cultural and geographic boundaries, brought a national marketplace to rural dooryards, and by their personal touch introduced fruit trees and ornamental plants to places and to people who had never cultivated them before. The general improvement of the countryside was so dependent on nursery agents that one commentator queried, "What would we have done without the aid of these very horticulturists, that are so often denounced as humbugs and swindlers?"[53]

While nurserymen and seed dealers used advertisements, catalogues, and the personal sales efforts of plant peddlers to bridge the gap between local markets and distant suppliers, they faced one last marketing hurdle. Their future prosperity hinged on expanding the market, and pushing the already firm demand for ornamental plants in new directions and among new customers. As they attempted to broaden and strengthen their customer base, commercial horticulturists developed a number of themes that surfaced repeatedly in their advertisements, catalogues, and other promotional efforts. In the process, they attached an array of new social meanings to ornamental plants that enhanced their desirability and importance, and insured that the call for trees, shrubs, and flowers could never be satiated.[54]

Among the most pervasive ideas promoted in horticultural advertising, the assertion that gardeners could achieve spectacular results with relatively little effort, and without special knowledge or skills, appeared frequently. Nurserymen and seedsmen often described plants or special collections of flowers as "showy and easy of culture," or as B. K. Bliss insisted about one of his offers, the plants were guaranteed to "give satisfaction to the purchaser" without a good deal of trouble. Although ostensibly offered as an assurance to inexperienced gardeners of all ranks, the emphasis on ease of plant culture provided a subtle yet important method of promoting horticultural interest among those with very little leisure time or who lacked the resources to hire gardening assistance. According to commercial nurserymen and seed dealers, ornamental plants belonged in every dooryard, not just around the homes of the well-to-do, and they insisted that they could supply beauty in a reasonable way.[55]

Cost, not only in terms of time and labor, but in the initial expenditure for plants, also surfaced frequently as an important theme in nursery and seed promotions. Illinois nurseryman F. K. Phoenix typified the emphasis in his newspaper advertisements, almost always calling attention to his low prices, large assortment, and quality stock. Most commercial horticulturists framed the issue of price very carefully, insisting on the one hand that their plants provided a good deal of beauty and pleasure at fair, but low prices, and on the other hand, that customers should expect to pay a reasonable price for quality. Like many others, D. M. Ferry warned against "cheap seeds," and noted that crop failures and disappointments in quality of bloom would make them "far dearer in the end."[56]

Commercial horticulturists performed another balancing act. While they advocated plants for the "million," stressing ease of culture and low cost, nurserymen and seed dealers also insisted that widespread availability did not lessen the value of the plants offered for sale. They consistently described their seeds or plants as "choice" or "select," and contended that they offered only the finest quality. Typically, commercial horticulturists defined quality in several ways. Most emphasized outcome for the gardener, and claimed that due to their careful cultural practices, customers could depend on consistent, pleasing results. In addition to the health and vigor of their plants and seeds, many nurserymen and seed growers also promoted the superlative ornamental qualities of their plants. Catalogues and advertisements were filled with descriptive terms designed to whet the appetite of even the most reluctant gardener. A sampling from Washburn's 1868 catalogue typifies the language and attributes that horticulturists routinely emphasized. Using the terms "bold-looking," "handsome foliage," "splendid profuse blooming," "extremely beautiful and interesting," "free-flowering, highly ornamental," and "graceful and magnificent," Washburn offered what amounted to a highly subjective appraisal of plant appearance, in most assertive and seemingly objective terms. These descriptions not only drew attention to the plant, but assured inexperienced or hesitant customers that their selections would provide an appropriate statement of taste or horticultural acumen. For more experienced gardeners or those more confident of their personal preferences, the superlatives opened the door to a huge range of ornamental possibilities. There was always something more beautiful, more desirable and more inviting to plant in the garden.[57]

While quality and exceptional ornamental value had importance, long-term interest in plants and their culture needed an additional stimulus. As

 they attempted to expand the market and insure its longevity, horticulturists turned to new introductions and the changing dictates of fashion as a way to pique on-going horticultural interest among all their customers. Frequently, commercial horticulturists used terms such as "new" or "rare" to emphasize the desirable characteristics of their plants, and emphasized their ability to obtain the latest fashionable varieties from Europe. Firms like the Toledo Nursery often advertised that they "spared no cost or expense to obtain everything new and valuable," while William Adair noted that he routinely "enriched" his collection with annual imports of fine European varieties. The Washburn Company, in turn, suggested that their long-standing relationship with growers in England, France, Germany, and Prussia enabled them to be the "*first* to possess every thing new whenever introduced." The Washburn Company was so eager to supply what they termed "novelties" quickly that they offered seeds without testing, but assured readers they imported their stock from reliable growers. Eventually, the Washburn Company offered a separate listing of novelties each season, and urged patrons to order early because stock was limited. In 1868, their listing included over a hundred varieties, with twenty-three "new and rare" gladioli alone. Others companies promoted novelties, but suggested that they evaluated each one, and would sell only the most exceptional new introductions. Ellwanger and Barry carefully screened new varieties each year, and added to their lists only those plants which were "real improvements." D. M. Ferry followed a similar policy and offered new introductions only after testing for quality.[58]

Along with simple "newness," commercial horticulturists frequently underscored the popularity of certain trees or flowers, recognizing that for many of their customers, acceptance by others constituted a very important selling point. In his advice to nursery agents, D. M. Dewey stressed the issue and advised peddlers to always emphasize the "most popular new varieties" for their customers. Horticulturists underscored the popularity of ornamental plants in several ways. Sometimes they simply pointed out the virtues of a well-known plant and described it as popular or in great demand. At other times, they indicated new or improved varieties of already established or well-accepted plants. The Washburn Company, for example, noted the "great improvement" that had been made in hollyhocks, a "fine old flower" that was now among the "most popular flowers of the day." Occasionally, horticulturists tried to establish new norms or popular categories by suggesting that because of their attributes or particular beauty, certain plants deserved a great deal more attention.

No matter how they manipulated the terms, their categorization of plants as popular or fashionable created a normative frame for their customers, assuring them that a particular selection fit comfortably within current ideas of good taste and style.⁵⁹

Fashion created its own necessities, and soon nurserymen and seedsmen combined designations of "popular" with other terms like "desirable," "indispensable" or "essential" for every garden. In an 1860 catalogue published in the *Michigan Farmer*, Hubbard and Davis used language that typified what increasingly appeared in other advertisements and catalogues. The firm offered customers over twenty-five varieties of fuchsias and insisted their plants were "unsurpassed and should be in every collection." They labeled delphiniums, in turn, "a great acquisition" for any interested gardener. B. K. Bliss offered very similar assessments in his 1872 catalogue, suggesting, for instance, that ornamental grasses "should occupy a prominent place in every garden," or that gardeners should consider planting the listed *Dianthus* varieties because they were "unsurpassed for effectiveness in beds or mixed borders."⁶⁰ Like the categorization of plants as popular, these seemingly objective assessments of ornamental plant value established standards of garden beauty and completeness against which customers might measure their own dooryards or flower gardens. The designations offered guidance, and the assurance that if customers selected the recommended plants, their gardens would reflect an appropriate degree of knowledge and sophistication. For nurserymen or seedsmen, promoting fashionable necessities offered endless opportunity. Despite the seemingly factual assertions, the standards were in fact arbitrary and ever-changing, enabling commercial horticulturists to establish changing norms and "necessary" new plants year after year.

As nurserymen and seed dealers went about creating new markets and expanding the old, they added layers of meaning to the carefully constructed public significance of trees, shrubs, and flowers. Commercial horticulturists certainly supported the claims of reformers who linked plants to properly embellished and morally influential homes, or who understood an ornamented dooryard as a sign of progressive tendencies. Often, they incorporated those ideas into their own promotional strategies. But in the competitive world of horticultural marketing, customers needed to link ornamental plants with newness, with fashion, and with the very powerful idea that individuals or families could, simply and economically, purchase refinement and good taste. In the hands of commer-

cial horticulturists, the right plants in the dooryard offered everyone the chance to obtain the symbols of respectability, transform their public image, and communicate their fashion acumen.

Speaking before members of the Michigan Pomological Society in 1872, Ohio nurseryman and horticultural advocate M. B. Bateham speculated about the reasons for the rising interest in horticulture, and offered an assessment of the role of commercial horticulturists that still rings true. Ornamental plant culture, according to Bateham, was indeed an activity "for the million," and was vastly encouraged by the advertising, promotions, and catalogues of nurserymen and seed dealers.[61] For the residents of Jackson, Lenawee, and Washtenaw Counties, and other Midwesterners, the efforts of commercial horticulturists added another piece to the horticultural puzzle. Whether in the inventories of small local nurseries, regional firms, or vast enterprises with a national clientele, nurserymen and seed dealers offered a huge array of trees, shrubs, and flowers. They encouraged horticultural consumers to think about ornamental plants as something to purchase rather than trade, urging them, at the same time, to move beyond local patterns of exchange and into a national marketplace. As commercial horticulturists diligently worked to perpetuate and expand their markets, they developed an egalitarian approach, welcoming as customers anyone who could read an advertisement or send for a catalogue, or who simply loved trees and flowers. Their promotional efforts overlooked disparities in economic rank or personal circumstances, and repeatedly argued that everyone could enjoy the simple, economical benefits and pleasures of ornamental plant culture. Most importantly, commercial horticulturists encouraged their customers to think of home embellishment as evidence of good taste and fashion consciousness. Ornamental plants and their arrangement were key variables in this new fashion interest, and their presence, arrangement, and culture reflected not only the moral virtue of homeowners, but their taste and sense of style.

Six
A Neighborly Nudge

Local Models and the Promotion of Horticultural Interest

The efforts of commercial horticulturists to develop markets and supply plants certainly extended interest in trees and flowers, but not all Midwesterners found their initial horticultural inspiration in the pages of nursery catalogues. In the mid-1870s, W. E. H. Sobes and J. Evarts Smith, both active members of the Ypsilanti, Michigan, chapter of a national agricultural reform organization called the Patrons of Husbandry or the Grange, reported on a new program that exemplified another, far more personal source of ornamental plant interest and enthusiasm. The men were part of a "visiting" committee that toured designated farms in the Ypsilanti region, and then described their findings to other Grange members. The committee was particularly concerned with "anything coming under our observation calculated to instruct or emulate us in our calling as husbandmen to the end that we may beautify and make comfortable our homes and add to the profits of our business."[1] As they looked over other Grange members' dooryards, outbuildings, crops, and livestock, committee members hoped to share in the first-hand experience of the area's progressive farmers. Their actions, however, underscored an important truth about the dissemination

of new ideas and innovative practices. For many Midwesterners, seeing was believing, and the examples of their neighbors or friends were far more influential in sparking horticultural enthusiasm than the most well-intentioned ad-vice of reformers or the innovative promotions of nurserymen and seed dealers.

Some horticultural advocates understood the importance of local examples in spreading their ideas, and frequently discussed the efficacy of "emulation" or "laudable rivalry" in promoting ornamental plant interest.[2] Local horticultural examples also had an immediacy and concreteness that made them particularly potent agents of change. While noting the new agricultural implements and improved livestock breeds exhibited at the first Jackson County Fair in 1853, Agricultural Society Secretary R. Lathrop articulated a widely understood benefit of firsthand observations. We "can see and examine for ourselves," Secretary Lathrop declared, and added that the experience offered opportunity for comparisons, self-appraisal, and eventually the "liberalizing" of attitudes and actions. Thomas Meehan, editor of the *Gardener's Monthly*, further emphasized the importance of local models for embellishment when he described the relationship between horticultural societies and the more esoteric and distant horticultural journals. "The magazine," Meehan insisted, "shows what should be done, the society reasons by results, and exhibits the sum to the public."[3]

The residents of Lenawee, Jackson, and Washtenaw Counties had a vast array of "magazines" and commercial resources at their disposal. But what of local opportunities to observe ornamental plant culture and use? This chapter examines three such resources, exploring first local models of home and farm embellishment designated by the agricultural press or societies, then considering the horticultural influence of fairs and exhibitions, and finally appraising how local "horticultural exemplars" aided in promoting ornamental plant interest. As they went about the business of planting and pruning, these local cultural mediators offered neighbors and friends a firsthand view of horticultural possibilities, and at the same time quietly provided additional reasons to plant trees and cultivate flowers.[4]

For some residents of southeastern Michigan, local horticultural models emerged from the pages of agricultural or horticultural journals. Like Thomas Meehan, many editors understood the effectiveness of concrete examples in carrying out their policies and programs, and frequently described what their readers could learn if they would only look around

their own neighborhoods. Editors or correspondents often traveled, sometimes to attend fairs or exhibitions, and at other times to evaluate specific agricultural or horticultural practices in their own region or in other parts of the country. As they moved about the countryside, these observers reported what they encountered along the way, noting unusual sights or simply describing farms, businesses, homes, or towns and villages. Frequently, the enterprises they highlighted exemplified some of the innovative or progressive practices that the agricultural or horticultural press consistently promoted, and the actual existence of a thriving example of deep plowing, a carefully cultivated orchard, or a well-ordered farmyard simply added fuel to their reforming fires. Although the practice was intended to stimulate interest and edify all readers, it also designated local models, setting a particular farm or business apart as an enterprise to notice and perhaps emulate. Distant subscribers received a broadened view of agricultural or horticultural progress, but for those nearby, the designated models established a touchstone of progressive tendencies, recognized beyond the region yet available for firsthand examination and comparisons. Often, ornamental plant embellishments emerged as an important part of these exemplary scenes.[5]

Lenawee, Jackson, and Washtenaw County families might read accounts of Michigan travels in newspapers such as the *Moore's Rural New Yorker* or the *Prairie Farmer*, but were most likely to encounter regional descriptions in the *Michigan Farmer*. In a typical example of the genre, horticultural editor S. B. Noble visited a number of Washtenaw County farms during his 1853 "Rambles" across the state, and pointed out innovative techniques of insect control, exemplary sheep flocks, and superior cultivation practices. In several instances, he also carefully described ornamented front yards, clearly linking shrubbery and shade trees with "good modern style" homes and well-managed farms. In 1856, the *Farmer* returned to Washtenaw County, visited the farm of William Burnett, and developed an even more definitive model of a progressive farm featuring both sound agricultural practices and horticultural improvements. The *Farmer* noted the quality of the farm buildings, the condition of the orchard, and Mr. Burnett's experiments with a new wheat variety, but then pointed out that the Burnett home was surrounded with a good selection of ornamental trees and shrubs. Mrs. Burnett, the *Michigan Farmer* added, was an "amateur cultivator of green house plants of the choicest kind." In addition, she nurtured roses and cultivated a wide variety of annual and perennial flowers. Her floriculture efforts, the *Farmer*

noted, were "worthy of emulation by others having equal advantages." Nearly twenty years later, *Michigan Farmer* correspondents still linked progressive agricultural practices with horticultural refinements. After examining the fencing, roads, drainage techniques, crop yield, and improved livestock breeds on the Ypsilanti farm of Danicl L. Quirk, the *Farmer* found all of those "improvements and changes" so exemplary that it declared the farm "almost a model." Right along with the more practical improvements, ornamental plants added to the progressive picture. The dooryard was a pleasing sight, displaying "some very fine evergreens, horse chestnuts, maples, and other shade trees, which are," the *Farmer* claimed, "in a very fine growing condition."[6]

In addition to farm landscapes, commentators also described village scenes, putting forth particular aspects of the economy like manufacturing firms, mills, or businesses as worthy of admiration and imitation. Here again, their descriptions often linked the town's appearance with its bustling industrial or commercial enterprises. As early as 1843, the *Michigan Farmer* urged its readers to look to Ann Arbor as an "excellent example for other villages," stressing its "order and neatness" and the practice of ornamenting streets and houses with "fine shade trees and ornamental shrubs." S. B. Noble's 1854 description of Ann Arbor offered an even more detailed appraisal of the desirable features of a prosperous, progressive city. The town's public buildings were "substantial brick edifices," the city enjoyed many educational resources, and its steam engine and stove factories, paper mills, and other manufactories and commercial enterprises were thriving. Noble described Ann Arbor as an attractive town, and pointed out that its well-built homes "with their beautiful gardens, its spacious streets ornamented with trees, [and] its well-stocked park" were unsurpassed in the state. For those who wanted to enhance their town or village, the examples put forth by the *Michigan Farmer* had a clear message. The most progressive villages and urban centers bustled with business, but also cultivated an ordered landscape embellished with appropriate trees, shrubs, and gardens. Practical horticultural refinements tempered the commercial activity with a note of repose, stability, and beauty, but were part of the same progressive mentality that fostered business and promoted industry.[7]

As editors designated local models for improvement, they also understood that an even more potent force for change existed in many towns, villages, and rural regions. Since its inception, the agricultural press firmly supported agricultural or horticultural societies, farmers' clubs, or any other organization that encouraged rural sociability, the exchange of information, and the sharing of progressive tendencies. The horticultural press also took up the call, and often described the effectiveness of horticultural societies in improving landscapes, increasing knowledge, and refining whole communities. Like journals and newspapers, many of these agricultural and horticultural associations conducted local or regional surveys, reported on the results to members, and in the process designated additional local models of progressive agricultural or horticultural practices. Generally, societies initiated their surveys for one of two purposes. In some instances, visiting committees evaluated a selected group of farms that were competing for recognition or prizes as a part of annual fairs and exhibitions. The society established a framework for the competition and solicited entries, but individuals or families put themselves forward as contestants for the premiums. In other cases, members of organizations like the Ypsilanti Grange simply wanted to learn from their neighbors and created a forum for communicating successful practices. Both processes set certain farms or homes apart as exemplars for others in the region.[8]

The records of the "Visiting Committees" of the Ypsilanti Patrons of Husbandry provide a particularly revealing account of both the process of defining local models, and the role that ornamental plants and horticultural embellishments played in a very localized vision of progressive agricultural practices. Along with other Midwestern states, Michigan experienced a real surge of interest in the Patrons of Husbandry in the early 1870s, and by 1874, residents had established nearly three hundred local Granges across the state. In August 1873, twenty-one men and eight women organized the Ypsilanti Grange of eastern Washtenaw County, and within the year, an additional twenty-six "brothers" and seventeen "sisters," as Grangers termed their members, joined the organization. Like other early Granges, the Ypsilanti Patrons of Husbandry addressed a broad range of agricultural concerns including the need for a general "increase of knowledge," boosting the "social position of farmers," and advancing farmers' familiarity with business and trade practices. In addition, they promoted activities that would "dispense" with middlemen and eventually "overthrow the credit system." The regional farm visits of the Ypsilanti Grange grew out of one final and more modest goal clearly articulated in the purposes of the organization. The Patrons of Husbandry vowed to "teach better culture of the soil and to surround our homes with beauty and comfort."[9]

Visiting Committees of the Ypsilanti Grange were normally composed of two or three members who traveled to a fellow member's farm, toured the premises, and commented on any practices they thought notable or worthy of imitation. Sometimes visits involved a sociable exchange among Grange members, and as they did after touring the farm of William Watling, the Committee occasionally capped the day with a "real grange feast" and a game of croquet. Visiting Committees often examined drainage and soil types, evaluated fencing, and noted the condition of barns, other outbuildings, and farm implements. The order and general neatness of the home, farm yard, and fields seemed a primary concern in most commentaries. In a typical appraisal, the Visiting Committee found Brother and Sister Miller to be "worthy patrons," commended their farm for having a "place for things and everything in place," and concluded the family was "worthy of our emulations as a modle [sic] in this regard."[10]

In most descriptions, horticultural embellishments formed an important part of the model farm, and appeared right along with good fences or well-tended tools as signs of superior management practices. At the B. D. Loomis farm, the Committee pointed out the quality of tillage, described the fencing, noted a "fair barn" and a "good comfortable house," and then recorded the "embellishing of the yard with shrubery [sic] and vines, giving the house a look of good taste and refinement." Brother Randall's well-managed Pittsfield Township farm, in turn, had a "neat and comfortable" home, surrounded by a "clean yard, nicely shaded by fruit trees on the east and a beautiful little grove on the west." On another visit, this time at the farm home of Brother Henry Preston, Committee members were particularly impressed with a "very handsome growth of sugar maples" ornamenting each side of a "beautiful avenue" leading to the farm buildings.[11]

Although their vantage points differed, the Visiting Committee of the Ypsilanti Grange and the agricultural press shared some common assumptions as they described regional model farms. In both circumstances, commentators cited ornamental plants as an important component of progressive farming practices, and seemed to consider them as integral to an appropriate model farm as good barns or well-tilled fields. Like improved livestock breeds or sturdy fences, ornamental plant culture represented a commitment to progressive management practices, agricultural innovation and the market economy. Generally, however, their comments accentuated certain types of horticultural ornaments and

described them within the context of the overall order and neatness of the farmscape. Refinements like shade trees or ornamental shrubs offered an element of beauty, but seldom required either undue expenditure or excessive labor. In addition, deciduous trees had the practical advantage of providing welcome summer shade, while judiciously planted evergreens could break the winter winds. For provident and wide-awake farmers, lawns, trees, and shrubs offered just the right level of embellishment, signaling an intelligent and productive refinement that was part and parcel of the progressive outlook, but never crossing the line into mere fashion or ostentatious display.

According to their proponents, designating local examples of improvement constituted only one of many ways that agricultural or horticultural societies worked their reforming magic. Ideally, farmers' clubs, Granges, horticultural associations, and agricultural societies all provided a forum for intellectual and social exchange among members, and promoted that positive competitive spirit that encouraged innovation and experimentation. More importantly, however, these organizations sponsored fairs and exhibitions, and mid-nineteenth-century commentators stressed their influence in offering examples of new ideas, technological innovation, and agricultural improvements. In his 1848 remarks before the Detroit Horticultural Society, Michigan nurseryman and agricultural promoter J. C. Holmes typified their enthusiastic appraisal. "Exhibitions are to Horticulture what museums are to painting and sculpture," Mr. Holmes declared. "It is publicity; It is example! Emulation, light, and sometimes it is glory! To create exhibitions of horticulture, is to give to that useful and graceful science, the benefit and the great principles of mutuality and of association." Over twenty years later, the *Jackson Daily Citizen* made a similar appraisal, noting it was well worth a farmer's time to participate in agricultural exhibitions, for there they could "learn the road to real progress and lasting improvement in their calling."[12]

By the mid-nineteenth century, the residents of Lenawee, Jackson, and Washtenaw Counties, like other Midwesterners, had an array of local, regional, and statewide agricultural and horticultural associations to choose from, and many of these organizations promoted their activities through annual fairs or exhibitions. Lenawee County residents organized their Agricultural and Horticultural Society in 1849, and held their first fair that same year. Officials pegged attendance at nearly five thousand visitors, and proclaimed their efforts successful. In 1851, fair officials estimated that between six thousand and eight thousand visitors enjoyed the

fair, and noted that the event had produced a "laudable spirit of emulation amongst all classes of producers." In addition to annual fairs, the Agricultural Society cooperated with the Adrian Horticultural Society to hold weekly horticultural exhibitions during the spring and summer months. The practice continued through the early 1850s, and although Society officials generally reported a good attendance, they sometimes lamented that farmers failed to show a "desirable" amount of interest in the meetings.[13]

While there is some evidence that Washtenaw County residents formed an agricultural and horticultural society in the early 1840s, the association drew up its constitution and was officially recognized by state officials in 1849. Like Lenawee County, Washtenaw County Agricultural Society members held their first fair in October 1849. Apparently, Washtenaw County officials had little trouble attracting members, for the Society's first president, William Finley, recorded the names of 243 county residents who paid their $1 membership fee in the Society's first year. By 1868, the Washtenaw County Agricultural Society had 182 life members while many additional residents continued to contribute $1 for an annual membership.[14] Jackson County residents also organized an agricultural society in the early 1840s, but did not receive official state recognition until 1852. The Society held its first fair in the Jackson public square in 1853, and briefly took over the court house for the exhibition of fruits, vegetables, and domestic manufactures. The Jackson County Agricultural Society flourished, quickly obtained land for a permanent fairgrounds, and by 1858 reported 1,211 members.[15]

In all three counties, annual fairs remained an important part of agricultural society activities for decades, and officials began to purchase and improve permanent fairgrounds. In the 1860s, Lenawee County officials reported that the "Society has very good grounds near the City of Adrian with a large and convenient hall and other buildings necessary for holding its exhibitions." A well-graded half-mile horse racing track and grandstand further improved the grounds by 1869. Despite opposition from other parts of the county, Washtenaw County Agricultural Society officials purchased a fairgrounds in Ann Arbor in 1859. The Society continued to make improvements, and by 1871 boasted "ample room and facilities to accommodate exhibitions," a good race track, and a new floral hall with four large wings and the center of the building devoted to flower display. By the late 1850s, the Jackson County Agricultural Society had also improved its fairgrounds with the planting of hundreds of shade trees, the grading of the race track and drives, the establishment of an eating house, and benches for fair visitors. When the State Fair came to Jackson in 1869, County Agricultural Society officials attempted even further improvements and purchased additional acreage to accommodate both fairs.¹⁶

Additional agricultural societies and farmers' organizations also flourished in the three-county area. In Lenawee County, for example, farmers in Franklin, Tecumseh, Ridgeway, and Raisin Townships established the Northern Lenawee County Agricultural Society in 1856. In 1859, residents in Medina, Hudson, Rollin, and Woodstock Townships joined their neighbors in Hillsdale County to form the Hillsdale and Lenawee Union Agricultural Society. Progressive Washtenaw County residents, along with western Wayne County citizens, founded the Washtenaw and Wayne Union Agricultural Society in 1856. A number of farmers' clubs also sprang up in the region. Like the Lenawee Junction Farmers' Club, formed in 1873, these organizations met regularly to discuss agricultural and horticultural topics.¹⁷

Some of these organizations supported their own fairs and exhibitions, and encouraged farmers in the region to attend. In a typical example, the Hillsdale and Lenawee Union Agricultural Society hosted an annual fair in Hudson, and in 1863 notified *Michigan Farmer* subscribers that they "hope[d] and expect[ed] a general attendance of the farmers and others." The first fair of the Northern Lenawee County Agricultural Society, held in Tecumseh in 1856, attracted nearly three thousand visitors and over five hundred entries for the one-day affair. Fair officials set up the exhibition in a "very beautiful grove" outside the village, and even constructed a building for the display of fruit, flowers, and domestic manufactures. The *Michigan Farmer* declared the event successful, and noted "we were highly pleased with the general spirit which pervaded all who visited or took part in the fair."¹⁸

While local residents busily organized their own associations, agricultural leaders like J. C. Holmes promoted a statewide agricultural society. Michigan's Territorial Legislature granted a charter for a regional agricultural association in 1833, but had no resources to support the organization. A second and more fruitful attempt at organization began in 1849. State officials put out a call for interested citizens to convene, and in March 1849 the newly formed group met to elect officers. The legislature quickly passed an act to incorporate the State Agricultural Society, and agreed to provide a $400 subsidy, if the Society could raise matching

funds. Over the years, the state gradually increased its appropriations, agreeing to $1,000 in 1851, and doubling that amount by 1858. Detroit nurseryman and horticultural advocate J. C. Holmes acted as secretary of the organization from its founding until 1857, and was followed by R. F. Johnstone, editor of the *Michigan Farmer*, who remained in office until 1875. A number of Lenawee, Jackson, and Washtenaw County residents were active in the organization of the State Society, and held important positions or maintained memberships for a number of years. Governor Ransom, for example, appointed Mr. Deming from Lenawee County and Mr. Salyes from Washtenaw as members of the organizational committee. In addition, residents like John Starkweather of Washtenaw County and A. J. Dean of Lenawee served on the State Agricultural Society's Executive Committee, while other county residents were active members of the organization.[19]

Like other agricultural associations, the State Agricultural Society depended on annual exhibitions to spread its message and stimulate interest. The Society held its first fair in 1849, and until 1862 conducted all but one of the early exhibitions in Detroit. In a letter to State Secretary J. C. Holmes, Washtenaw County nurseryman E. D. Lay described the interest and support of area farmers for the first fair, and asked for "some half a dozen copies of the regulations of the State Fair as there is quite a number enquiring about it." Lay assured Mr. Holmes that with a little promotion in the region, at least fifty local residents might provide entries for the State Fair competition.[20] Beginning in 1863, State Fair officials hoped to improve participation and attendance by moving the Fair around the state. Kalamazoo hosted the festivities in 1863 and 1864, Adrian in 1865, and Jackson in 1869 and 1870. In 1871, an association called the Northern Michigan Agricultural and Mechanical Society, organized to meet the needs of northern agriculturists, held a competing and very successful fair in Grand Rapids. Finally recognizing the dangers of too much competition, the two organizations joined forces and together held another well-attended Grand Rapids exhibition in 1873.[21]

In addition to agricultural societies, the residents of Lenawee, Jackson, and Washtenaw Counties enjoyed the benevolent influence of a number of horticultural associations. In 1846, a small group of plant enthusiasts organized the Detroit Horticultural Society, probably the first official horticultural society in the state. The organization began with ten members, but during its relatively short life attracted about a hundred supporters, including many of Detroit's commercial nurserymen like

William Adair, E. G. Mixer, J. C. Holmes, and Canadian nurseryman James Dougall. A number of professional and amateur fruit growers and floriculturists from the eastern side of the state also participated. From the three-county area, Ypsilanti nurseryman E. D. Lay and Mark Norris, an amateur Washtenaw County horticulturist, were active Detroit Horticultural Society members. Unlike most agricultural societies that charged a $1 "yearly subscription," members of the Detroit Horticultural Society paid a $2 induction fee in addition to the yearly subscription. The higher cost may at least partially explain why membership was always somewhat limited.[22]

Although lacking in numbers, members of the Detroit Horticultural Society took up the horticultural cause with enthusiasm. They were convinced that the activities of horticultural societies had a very positive influence on "mental culture and enterprise," and promoted their ideas through an ambitious program of horticultural exhibitions. Society officials established committees to consider fruits, vegetables, trees and shrubs, indigenous plants, greenhouse plants, and florists' flowers, and urged horticulturists from across Michigan to send specimens for exhibition. Using the Odd Fellows Hall and several other sites around Detroit, the Society held three exhibitions in 1846 and four in 1847. The *Michigan Farmer* reported that the events were well-attended and featured good displays of fruits and vegetables. Visitors could also enjoy modest, but often amazingly varied, exhibits of greenhouse plants and cut flowers. By 1852, interest in the Horticultural Society declined. The Society attempted to broaden its membership base, and the *Michigan Farmer* described it as a "substantial and we trust permanent organization," but its 1852 exhibitions generated little enthusiasm, and members soon turned to new interests and organizations.[23]

With the decline of the Detroit Horticultural Society, horticultural advocates from other parts of the state began clamoring for an new organization. In September 1853, a convention of southern Michigan nurserymen met at Jonesville, in Hillsdale County, to discuss their needs, and called a second meeting for February 1854, in Adrian, to elect officers and adopt a constitution. The Michigan Nurseryman's and Fruit Growers' Association was primarily interested in regularizing the nursery trade, in controlling competition especially from eastern plant peddlers, and in sharing horticultural information, as members hoped to "advance generally horticultural interests of this and neighboring states." A number of horticulturists from the three counties were active in the Association, including

nurserymen B. W. Steere and D. K. Underwood of Adrian, John Merritt of Tecumseh, and S. B. Noble of Ann Arbor. Well-known amateur horticulturist S. O. Knapp of Jackson also participated. The Association met for several years in Adrian, and in 1856 held a large exhibition in Jackson, but once more interest and support waned.[24]

Horticulturists put out another call for a new organization, this time advocating a group with broader interests and less commercial emphasis, and after a September 1857 meeting in Jackson to draft a constitution and by-laws, organized the Michigan State Horticultural Society. In January 1858, the Society held a very successful meeting and exhibition in Kalamazoo, but a June exhibition in Detroit attracted few exhibitors or visitors. Again, a number of residents from the three-county area supported this new state society, with B. W. Steere serving as a vice president, D. K. Underwood as a director, and others like Daniel Cook of Jackson exhibiting horticultural products. Unfortunately, the Michigan State Horticultural Society was, like its predecessors, short-lived, and its members soon needed to look elsewhere for horticultural stimulation.[25]

In 1870, largely inspired by a growing interest in both commercial and home fruit culture, Michigan horticulturists finally managed to form a more permanent statewide organization. Meeting in Grand Rapids, organizers agreed to hold monthly meetings and an annual exhibition, and to call their new organization the State Pomological Society of Michigan. A committee worked to secure state incorporation for the society, but finding that state law did not cover horticultural societies, turned their attention to lobbying for appropriate legislation. The state legislature complied, and in 1871 passed a law giving horticultural societies the same rights and privileges of agricultural societies. The law stipulated that the transactions of a state horticultural society, like those of the agricultural society, should be printed and distributed annually. Over the years, the state subsidized printing of the Pomological Society transactions, but provided little additional financial support for the organization.[26]

The State Pomological Society held its first fair, in conjunction with the Kent County Agricultural Society, in 1870 and its second (1871) and third (1872) in cooperation with the newly formed Northern Michigan Agricultural and Mechanical Society, also in Grand Rapids. Beginning in 1873, the Pomological Society joined forces with the State Agricultural Society to hold joint annual exhibitions. Under the guidance of experienced and enthusiastic horticulturists, the fruit, vegetable, and flower departments of the State Fair received a considerable boost, and by 1875

fair organizers proudly reported 1,192 horticultural entries. In 1874, the Society voted to hold quarterly rather than monthly meetings, and continued its policy of convening meetings around the state, often joining with local societies to share information, reports, or to host small exhibitions. Initially, southwestern Michigan horticulturists dominated the state organization, but gradually membership broadened, and a number of residents from the three-county area became actively involved. Jackson's S. O. Knapp, for example, served on the executive committee in 1873, and Lenawee County residents Artemus Sigler and J. W. Helme presented papers at the January 1874 Adrian meeting. As the years went by, a number of other residents of the region read papers, participated in discussions, or served as Pomological Society officers.[27]

The activities of the State Pomological Society seemed to encourage local societies, and in the late 1870s and early 1880s, a number of counties formed horticultural societies as auxiliaries of the State Society. Washtenaw County, for example, organized a Pomological Society in 1878, while Jackson County horticulturists organized their own society in 1880. Membership numbers never matched those of agricultural societies, but these local organizations boosted interest and the influence of the State Pomological Society. In the mid-1880s, Charles Garfield, long-time secretary of the State Pomological Society, reported that the Society had 165 life members, 150 annual memberships, and an additional 670 members drawn from the local auxiliaries who also enjoyed membership in the statewide Society. Local societies, in turn, often attempted to promote horticultural interest among friends and neighbors. The Jackson County Horticultural Society noted that it had only seventeen members, but that the Society regularly published meeting reports in local newspapers, and distributed horticultural information at the Jackson County Agricultural Fair in an attempt to broaden its reach.[28]

The residents of Lenawee County enjoyed the benefits of a local horticultural society longer than any other region in the state. Established in 1851, the Adrian Horticultural Society maintained its independent and very active status until 1881, when it, like other local societies, became an auxiliary of the State Society. The Adrian Society encouraged both men and women to join as voting members, charged an annual fee of $1, and attempted to offer members access to the latest horticultural information. The Society subscribed to all the prominent horticultural journals, and by 1877 reported a library of "upwards of two hundred volumes." Notices in the local newspaper often announced meeting times and location, and

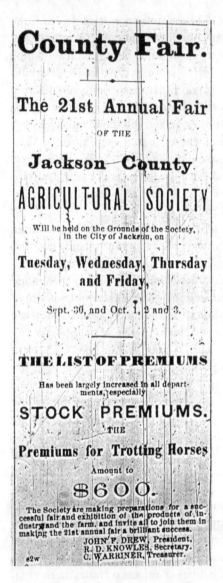

Figure 6.1. To encourage participation, most agricultural societies advertised their fairs, and offered premiums for those who entered items for judging. From *Jackson Weekly Citizen* (September 1873): 6.

encouraged all to attend. Fruit and flower exhibitions were a popular feature of the Adrian Horticultural Society activities. In 1869, the *Adrian Times and Expositor* reported that the annual exhibition took place in an unoccupied store several doors down from the newspaper office, and that the flower display was particularly eye-catching.[29]

Members of all of these organizations enjoyed access to horticultural information through lectures, discussions, and the sharing of printed materials, but annual fairs and exhibitions extended the reach of the associations into the broader community. These gatherings quickly became major social as well as educational events, and drew visitors from near and far to examine the entries, enjoy the competition, or simply to mingle with other fairgoers. When the first State Fair opened in Detroit in 1849, crowds abounded, and the *Michigan Farmer* could hardly contain its excitement. "The multitudes came pouring in from every quarter, and through every channel," the *Farmer* gloated, "by railroad, steamboat and stage—in wagons, on horseback and on foot, until our streets swarmed with human beings." Even smaller regional fairs attracted many visitors, and on fair days towns like Adrian or Jackson bustled as farm families mingled with city residents to enjoy the festivities. The *Adrian Times* described a typical scene for the 1869 Lenawee County Fair, declaring that "the city was crowded with teams and wagons bearing full freights of passengers to the fair." Several years later, the *Times* again noted that as the County Fair began, "Teams filled and crowded the streets, and foot passengers jostled each other on the side walks, and every portion of the county seemed to be represented."[30]

Thousands of visitors per day created all the hubbub. At their 1873 County Fair, for instance, Jackson County Agricultural Society officials registered an attendance of seven thousand on a single day, and thought it a record among county fairs for that season. (See Figure 6.1.) The State Fair sometimes drew even larger crowds. Officials sold an estimated thirty thousand tickets on opening day when they held the 1870 State Fair at Jackson. Fair attendance on that day nearly quadrupled the population of the city, and was only five thousand short of the entire population of Jackson County. Needless to say, the event put tremendous pressure on the city to provide food and accommodations for all the visitors, attracted a contingent of "roughs, gamblers, and prostitutes," and generated a "fearful" amount of dust from all the comings and goings.[31]

Many visitors traveled to fairs by rail. Railroad companies, in turn, were very willing to accommodate the crowds, often adding special excursion trains or reduced rates for fair visitors. For a number of years, fair exhibitors received an even greater boost, for railroads transported "stock, implements, and goods" free of charge, and offered exhibitors themselves reduced rates. Their assistance in carrying livestock or farm implements was particularly important in encouraging those distant from the

fair site to enter the competition, but the railroads knew a good investment when they saw it. They practiced their philanthropy with the clear understanding that a fine exhibition would draw good crowds and more rail passengers from around the state.[32]

Once they arrived at their destination, fair visitors had many opportunities to expand their horticultural horizons. A quick trip down Ann Arbor's tree-lined streets or a walk through Jackson's newly developed rural cemetery, complete with rolling hills and choice evergreen plantings, offered a glimpse of progressive new landscape practices.[33] For most visitors, however, the real focus of their trip was the fair itself and they turned to the exhibitions as a source of both new ideas and horticultural inspiration. Fair officials hoped that the fruit, vegetables, and flowers, tucked among the livestock, plows, and domestic manufactures, represented the best the region could offer, and in their abundance and beauty would encourage even their most reluctant fairgoers to think about horticultural improvement. Local or state fairs were indeed an important source of both horticultural information and inspiration, and because of their popularity, the messages they transmitted deserve careful appraisal.

Readers of mid-nineteenth-century horticultural advice literature were in for a surprise when they visited exhibits across the region. Even though professional horticulturists continually emphasized tree and shrub planting or lawn maintenance as fundamental to properly embellished landscapes, exhibitors could hardly bring their Norway spruce or favorite shrubs to the fair. To a large extent, fairgoers saw what horticultural exhibitors could readily transport to the exhibition grounds, usually cut flowers displayed singly or in collections, bouquets of various types, or potted greenhouse or parlor plants. Fair officials were well aware of this bias and over the years tried to remedy it by establishing premium categories that included shade trees, ornamental grounds, suburban yards, or flower gardens, and scheduling on-site visits to evaluate examples and award prizes. In 1851, the Lenawee County Agricultural Society established two premiums, the first rewarding the "best Ornamental Forest Trees" planted along roadsides, and a second for the "greatest number of Ornamental Trees in or near the road, so as to shade it, saved on any Farm in the County, and trimmed this Spring." That same year, the Agricultural Society also offered premiums for the best nursery in the county and for superior flower gardens. Several years later, the Lenawee County Society added a category for the "best ornamental trees, saved and trimmed," on

a farm. The State Pomological Society, in turn, established prizes for the best suburban ornamental grounds, the best village or city garden, and the best private planthouse. All of these measures rewarded society members for diversified interests and may have encouraged them to broaden their horticultural perspectives, but did little to expand the ornamental plant offerings available to the average exhibition visitor.[34]

In addition to the limits imposed by the exhibition structure, the premium system strongly influenced the types of entries that exhibitors normally displayed at fairs. Most agricultural society officials assumed that individuals needed the encouragement of a monetary prize if they were to put forth their best efforts, and, in order to judge the entries and award prizes, designated the competitive categories open to participants. For those exhibitors thinking only of the competition, fair categories might influence the kinds of flowers they grew. For fair visitors, in turn, premium categories not only defined the types of plants they could see, but underscored a particular set of floral characteristics as most beautiful or desirable. Mid-nineteenth-century nursery and seed catalogues offered hundreds of varieties of greenhouse plants and garden flowers, yet out of a myriad of horticultural possibilities, fair officials often selected very few. At the 1850 Michigan State Fair, for example, premium categories included dahlias; the best variety of cut flowers, roses, and indigenous flowers; the best floral design; the best hand, flat and round bouquets; and the best design for a flower garden. By 1855, categories grew to include the best variety of cut flowers; greatest variety of dahlias, roses, phlox, verbenas, petunias, pansies, German asters, indigenous plants; and the best collection of greenhouse plants. Officials awarded additional premiums for the best bouquets and baskets of flowers. A comparison of the total list of entries with those that actually won premiums confirmed that most exhibitors entered plants that corresponded to established categories. County fair officials devised similar categories. In 1852, the Lenawee County Agricultural Society offered prizes for the greatest collection of hardy June roses, perpetual roses, annual flowers, perennial flowers, greenhouse plants, and the best bouquet.[35]

There were several loopholes in the premium system that promoted at least some additional diversity in horticultural displays. Prizes for "best collections," whether of a particular kind of flower or for the greatest variety of blooms, broadened the range of plants exhibited. The entries of Mrs. Mark Norris, an amateur horticulturist from Ypsilanti illustrated the possibilities for these inclusive categories at the 1850 Michigan State

Fair. Mrs. Norris competed for a prize labeled "best variety of cut flowers," and brought a collection of eighty-two different flowers to display. Fairgoers could admire perennial and annual hibiscus, ageratums, alyssums, amaranthus, asters, verbenas, zinnias, Chinese primroses, scabiosas, heliotropes, dahlias, dwarf lupines, portulacas, and a number of other flowers, all from the garden of a single exhibitor.[36]

Although influential, established categories and the premium system did not wholly determine what plants came to the fair. Whether eligible for prizes or not, a few participants brought along valued specimens, unusual collections, or unique varieties. If resources were available, fair officials sometimes recognized their efforts by awarding discretionary premiums. In 1851, Lenawee County officials offered premiums for collections of hardy perennial flowers, annuals, dahlias, and greenhouse plants, but noted that "discretionary premiums not exceeding $30 may be awarded on such fruit and flowers as are not provided for in the foregoing list, but which from their merit, may be entitled to premiums." Twenty years later, the Washtenaw County Agricultural Society followed a similar policy. Of the thirteen premiums awarded, eleven fell into previously articulated categories, but two entries, the "best fig tree" and the "best oleander," did not match the premium list. Recognizing their merit, officials awarded the entries discretionary premiums.[37]

The season when most agricultural societies conducted their exhibitions also influenced the range of horticultural entries. Designed to showcase products of the agricultural harvest, an exhibition held in late September or early October missed the prime flowering time for many outdoor ornamental plants. The season was particularly significant since most fair officials emphasized that plants should be in flower at the time of the exhibition. Fair officials hoped for a brilliant and attractive display, but this emphasis on the flower rather than the quality of the plant's culture, its uniqueness, or its vegetative qualities tended to skew the available horticultural examples. The range of houseplants available for exhibition was less seasonally dependent, but similar preference for flowers again limited the possibilities. Michigan State Pomological Society officials underscored the point, commenting during a state exhibition that they could see little "propriety in offering premiums for flowering plants that cannot be in blossom during the period of our state fairs." Their decision removed ornamental favorites like camellias, azaleas, and even some geraniums from the specimens available to visitors, and did nothing to enhance the floriculture possibilities for fair visitors.[38]

Although their scope varied tremendously, horticultural societies tended to offer a more diversified array of ornamental plants than did their counterparts in agricultural societies. The broadened perspective stemmed from several sources. Horticultural societies generally had a much more specific focus, and often drew exhibitors from a group of skilled and intensely interested amateurs who specialized in unusual plant types or who cultivated large collections. Small local or regional horticultural societies were far more likely to hold exhibitions without the guiding structure of the premium system, and as a result often showcased an amazingly diverse, if not disparate, collection of ornamentals. Occasionally, horticultural societies expanded their offerings by hosting several smaller exhibitions throughout the growing season, or by presenting their primary exhibition earlier in the summer. This practice enabled exhibitors to display a much broader sampling of their collection, and offered repeat visitors the opportunity to see a changing selection of plants as the season progressed. In 1848, for example, the Detroit Horticultural Society held exhibitions in late May and in late June. At the first show, visitors could enjoy tulips, lilacs, peonies, phlox, and a vast array of greenhouse plants. At the second exhibition, hardy roses, nearly twenty varieties of pelargoniums, petunias, dusty millers, and veronicas were among the many featured flowers. Taken together, both exhibitions offered a far better sense of floral diversity than was typically available at a single autumn agricultural fair.[39]

No matter how fair officials defined exhibition categories, commercial nurserymen and florists often provided the most comprehensive exhibits and offered visitors the best opportunity to see horticultural diversity. These professional growers could use their greenhouses and large acreage devoted to flower culture to overcome at least some of the seasonal limitations experienced by amateur growers. Sometimes local commercial interests, like Adrian florists Loud and Trask, broadened the offerings at agricultural fairs. In 1870, the firm contributed twenty-four entries to the Lenawee County Fair, and "render[ed] their portion of Floral Hall very attractive." Several years later, the *Adrian Times and Expositor* pointed out that James Vick, the well-known Rochester, New York, seedsman, had made a very extensive addition to the floral display at the State Fair. The exhibit was "immense" and so diverse that the commentator thought a "Florist's Guide" necessary to enumerate all the offerings. Other commercial seed houses routinely exhibited as well. In 1871, the *Jackson Weekly Citizen* insisted that Briggs and Brothers, another Rochester firm,

presented the best display of flowers at the State Fair, noting it was "really a very fine one."[40]

At many state and local fairs, the methods of displaying flower and plant entries also offered fairgoers a range of horticultural views, and added another dimension to ornamental plant significance. With the exception of the Michigan State Pomological Society, most of the regional horticultural societies were small organizations and used whatever space was conveniently available for their exhibitions. Well-established county agricultural societies, however, usually exhibited their horticultural entries in a building called "floral hall." Even small agricultural societies, lacking permanent fairgrounds, normally constructed a temporary structure to shelter the horticultural or other "ornamental" entries. The horticultural embellishment of these floral halls, apart from the entries, became a focal point of the fair, and decoration committees, normally comprised of area women, went to considerable effort to create a luxuriant and lovely scene. Festoons of evergreens, floral pyramids, dried grass bouquets in "mossy vases," moss-covered display shelves, "hanging moss baskets brimming over with flowers and fruit," and other horticultural accessories often turned floral halls into a "perfect bower." Some commentators claimed the decorations enhanced the exhibits, while others emphasized that the display added a note of taste, elegance, and refinement to the entire fair proceedings. Visitors could see examples of horticultural embellishment at work, and perhaps draw lessons from its effects to use at home.[41]

To cultivate interest and draw visitors, many societies featured additional amenities in floral halls that further emphasized the air of luxury and refinement. Some fair officials favored fountains, arguing that they added beauty and the soothing sound of water, but also because they showcased the community's progressive attitudes. Fair officials or the hosting town or village needed to make a significant financial commitment to supply water to the fair buildings, and its presence often signified that the town enjoyed the convenience of a municipal water system. When the 1863 State Fair moved to Kalamazoo, city officials "at an expense of $1,000 ... laid pipes through which water is conducted to the hall." The City also "ordered a beautiful fountain from New York," and placed it in "a bazin [sic] fifteen feet in diameter, fed from the pipes, while multitudes of fish from the adjoining lakes will sport in the basin, for the amusement of the spectators." With Kalamazoo's progressive and beautiful display providing a model, subsequent State Fairs almost al-

ways featured a fountain. In addition, cages of canaries and even parrots sometimes graced floral halls, their calls adding to the excitement and giving a special quality to the exhibition.[42]

Despite the name, floral halls seldom housed only flowers. State Fair officials eventually constructed separate buildings for the display of manufactured items or even musical instruments, but most smaller agricultural societies exhibited domestic manufactures, art work, and fancy needlework as well as many commercial exhibits right along with all the fruits, vegetables, and flowers. The floral hall at the 1870 Jackson County Fair typified the arrangements and suggested the diversity of goods that fairgoers encountered as they examined the horticultural entries. The center of the hall was filled with a "fine collection of flowers and plants," but nearby local businessmen displayed sewing and knitting machines, chromolithographs, bound books, stereoscopic views, organs, paintings, cases of boots and shoes, and an array of shells and corals. In one wing of the hall, visitors could enjoy another display of plants and flowers along with an exhibit from the business college, a private conchological collection, a display of sewer pipes, household furniture, a sash and door exhibit, and numerous fancy needlework entries. Fruits, vegetables, grains, and the bread, butter, and canned good entries took up yet another wing of the floral hall. The jumble of goods was not so disparate in the eyes of fair officials, who considered the floral hall an "ornamental building" that might reasonably house various exhibits "designed for ornament and to gratify educated tastes." Flowers, ornamental needlework, paintings, and commercial displays of chromolithographs or musical instruments fell logically within those parameters, but other commercial or manufactured items had a less obvious relationship to ornament. Considering the need for some structure for proper display and protection from the elements, smaller societies had little choice but to display them in what was frequently the only available enclosed space.[43]

Expediency may have shaped exhibit layout, but the fact that potted geraniums or bouquets rubbed shoulders with the latest sewing machines, fashionable attire, or colorful chromolithographs was not lost on fair visitors. The exhibition practice moved ornamental plants out of the realm of the routine and practical, into a new field of fashion, progress, and commercial enterprise. The habit of adding exotic or unusual embellishments to floral halls further emphasized the links between ornamental plants and broader issues of advancing taste or refinement. Fair officials hoped to create another world for visitors when they stepped into

the floral hall, one in which the mundane experience of farm and village life could be momentarily forgotten. The vision they assembled was a marketplace of new ideas, and a curious blend of the homegrown, the exotic, the exciting, the fashionable, and the commercial. The vision clearly transcended regional boundaries, linking Midwestern fairgoers with their counterparts across the nation.

Inadvertently perhaps, fair officials created a horticultural model significantly different from the practical, ordered refinements they so frequently advocated in print or as they designated exemplary farms, homes, and villages in the region. For fair visitors who had little additional contact with professional horticultural advice or with the more general concerns of agricultural or horticultural societies, exhibitions clearly placed floral ornaments within a new world of commercial enterprise and expanding materialism. The large number of entries from commercial firms made the exhibition visually attractive, but at the same time offered the subtle suggestion that successful plant culture required professional know-how, or at minimum, the assistance of commercial horticulturists in establishing fine home gardens. In most exhibitions, garden flowers or indoor houseplants dominated the entries, placing considerable emphasis on those horticultural embellishment at the expense of less labor-intensive tree or shrub plantings. Many prize-winning plants had large showy flowers, a familiar attribute of fashionable flowers at mid-century, but one that again emphasized the importance of floral display over other botanical attributes. The casual visitor or inexperienced gardener might assume that fashionable, brightly colored flowers were far more desirable than a perennial border, well-placed flowering shrubs, or a simple, well-tended lawn embellished with shade trees. They might also conclude that the best horticultural embellishments required economic resources and a significant commitment of time and labor. Many of the frequently exhibited plants required special care. Dahlia tubers and gladiolus corms, for example, needed winter protection, and if they were to survive had to be dug in the fall, carefully stored to avoid freezing, and then replanted in the spring. Purchased as bedding plants, verbenas, in turn, created a colorful display during the summer months, but could not withstand Michigan winters. They required overwintering as indoor potted plants, storage in a cellar, or had to be repurchased the following spring.

For those who cared to receive the message, fairs suggested they might purchase or plant refinement and obtain a new social image. Some professional horticulturists, sensing the implications, worried that combining exhibitions of fruits, flowers, and decorative arts took away the educational value of the exhibit, and made purposeful study of the entries or careful comparisons of horticultural qualities almost impossible. Others clearly understood the juxtaposition as another way to encourage progressive practices and new ideas without significantly disturbing traditional social values.[44]

Although agricultural society fairs and horticultural exhibitions drew visitors and created a range of horticultural examples, many farm families or village residents turned to horticultural models even closer to home. As they went about their daily lives, the residents of Lenawee, Jackson, and Washtenaw Counties simply had to look around them for horticultural inspiration and ideas, noting as they did what their friends and neighbors had planted in the dooryard or how they cultivated their flower beds. A number of horticultural commentators documented the influence of neighbors in spreading horticultural interest. In 1853, William H. Scott, a Lenawee County resident, noted the reality of local horticultural observations in a tongue-in-cheek *Horticulturist* article. Mr. Scott pointed out that "Johnny Slattern" and "Bill Carenought" gave a "passing look" to horticultural embellishments as they traveled "towns streets and suburban roads" on their way to market. They needed, however, "example[s] of their own class," and Mr. Scott urged progressive farmers to lead the horticultural parade. Others echoed the theme. "No infectious disease is more certain to spread than the spirit of planting trees and cultivating flowers," Robert Allyn asserted in his prize-winning Illinois State Agricultural Society essay, "when you set one man at work with a will around his own home, you may rely on it, he will inoculate a hundred with his own blessed enthusiasm." The *Prairie Farmer*, too, had every confidence that a neighborly "nudge" was the most effective way to promote horticultural improvements. "You may talk to your neighbors of progress and improvements," the *Farmer* declared, "but your example is worth all and vastly more than your talk." Plant a tree, shrub or flower, the *Farmer* urged, and your neighbors would soon follow.[45]

Who among the many farm, village, and city residents of Jackson, Lenawee, and Washtenaw Counties functioned as horticultural exemplars for others in the region? Certainly, county residents might look to the homegrounds of a number of prosperous or prominent neighbors as they shaped their own horticultural preferences and opinions. Progressive farmers like John Starkweather of Washtenaw County provide a typical example. Agricultural leaders recognized Starkweather as one of the

Figure 6.2. Home of small-town businessman and banker J. K. Boies. From *Combination Atlas Map of Lenawee County,* Michigan (Chicago: Everts and Stewart, 1874), 85.

state's leading farmers, and he received accolades for his orchards, Spanish Merino sheep, and full-blooded Durham cattle. Mr. Starkweather's horticultural interests were modest, however, and his neatly tended but simple dooryard, dominated by deciduous shade trees, a few shrubs, and a long, narrow garden, exemplified the restrained embellishment so often promoted in the agricultural press. Mrs. Starkweather appeared to take the most active interest in floriculture, and her petunias, houseplants, ornamental foliage plants, hanging baskets, sweet-scented geraniums, and gladioli frequently won prizes at the Washtenaw County Fair.[46]

Prosperous merchant, banker, and politician John K. Boies of Hudson in Lenawee County offered a contrasting example of horticultural display. Mr. Boies typified successful businessmen in the region, and seemed to take pride in well-ornamented homegrounds that reflected his prosperity and position in the community. Mr. Boies lined the driveway to his spacious home with a well-trimmed hedge. Several weeping trees graced the front yard, and a number of conifers, deciduous trees, and flowering

A Neighborly Nudge 171

RES. OF JAMES O'DONNELL

Figure 6.3. Home of James O'Donnell, editor and publisher of the *Jackson Daily and Weekly Citizen*. From *Combination Atlas Map of Jackson County, Michigan* (Chicago: Everts and Stewart, 1874), 116.

shrubs dotted the lawn. A single small cut-in garden bed, surrounding a shrub, appeared directly in front of the house.[47] (See Figure 6.2.) In contrast to the Boieses' home, another urban resident, James O'Donnell, had few ornamental plants around his Jackson city home. Mr. O'Donnell, who was editor and publisher of the *Jackson Daily and Weekly Citizen*, ornamented his lawn with a single line of deciduous shade trees bordering one side of the property. The O'Donnell family also displayed potted plants in their bay window, but enjoyed few other horticultural amenities.[48] (See Figure 6.3.)

If they knew these homeowners or their stature in the community, observers might conclude that the homes of upstanding citizens required at least some horticultural embellishment. With a little more thought about the varied examples, however, residents might reason that a wide range of plantings was acceptable. Mr. Starkweather, for instance, was a successful farmer, yet had only modest embellishments around his farm

home. Small-town businessman J. K. Boies reflected his success with slightly more complex plantings, yet city resident James O'Donnell, prominent because of his position as newspaper editor and publisher, had few horticultural amenities in his small yard. Hundreds of similar examples among residences pictured in the *Atlas Maps* of Lenawee, Jackson, and Washtenaw Counties corroborated the importance of trees, shrubs, and flowers, but simultaneously emphasized the diversity of display possibilities.

In addition to the homes of prominent citizens, the residents of southeastern Michigan could turn to a number of well-respected horticultural enthusiasts who offered even more comprehensive models for ornamental plant culture. Two individuals from Lenawee County were particularly influential and, because of the level of their horticultural interest and their varying circumstances, deserve an extended appraisal. The first, Dr. Woodland Owen of Adrian, was born in England in 1819. As a young man, Owen trained as a pharmacist, but eventually took up dentistry, and in 1842 came to America, opening a dental practice in Rochester, New York. Still restless after several years, he headed west to explore Wisconsin, Illinois, Indiana, and Michigan. He liked the looks of southeastern Michigan, and in 1848 moved to Adrian and again established a dental practice. In his new home, Woodland Owen was active in public affairs. An ardent abolitionist, Owen supported the Michigan Anti-Slavery Society, and often hosted abolitionist leaders in his home. In the early 1860s, he served as Adrian City Alderman, and made an unsuccessful bid for mayor in the early 1870s. Dr. Owen also served as an Adrian College trustee and on the board of Adrian's Plymouth Church.[49]

Owen's horticultural enthusiasm manifested itself in both his activities and his homegrounds. Whether influenced by his father's progressive agricultural interests, his years amidst the burgeoning Rochester nursery industry, or his natural proclivities, Woodland Owen simply loved plants, and devoted much of his leisure time to their culture. He was active in the Adrian Horticultural Society from its founding in 1851, often hosting meetings at his dental office and contributing to the society's exhibitions. In addition, Dr. Owen housed the Horticultural Society's library in his "Dental Rooms." He also supported the Lenawee County Agricultural Society, and frequently entered his flowers in the county fair competition.[50]

By the mid-1870s, Dr. Owen had filled the grounds surrounding his spacious home with a variety of horticultural embellishments. He inter-

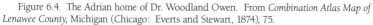

Figure 6.4. The Adrian home of Dr. Woodland Owen. From *Combination Atlas Map of Lenawee County,* Michigan (Chicago: Everts and Stewart, 1874), 75.

spersed at least nine evergreen trees with deciduous ornamental trees in a seemingly random design around the house. A number of flowering shrubs also dotted the lawn, while rows of shade trees lined the streets adjoining the yard. Vines ornamented at least one side of the house, and additional vining plants grew on free-standing trellises in the yard. A large, elongated vine-covered arbor or summer house graced one side of the homegrounds, and Dr. Owen formed a small rockery in front and slightly to the left of the house. The exposed position of the rock-work suggested that he probably cultivated succulent or alpine plants among the rocks rather than ferns or woodland species, which advisers frequently recommended for more sheltered situations. Several plant stands and flower vases ornamented the Owens' side yard. The Owens also enjoyed indoor plant culture. A large bay window and a small lean-to greenhouse attached to an outbuilding housed a varied selection of tender plants. Over the years, Dr. Owen or his wife entered a diverse array of plants in local exhibitions and fairs including pansies, geraniums, roses, fuchsias, verbenas and other bedding plants, gladioli, and even at one point a cotton plant in "full blooming condition."[51] (See Figure 6.4.)

The size of the embellished area surrounding the Owen home was not radically different from other farm, village, or city homegrounds in the region. Dr. Owen also pursued his horticultural interests with relatively

modest financial resources. In 1870, Dr. Owen's economic worth totaled $10,500. While far from poor, the value of Dr. Owen's personal and real property was well below that of many farm families in the region. In addition, Dr. Owen maintained a busy dental practice, caring for his trees and flowers during his leisure time. Both Mrs. Owen and their son Henry expressed a strong interest in ornamental plants, and may have helped in homeground maintenance and plant culture, but there was no evidence that the Owens routinely hired help to assist in outdoor chores. The Owens' homegrounds provided many diverse examples of typically advocated horticultural improvements, but did so on a scale that was within reach of many other aspiring horticulturists in the region. A farm family passing by might not choose to add a greenhouse to their barn, but could find the appearance of vines bordering the bay window attractive and think about such a practice at their own home.[52]

Another avid Adrian horticulturist, Dr. Owen's friend and fellow member of the Adrian Horticultural Society Edwin P. Powell, pursued his ornamental plant interests under very different circumstances. Born in New York, Powell came to Adrian in 1861, and as a twenty-eight-year-old bachelor, took up his post as the new pastor of the Plymouth Congregational Church. Rev. Powell found lodging with Kelly Beals and his family, a grocer living in a modest neighborhood of artisans and shopkeepers. Powell also rented a room at the Masonic Temple to use as a study and meeting place for his parishioners. Almost immediately, Powell purchased some acreage about a mile from his rented rooms, and set about developing his gardens. Presumably to ready himself for spring planting, Rev. Powell placed the first of several plant orders with Rochester nurserymen Ellwanger and Barry in December 1862. Several years later, Powell reported that his grounds included an acre and a half of lawn and nearly sixty separate flower beds. By 1871, Rev. Powell owned over five acres of gardens and orchards in one spot, and an additional two acres in another location in the city. He covered his ornamental grounds with over forty varieties of fruit trees and small fruits, a selection of evergreens, and a "nearly perfect hedge along the roadside." Powell even designed and cultivated parterres to accentuate the colors and contrasts of his many flowers.[53]

To satisfy his passion for ornamental plants year-round, Powell developed an innovative system of indoor plant culture that was suitable even in his rented rooms. With a conventional window forming one side, Powell built a framework surrounding the window and extending into the room, and then lined the frame with window sash and glass doors. He covered the floor of his inverted "bay window" with a zinc lining to collect moisture and dirt, and mounted a series of shelves inside the small conservatory to display his plants. Hooks in the ceiling enabled him to suspend baskets of plants in the center of the glass enclosure, which further increased display space. This inexpensive and rather modest arrangement created a fine environment for tropical plant culture, and Powell was able to grow camellias, geraniums, azaleas, fuchsias, heliotropes, abutilons, pansies, mignonettes, verbenas, petunias, and many other varieties that required greenhouse conditions during the winter months. For plants that needed additional moisture but could tolerate low light levels, Powell established a large Wardian case, and in it cultured many varieties including begonias, caladiums, and even lycopodiums. Powell provided a detailed description of his innovative bay window and Wardian case in one of several articles he wrote for the *Gardener's Monthly* in the mid-1860s.[54]

Like Woodland Owen, E. P. Powell frequently displayed his plants at the Adrian Horticultural Society exhibitions and the Lenawee County Fair, but in addition often entered his collections in Michigan State Fair competitions. At the 1868 State Fair, for example, Rev. Powell took first prize for the greatest variety of roses, the best six ornamental-leafed plants, the greatest variety of dahlias, best seedling phlox, best petunias, tropaeolums, zinnias, pinks, hardy annuals, and German asters. He exhibited a comparable variety at many other exhibitions, and at one fair he even caught the eye of visitors with a green petunia.[55]

While horticulture was an avocation for Rev. Powell, he could not resist weaving it into his ministerial duties. In 1869, Powell delivered a special floral sermon at the Plymouth Church, telling parishioners that a love of flowers beautified the soul. The church was filled with wreaths, bouquets, and evergreen boughs for the occasion. Powell routinely enhanced visits to the sick or elderly with floral bouquets, freshly cut from his own gardens, and even urged his fellow clergymen to use floriculture as a sure way to ease their own minds and calm their spirit. The practice, Powell argued, would result in a "healthy body, a mind under control ... and an inestimable amount of joy and comfort."[56]

Powell's efforts provided a plethora of horticultural examples for members of the Plymouth Church, for those who received his floral gifts, and for community members that happened to pass his ornamented grounds, but sent several contrasting messages when compared to the example of

Woodland Owen. Most obviously, Powell's love of plants was closely tied to his calling as a minister, and he clearly associated their appreciation and culture with a love of God. With that linkage, Powell offered a living model of what domestic reformers, social critics, and some horticultural advisers had argued for years: there was a connection between the love of natural beauty and an elevated spirit. In addition, Powell's horticultural interest extended far beyond the region, and he offered local residents a very active connection to contemporary horticultural practice. His plant orders from Rochester nurserymen Ellwanger and Barry underscored his horticultural savvy, for they, at the time, operated the largest nursery in the country and offered their customers unprecedented selections of plants.[57] Powell kept abreast of current horticultural literature, and in the mid-1860s and early 1870s, published several commentaries in the well-respected *Gardener's Monthly* that confirmed his horticultural competency. In 1871, the *New York Independent* acknowledged his expertise, describing Powell as the "most successful floriculturist in Michigan."[58] Finally, E. P. Powell achieved his horticultural reputation and nurtured his vast array of plants with rather limited means. He did not own a home, yet cultivated huge gardens. His ministerial duties placed tremendous demands on his time, but he routinely set aside hours each day for plant culture. His economic resources were also modest when compared to most other homeowners in the region, yet he skillfully managed what he had to maximize his horticultural opportunities. He clearly valued his plants enough to shape the pattern of his life around them.[59]

For all its positive elements, Rev. Powell's role as a horticultural exemplar had another side. The *New York Independent* described Powell as a successful clergyman, but an "eccentric." Some of his parishioners apparently agreed, for discord and declining membership marked his years at Plymouth Church, and in 1871 Rev. Powell resigned. The horticultural implications of the move were graphically illustrated when Powell advertised his land and household goods for sale. The land, Powell claimed, was beautifully ornamented, and all was in order "for a first class home." Amidst a meager list of household furnishings appeared a "Wardian-case filled with plants" and "one thousand gladiolus bulbs of 30 mixed varieties, superb." Powell had offered the residents of the area a whole new perspective on horticultural diversity, but his personal difficulties may well have tempered the efficacy of the message.[60]

Despite their differences, Woodland Owen and E. P. Powell acted as very important resources for those who contemplated horticultural improvement. They both functioned as sophisticated cultural mediators, translating commonly articulated and nationally disseminated horticultural rhetoric, standards, and trends into local realities. Their level of expertise and knowledge compared favorably to other skilled horticulturists nationwide, and the examples they offered communicated no taint of parochialism. These avid horticulturists delivered yet another message to friends and neighbors who viewed their homegrounds and gardens. Both the diversity of their personal circumstances and the range of horticultural embellishments they displayed confirmed that homeowners had many options.

What lessons did the residents of Lenawee, Jackson, and Washtenaw Counties extract from this array of horticultural models? Did exhibits at the county fair, designated model farms or villages, and exemplary neighbors speak with a unified voice? In one sense the message did have a common theme, for all promoted an interest in horticulture and associated plant culture with new ideas, a progressive outlook, and an "improving spirit" that extended far beyond the region and connected local residents to a developing national landscape and culture.[61] For those who considered shade trees and brilliant flowers more deeply, however, the central theme was nuanced, sometimes associating horticultural display with neatness and order, at other times with fashionable material goods, or finally with a personal expression of social position or horticultural interest and knowledge.

Beyond their symbolic significance, these local examples provided concrete opportunities to view often contrasting horticultural possibilities. Fairs and exhibition entries tended to encourage bright, showy floral displays, while designated models of farm or village homes showcased modest, well-tended embellishments like shade trees and flowering shrubs. With their varied yet comprehensive horticultural interests and backgrounds, local horticultural exemplars offered county residents a kind of smorgasbord of horticultural choices ranging from tree-shaded lawns to rock-work, cut-in gardens, and greenhouses.

The diversity represented in these local models created a more complex physical and symbolic reality than the picture routinely painted by professional horticulturists and commercial nurserymen. The residents of Lenawee, Jackson, and Washtenaw Counties who learned by looking encountered a varied horticultural landscape shaped by individual circumstances, tastes, and choices. When planted in regional village and farm dooryards or exhibited at the county fair, ornamental plants meshed with

 a myriad of local realities, and communicated a number of contrasting and sometimes conflicting cultural messages. The nuances were in themselves eloquent, for they suggested that the shade trees and flowering shrubs dotting Midwestern domestic landscapes melded widespread public meanings with personal and very individualized significance.

Seven

Bergamot Balm and Verbenas

A Personal Perspective on Ornamental Plant Culture

A glance through contemporary domestic reform literature, horticultural and agricultural journals, nursery and seed catalogues, or county fair records suggests many reasons why Midwesterners may have turned to ornamental plant culture in the middle decades of the nineteenth century. Reasoned arguments, commercial enticements, and the examples of progressive neighbors and friends abounded, and all pointed out the possibilities and propriety of horticultural embellishments. Advocates promoted their cause by assigning a range of very public meanings to ornamental plants. Some argued that trees and flowers signaled good taste and refinement or broadcast a family's progressive outlook and attitudes. Others insisted that a well-tended flower garden communicated the resident's "pure-mindedness" and intelligence, while disorder in the dooryard suggested moral lassitude or worse. In the hands of their promoters, flourishing flower beds or stately shade trees, along with well-kept dooryards and neatly mown lawns, became quiet yet very public reminders of respectability and shared cultural values.[1]

Although commentators attached easily translated social messages to ornamental plant culture, their significance in Midwestern domestic landscapes remains neither clear nor simple. Horticultural proponents themselves perceived trees and flowers as a way to not only communicate, but to inculcate social values, and to embody those values in domestic landscapes across the nation. Commercial nurserymen and seed dealers, in turn, understood the economic realities of a flourishing horticultural trade and, at least in part, saw their livelihood enhanced by the spread of flower gardens and shade trees. But what of the Midwesterners who purchased, planted, and nurtured all those trees and flowers? Did they perceive ornamental plants as simply public signs of their fashion awareness, respectability, or progressive tendencies? Or did they establish new roles for ornamental plants, imbuing them with additional meaning as they wove them into the circumstances and realities of their individual lives?

Hints of a complex and deeply personal significance surfaced as farm families and city dwellers wrote to Illinois nurseryman John Kennicott about their trees and flowers. Most letters inquired about plant orders or sought horticultural advice, but interspersed with the business-at-hand was another perspective. Some Midwesterners turned to plants to overcome their grief and loneliness. Morilla Cates of Garden Prairie, Illinois, for example, ordered a Norway spruce to mark her young son's grave, and promised to write for bulbs in the fall if she could put aside enough money from her sewing. Others hoped that ornamental plants might soften and shape an unfamiliar landscape. Daniel McCarthy longed for fruit and ornamental trees to add life and interest to the "bare and naked prairies" of his new Davenport, Iowa, home. Some Midwesterners found the simple love of flowers compelling, no matter what their circumstances. Chicago resident L. L. Klemin rented his home, but wanted to plant flowers anyway. He told John Kennicott, "I am very fond of flowers and want to gratify that love with as little expense as I can." Mary Rowells, a Joliet, Illinois, farm wife, sent Kennicott a very modest order, but revealed that she planned to divide and propagate the plants to make herself a "longlooked [sic] for new home garden." Another farm wife described the pride and pleasure she found in her gardens. "I have had a fine show of flowers all summer, for a farmers wife [sic]," Margaret Carle of Urbana, Illinois, wrote, "I took first premium on cut flowers at our county fair." R. W. Arnold summed up the sentiments of many others when he observed, "poverty does not crush out the love of the beautiful altogether." Mr. Arnold ordered a few roses for spring planting in his farm dooryard, but asked Kennicott to accept payment after the fall harvest. Clearly, for Mr. Kennicott's correspondents, roses in the dooryard or a single Norway spruce told a very private story. In large measure, these Midwesterners appeared to take up ornamental plant culture in their own way and for their own reasons, and in the process imbued dooryard trees and flowers with new layers of meaning.[2]

This tantalizing glimpse of personal meaning calls for a more sustained appraisal of horticulture in the lives of individual Midwestern families. An unusual collection of personal writings from several intertwined southern Michigan farm families offers just such an opportunity. Three members of the Lawrence, Buell, and Copley families of Volinia Township, Cass County, Michigan, detailed their lives and horticultural interests in diaries. Beginning in 1855, and continuing through much of the century, farm wife Esther Copley Lawrence recorded daily routines, often commenting on her husband, their three sons, farm chores, community events, and visits with neighbors. Barber Grinnel Buell, Esther Lawrence's brother-in-law and neighbor, kept a similar comprehensive account of farm and community activities extending from 1869 until 1898. Charlotte Copley, Esther's oldest sister, also left behind a brief, but very significant written record, this time describing a portion of her seventeenth year, and her family's journey from Ohio to Michigan in 1833. The Lawrence, Buell, and Copley families incorporated trees and flowers into the fabric of their daily lives, and as they cultivated their gardens, or simply enjoyed the beauty of verbenas and bergamot balm, they also added layers of varied and sometimes very private significance to what reformers had defined as public signs of respectability and good character. This chapter looks at ornamental plant culture from this very private perspective, exploring first the acceptance of horticultural norms, the significance of obtaining plants, and the links that horticultural interest fostered to new ideas and progressive practices. The chapter then plumbs the personal, investigating how trees and flowers altered work routines, enhanced family ties, and finally offered emotional comfort to these rural Midwesterners.[3]

Following the pattern of many other southern Michigan settlers, the Copley, Lawrence, and Buell families came to Michigan as the last stop on a gradual western migration. Esther Copley Lawrence's father, Alexander Copley, was born in Connecticut, but as a young man moved to New York State. He married Esther Nott in 1814, and the young couple began a series of moves to various towns and villages in western New

York. Copley was trained as a cabinet maker, but with the modernizing textile industry creating new demand, developed particular skill as a "textile machinist," producing spindles and looms used in the production of cotton cloth. In 1829, the Copleys and their five children moved to Dayton, Ohio, where Mr. Copley superintended a newly built cotton mill. The family established a new home on fifteen acres, and lovingly christened the spot "Delightful Hill." In the four years they remained in Ohio, Esther Copley bore two more children. Alexander Copley, in turn, used his hours away from the mill to experiment with mulberry tree and silk worm culture, and dreamed of establishing an integrated spinning and weaving mill to produce silk cloth. In 1832, the still restless Copley explored a fertile prairie region in what is now Volinia Township, Cass County, Michigan. He liked what he saw, and in June 1833, the Copley family set off for Michigan Territory. At the time of the move, their oldest daughter, Charlotte, was eighteen, and her sister Esther Copley was nine years old. The family made the 234-mile journey from Ohio to Michigan in twenty-three days, traveling in two ox-drawn wagons, and bringing with them cows, chickens, a span of horses, grape vine cuttings, rhubarb roots, and their household goods.[4]

The Copleys settled near the present site of Nicholsville in Volinia Township, and Alexander Copley established the area's first saw mill and cabinet shop. Eventually, they procured a 273-acre farm about a mile north of Nicholsville and bordering rich farm lands known as Little Prairie Ronde. Over the years, the Copley family fondly called their home the "old homestead" or "Locust Grove" because of the cluster of black locust trees in the front yard. Mrs. Copley had two more children after the family moved to Michigan, a son, Napoleon, in 1834; and Harriet, the last of the Copley children, in 1836. Upon her arrival in Michigan, Charlotte Copley taught school, first at the family home and later at the newly built Volinia Township schoolhouse. She eventually married Samuel S. Ward of Niles, Michigan, bore one daughter, and died in 1845 at age thirty.[5]

Levi Lawrence, his wife, and seven children also arrived in Michigan Territory in 1833, and built a home near the Copleys at Nicholsville. Born in Massachusetts, Levi Lawrence was a skillful blacksmith and "natural mechanic," and had plied his trade in Massachusetts, New York, Pennsylvania, and Ohio before coming to Michigan. Levi Lawrence eventually moved further west, but his son Levi B. Lawrence recognized the great agricultural potential of Cass County's fertile prairies, and stayed

Figure 7.1. The Buell family posed in front of their Cass County farm home. Family members include (l to r): Lincoln, Flora, and Frank Buell; B. G. Buell and Hattie Buell; Mrs. Lincoln Buell, Bertha and Jennie Buell; Everett, Hal, and Josiah Buell. Original in the Bentley Historical Library, University of Michigan, Ann Arbor, Michigan.

behind to begin his long and productive career as a Volinia Township farmer. In 1849 at the age of thirty, Levi B. Lawrence married his twenty-five-year-old neighbor and friend Esther Copley, and together they established a new home adjacent to the Copley homestead north of Nicholsville. Over the years, Levi and Esther worked to expand and improve their farm, and in the process raised three sons, Linneaus, Archie, and Austin Lawrence.[6]

Barber Grinnel Buell arrived in Volinia Township nearly twenty years after the Lawrences and Copleys. B. G. Buell was born in New Hampshire in 1823, but spent his boyhood in western New York. In 1854, accompanied by his mother, his sister Eunice, and his brother Emmons, Buell moved to Michigan, and purchased a partially completed house and farm land bordered by the Copley and Lawrence properties to the north and east. Buell gradually completed his home, and in 1862, married his twenty-six-year-old neighbor Harriet Copley at the home of her sister Es-

Figure 7.2. The Lawrence, Buell and Copley families all settled in Cass County, Michigan, near Little Prairie Ronde. From State of Michigan Specialized Technology/Mapping Unit, Lansing, Michigan.

ther Lawrence. B. G. and Harriet Buell raised a family of three daughters and three sons on their "Greenwood Stock Farm." Their oldest daughter, Jennie, carried on the family tradition of agricultural activism by carving out a long and distinguished career in the Michigan State Grange.[7] (See Figure 7.1.)

As residents of Cass County, the Lawrences, Buells, and Copleys enjoyed environmental, economic, and cultural circumstances similar to those of farm families in other southern Michigan counties and throughout the productive agricultural regions of the Old Northwest. In the mid-nineteenth century, agriculture dominated the region's economy, and several prairies, including Little Prairie Ronde in Volinia Township, offered farmers especially rich and productive soil. In addition, the tempering effects of Lake Michigan enhanced growing conditions, and made Cass County particularly well-suited to fruit culture. Wheat and corn led Cass County production, but many farmers raised livestock and cultivated large orchards in addition to their typical field crops. Like other Michigan counties, Cass drew its population from diverse regions. Many early settlers came from Ohio and Indiana, but New York and New England also contributed sizable numbers. Beginning with the settlement of freed slaves in the 1830s and 1840s, Cass County's Calvin Township was home to the largest African-American population in the state outside of Wayne County. The total population of Cass County, however, remained less than some neighboring counties to the east, reaching just over 21,000 citizens in 1870.[8] (See Figure 7.2.)

In contrast to Jackson, Washtenaw, and Lenawee Counties, Cass County boasted no large towns. The county seat at Cassopolis had a population of slightly over seven hundred residents in 1870, while Dowagiac, the largest town, claimed just over nineteen hundred citizens. The Michigan Central Railroad passed through Dowagiac in 1848, and a number of industries including a foundry, planing mill, farm implement company, and basket factory grew up in its wake. The "Air Line" branch of the Michigan Central also ran through Cass County, and in the 1870s, the Peninsular Railroad established a line diagonally across the county linking Chicago and Lansing, Michigan. Although there were few large towns nearby, county residents enjoyed ready access to distant markets, and with regular passenger service could easily visit Kalamazoo, Niles, or other commercial centers in the region.[9]

Like other Midwesterners, some Cass County citizens supported agricultural associations. Residents organized the Cass County Agricultural Society in 1851, and the Society held its first fair in the fall of that year. The Society purchased fairgrounds in the mid-1850s, but because of repeated conflicts with the railroads, changed sites twice over the next thirty years. Cass County was also home to one of the state's most active and progressive farmers' clubs. Organized in 1865, the Volinia Farmers' Club met monthly to discuss pertinent agricultural and horticultural topics, and held a popular and widely attended two-day fair every autumn.[10]

In many ways, the Copleys, Lawrences, and Buells typified progressive rural families, and the "improving spirit" so ardently promoted by the mid-nineteenth-century agricultural press. The three families were active readers. They subscribed to and exchanged books and journals

like *Harper's*, the *Atlantic Monthly*, *Moore's Rural New Yorker*, the *American Agriculturist*, *Hearth and Home*, the *Country Gentlemen*, the *Western Rural*, and the *Michigan Farmer*. In addition, B. G. Buell read several national, regional, and local general-interest newspapers including the *New York Tribune*, the *Detroit Tribune*, and the *Cassopolis Vigilant*. Esther Lawrence, in turn, read the *Women's Advocate*, a paper that she described as "owned, printed and edited by women," and especially enjoyed *Arthur's Home Magazine*. The families were also acquainted with several agricultural journal editors. When H. F. Johnstone, the long-time *Michigan Farmer* editor, was in the area, he took his dinner at the Lawrence home, as did Mr. Wood, editor of the *Western Rural*. Like many of his fellow Midwesterners, Buell liked to consider regional progress, and noted when he purchased the *History of Cass County*, a work memorializing the county's settlement and advancement.[11]

The Buells and Lawrences often attended lectures in area towns, sometimes riding together and frequently commenting on the topic and the quality of the speaker. The subjects varied widely, ranging from reform issues like temperance and women's suffrage to philosophical and scientific topics like spiritualism and phrenology. On one occasion, both families enjoyed an informational program on Egypt and its pyramids. The Buells and Lawrences also attended Farmers' Institutes held in the nearby village of Decatur and sponsored by the Michigan Agricultural College. At a typical institute in 1876, Buell and a full contingent of other area farmers heard lectures on insect problems, corn culture, and fruit growing.[12]

Both families belonged to the Cass County Agricultural Society, sometimes provided leadership for Society activities, and usually supported the Cass County Fair with their attendance and entries. B. G. Buell and Levi Lawrence were founding members of the Volinia Farmers' Club, and often attended meetings accompanied by Hattie and Esther. Like other southern Michigan residents, some Cass County citizens, recognizing the on-going need for agricultural and political reform, organized a local chapter of the Patrons of Husbandry or Grange in the early 1870s. B. G. and Hattie Buell regularly participated in Grange meetings, and Buell often commented on writing and sending local Grange reports to state headquarters.[13] B. G. Buell was also active in the State Agricultural and Pomological Societies. Esther Lawrence and Hattie Buell particularly enjoyed the Home Culture Society, and frequently met with the group to discuss such topics as "systematic housekeeping" or floriculture. In the late 1870s, Esther Lawrence began a regimen of diligent study and reading through the Chautauqua Library Study Course.[14] Both families also eagerly embraced labor-saving machinery and new ideas. They regularly visited neighbors to inspect and discuss any innovations in agricultural or home management practices, and by the late 1870s and early 1880s, enjoyed the benefits of new farm equipment and domestic tools like a sewing machine, a clothes wringer, a furnace, and even a telephone.[15]

Their progressive tendencies apparently paid off, for the Lawrence and Buell families prospered. They raised wheat, corn, oats, and hay, and prided themselves on their livestock, including dairy cows, hogs, and sheep. By 1870, both families produced hundreds of pounds of wool each year. B. G. Buell had a thirty-nine-acre commercial orchard, and regularly sold his fruit to regional merchants or sent it off to lucrative markets in New York or Chicago. In addition, he supplied his neighbors with fruit trees and sold some ornamental trees and shrubs.[16] The Lawrence family also sold apples along with honey and beeswax. Both families grew potatoes and produced hundreds of pounds of butter yearly. The Lawrence family's advance was particularly striking. In 1850, Levi B. Lawrence farmed sixty-five acres and reported the value of his property and farm machinery at $1,800. By 1870, the Lawrences owned a total of five hundred acres, valued along with personal property at $38,000. B. G. Buell recorded a similar increase in acreage and assets, and both families were eventually among the most prosperous in Volinia Township. In 1874, the Lawrence family celebrated their good fortune by hiring an architect to plan and oversee the construction of a new brick house complete with bay windows, a slate roof, a furnace, and a stone walkway.[17]

What did ornamental plants mean in the lives of these busy, successful rural families? Appearances offer preliminary clues, and as one might expect from the chorus of reform literature, the dooryard of the Lawrence home and the surroundings of the Buells' Greenwood Stock Farm boasted mown lawns, shade and ornamental trees, a smattering of flowering shrubs, and carefully tended flower beds. A slightly more thoughtful appraisal of specific plantings and practices revealed that the Lawrences and Buells were well aware of fashionable trends and prescribed horticultural norms, and although never articulating their motives, occasionally chose to incorporate some of those elements into their own domestic landscape plans. In one instance, B. G. Buell indicated his interest in a fashionable new leisure activity when he purchased a "set of crokay [*sic*] balls and bats" for $2, and then encouraged his children to invite neigh-

borhood youngsters for a day of croquet, swinging, and lemonade and sugar.[18] Buell's actions may have signaled a simple interest in his children's happiness, but on another level indicated an awareness and implicit acceptance of several widely promoted social norms. As so many commentators advocated, Buell apparently felt an obligation to enhance his family's well-being by providing more than adequate food and appropriate shelter. His decision also underscored a willingness to devote potentially productive agricultural lands to nonproductive use, and to dedicate the time and labor required to maintain a smooth mown lawn as a playing surface for games like croquet. Most importantly, the purchase of a croquet set, and his encouragement in using it, communicated to his children and their friends the benefits and value of a widely touted leisure activity. Whether intended or not, a croquet set on a smoothly mown lawn also told neighbors, friends, and passersby that the Buell family had the resources and the inclinations to enjoy an afternoon of leisure.

In another instance, B. G. Buell planted a line of Norway spruce trees to protect his strawberry beds, and then selected a variety of other evergreen trees to ornament his homegrounds. Again, his choices were in accord with many mid-nineteenth-century horticultural commentators who frequently described the virtues of coniferous trees for ornamental plantings, and especially praised the Norway spruce as attractive, hardy, and adaptable. Evergreens, these advisers pointed out, beautified dooryards year round, and enlivened otherwise desolate winter landscapes with a welcome spot of color. According to proponents, evergreens also provided particularly effective windbreaks, offering protection for more delicate plants no matter what the season. Buell's plantings suggest that he recognized and accepted those advantages, and his own writings corroborated that view. In an unpublished "Essay on Orchards," Buell urged other residents to plant rows of trees like "Lombardy Poplar, Cottonwood, Norway Spruce or White Pine" as windbreaks for a young orchard, recommending those species because they could withstand harsh conditions and grew quickly. He also noted that the Norway spruce was particularly useful because it was easily propagated, pruned well, and maintained its lower branches to make a sturdy shield against the wind. Like the croquet set, evergreen trees required purchase, and their presence as both windbreak and ornament testified to the economic well-being of the Buell household.[19]

Esther Lawrence also indicated an awareness of widespread horticultural advice and fashion trends as she selected and cultivated her garden

Bergamot Balm and Verbenas 189

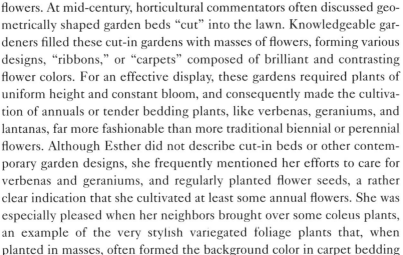

flowers. At mid-century, horticultural commentators often discussed geometrically shaped garden beds "cut" into the lawn. Knowledgeable gardeners filled these cut-in gardens with masses of flowers, forming various designs, "ribbons," or "carpets" composed of brilliant and contrasting flower colors. For an effective display, these gardens required plants of uniform height and constant bloom, and consequently made the cultivation of annuals or tender bedding plants, like verbenas, geraniums, and lantanas, far more fashionable than more traditional biennial or perennial flowers. Although Esther did not describe cut-in beds or other contemporary garden designs, she frequently mentioned her efforts to care for verbenas and geraniums, and regularly planted flower seeds, a rather clear indication that she cultivated at least some annual flowers. She was especially pleased when her neighbors brought over some coleus plants, an example of the very stylish variegated foliage plants that, when planted in masses, often formed the background color in carpet bedding designs. In accord with frequently repeated horticultural concepts, Esther also cultivated ornamental vines, and especially favored Madeira vine, one of the most popular varieties at mid-century. Her pleasure in dahlias, her love of roses, her pride in her chrysanthemums, and her cultivation of gladioli and fuchsias all reflected an interest in very fashionable flowering plants, and at least some willingness to invest in horticultural display.[20]

While B. G. Buell and Esther Lawrence selected some widely touted horticultural embellishments for their dooryards, they also expressed their ornamental plant proclivities by what they failed to establish or cultivate. Certainly within their economic means, they passed up more elaborate and time-consuming embellishments like rockeries, summer houses, or plant conservatories, despite the fact that horticultural commentators described all of them as signs of refinement and horticultural good taste. Even in her choice of fashionable plant varieties, Esther Lawrence appeared to temper her selections with practicality. She routinely cultivated a wide range of perennial and biennial flowers, and a variety of spring and summer bulbs as well as more labor-intensive bedding plants, tender perennials, and annuals. Although she never described actual garden layouts, she may have added bedding plants and other popular varieties as accents in a mixed bed of her own design. The combination of flowers satisfied her desire for a bright and colorful floral display and communicated her fashion awareness, but at the same time, conformed to her own ideas of appropriate and pleasing embellishments.[21]

The Buells and Lawrences also signified their awareness of and adherence to broad, altruistic community goals and ideals through their use of ornamental plants. Like many other Midwesterners, B. G. Buell planted rows of deciduous trees along the roadsides adjoining his farm, selecting butternut and whitewood trees from a nearby woods to complete the task. Although the trees enhanced Buell's holdings, they also indicated his responsiveness to a widespread mid-nineteenth-century movement to improve the rural landscape. The horticultural and agricultural press frequently touted the benefits of roadside trees, arguing that they provided shade and comfort for travelers and enhanced rural beauty. Individual farm families, commentators insisted, should take responsibility for tree planting along roadsides bordering their property, and if they did so, they signified their status as public-spirited citizens. Some county agricultural societies added their influence to the movement by appointing committees to inspect roadsides, and offering premiums to members who planted the most trees. The state of Michigan supported the improvement of roadsides as well. An 1857 law encouraged roadside tree planting, and ten years later an amendment offered a tax break to farmers who took up the call. Not every farm family, however, accepted the challenge. Well-tended, tree-shaded roadsides remained a mark of distinction for much of the century, providing not only welcome shade, but a clear sign of an individual farmer's initiative and progressive tendencies.[22]

In addition to roadsides, Buell devoted some resources and influence to the embellishment of the nearby schoolyard. Although the issue appeared less frequently than comments on roadside planting, the agricultural press had, for decades, urged readers to consider school environments and their influence on the sensibilities of young scholars. In a typical commentary, the *American Agriculturist* acknowledged that critics might find the concept "unimportant, useless, foolish even," but then pointed out that most people would prefer to send their children to a school where "some little attention is given to the cultivation of a taste for the good and beautiful, rather than where dilapidated fixtures and disgustingly dirty grounds are the most prominent features." The citizens of Nicholsville apparently agreed, for on April 12, 1871, Buell, along with a committee appointed to consider schoolyard ornamentation, met to discuss the possibilities, and then adjourned to the nursery of local farmer Benjamin Hathaway to purchase $25 worth of evergreens of "various kinds." Both Buell and Hathaway agreed to donate an additional four pine trees to complement the purchased conifers. Two days later, Buell loaded his contribution on his stone boat, hauled the trees to the schoolyard, and planted them, along with Hathaway's donations, as a windbreak on the west side of the school property. The committee's actions bore several fruits. The children of the Nicholsville area had a more pleasing schoolhouse, and area residents, led by Buell, had the opportunity to define their position and express their willingness to support refinement and the public good with their time and money.[23]

In a slightly different indication of public sentiment, Levi Lawrence joined with many of his fellow citizens in a ceremonial tree planting. In 1876, Michigan Governor John J. Bagley issued a proclamation honoring the nation's centennial celebration, and urging all residents to express their patriotism by planting special centennial trees. He acknowledged that trees were only symbols, an "external show," but hoped that they might serve as a very necessary reminder of the nation's founding principles. Bagley set aside April 15, 1876, as the official day for tree planting, and on that day, Levi Lawrence, along with thousands of others across the state, took action. "Pa," Esther Lawrence wrote in her diary, "set out a maple and two chestnut trees for the Centennial Trees."[24] Whatever their personal motivation, B. G. Buell's and Levi Lawrence's simple actions transformed ornamental plants into public statements of civic virtue, and communicated their status as progressive, public-spirited citizens. Appearances spoke loudly to passing residents, and along with neatly mown lawns and well-tended flower beds, shaded roadsides, ornamented school grounds, and centennial trees let them know that up-to-date, progressive families farmed the region.

The appearance of the Lawrence and Buell homegrounds indicated a selective use of prescriptive horticultural norms and fashionable trends, but the consideration of the daily realities of ornamental plant culture creates a far more complex picture. No matter what kind of fashion statement they wished to make or practical advice they planned to pursue, Midwesterners had to first obtain ornamental plants, perhaps transplanting native trees and shrubs into their dooryards, trading roots, cuttings or seeds with friends and neighbors, or purchasing plants from an expanding array of eager commercial horticulturists. The process sometimes drew Midwesterners into a new national marketplace and at other times enhanced local, long-established networks of exchange, but in either case added new layers of meaning to dooryard trees and flowers.

With their interest in horticultural matters, their progressive tendencies, and their economic resources, B. G. Buell and Esther Lawrence

 would seem ideal customers for the booming mid-nineteenth-century horticultural trade. They certainly encountered many advertisements for seed companies or large regional or national nurseries over their years of reading the *Michigan Farmer*, the *American Agriculturist*, or even local newspapers. In addition, their good friend and neighbor Benjamin Hathaway operated an extensive nursery on his nearby farm, providing a convenient local source for plant purchases and information, and offering an impressive selection of ornamental trees and shrubs. Hathaway was particularly noted for his roses, and even cultivated some tender plants in a greenhouse.[25]

Conforming to expectations, both B. G. Buell and Esther Lawrence did buy plants, but the diaries indicated only occasional purchases, suggesting that commercial transactions constituted one of several methods for increasing their plant collections. In addition, both appeared to divide their transactions between distant firms and local nurserymen. In January 1861, for example, Esther Lawrence visited a neighbor to get information about sending off for flower seeds. On another occasion, she ordered rose bushes, and was pleased when they arrived safely only ten days later. Buell, in turn, wrote to Storrs, Harrison and Company, a large regional nursery in Painesville, Ohio, for their catalogue and "flyer" on chestnuts, placed his order, and eventually noted that he had paid the $.88 bill. In another transaction, Buell ordered and received hedge plants from a distant nurseryman, and when they arrived by train, enlisted the help of his neighbor Levi Lawrence to bring them home from the depot at Decatur.[26] When they purchased plants locally, both families seemed to most frequently favor Benjamin Hathaway with their business, and over the years recorded a number of purchases from the Hathaway nursery including evergreens, willows, and roses.[27]

Their intermittent purchases suggested that Esther Lawrence and B. G. Buell turned to commercial suppliers when they sought specific horticultural varieties, and had compared the selection of one of the large national or regional firms to that offered by a nearby nurseryman. In the end, they made their purchase from the source that could best meet their particular need. The process and their selections maintained a balance between local purchases and the national marketplace. Each source had certain advantages, and rather than turning away from one to embrace the other, Mrs. Lawrence and Buell seemed to value both contributions. When they ordered from well-known, but distant commercial horticulturists, they enjoyed the benefits of a large selection, the excitement of re-

ceiving plants and seeds through the mail or by express, and the sense that their homegrounds displayed the best the nation had to offer. These transactions, however, typically lacked the personal interaction they were accustomed to, and afforded no opportunity to look over and discuss particular plants before they made their purchases.

When they purchased from local sources like the Hathaway nursery, they may have had a diminished selection, but could enjoy the personal attention and advice of their friend and neighbor. Esther Lawrence often noted her pleasure in such exchanges, and because of Hathaway's generosity, frequently came away from the transactions with additional plants. On a visit in 1858, for example, she brought home "five varieties of chrysanthemums in full bloom, and a slip of rose geranium" as gifts that underscored their friendship as well as their commercial relationship. Nearly twenty years later, she expressed particular delight when the Hathaways gave her some special coleus plants to enjoy.[28]

Even when they were involved in transactions with distant suppliers, Lawrence, Buell, and their Volinia Township neighbors managed to transform the occasion into an opportunity for community interaction. In the early spring of 1866, for example, Esther Lawrence, Mrs. Hathaway, and several other neighbors met at Hattie Buell's home, and "selected some seeds to send for." The ladies were forming what commercial seedsmen like James Vick termed a "club," and by sending their orders in a single letter, enjoyed the reduced price offered for large orders. Economy was not the only motivating factor, however, for although some seed dealers volunteered to send club orders to individual homes, Esther and her friends again assembled to divide up the newly arrived seeds. "The ladies met to distribute flower seeds," Esther wrote of the event, "we had quite a company." Both meetings provided a pleasant social exchange, an opportunity for women to interact with friends and neighbors, and the shared excitement of receiving seeds from a distant source.[29]

In 1870, B. G. Buell described a very similar transaction, this time involving a shipment of seeds from the U. S. Department of Agriculture. Beginning in 1849 and continuing through much of the century, the Patent Office, the predecessor of the Department of Agriculture, and the Department itself distributed packets of free vegetable, field crop, and ornamental plant seeds to individuals, agricultural societies, or other local organizations. On April 10, 1870, Buell noted that the secretary of the Volinia Farmers' Club had received a large shipment of government seeds, and had called a meeting the next evening for distribution. Buell,

Levi Lawrence, and an unusually large group of twenty other members attended the meeting, voted to give the four bags of oats to Benjamin Hathaway, and then distributed the "balance ... among all members present." Like the women of Volinia Township, Farmers' Club members understood the seed distribution as a way to improve their gardens, and as a way to enjoy the company of friends and neighbors.[30]

Although the diaries recorded an occasional purchase of seeds and plants, Esther Lawrence and B. G. Buell obtained ornamental plants far more frequently through a complex system of trade and gifts among friends, neighbors, and family members. These transactions usually involved excursions and visits, and were woven into the fabric of community social life. Like her neighbors, Esther Lawrence often went visiting, sometimes accompanied by her husband, one of her sons, or her sister Hattie. The group walked or took the buggy to a neighbor's home, chatted, and frequently came home with new plants to enjoy. Sometimes these plant acquisitions appeared to be the by-product of a social visit, and at other times actually sparked the social interaction. The exchanges themselves also varied a good deal. In some instances, both parties traded plants, and at other times Esther called on her neighbors to get specific plants, but did not mention bringing anything in trade. Other community members went home with cuttings or seeds after a visit with the Lawrence family. Typically, neighbors and friends exchanged a wide variety of plants, including dahlia tubers, hyacinths, lilies, roses, and perennial flowers that had grown large and needed division.[31] B. G. Buell also gave ornamental plants to others in the community, but tended to offer trees and shrubs rather than flowers. When neighbor William Lyon came to purchase seedling apple trees, for instance, Buell freely added a horse chestnut tree and a white lilac to the plants he had ordered. On another occasion, Buell divided the seeds he received at the Farmers' Club meeting with a neighbor and included a white lilac for the dooryard with his gift.[32]

These ornamental plant transactions also provided links of on-going common interest among somewhat disparate community members. Many of the transactions that Esther Lawrence and B. G. Buell recorded occurred among their immediate farm neighbors, but these families varied considerably in age and economic rank. Older neighborhood women like Benjamin Hathaway's mother, Naomi, or Sarah Goodspeed, a close friend of Mrs. Copley's, routinely exchanged plants with Esther. The Goodspeeds were prosperous farmers, and by 1870 the Goodspeed brothers

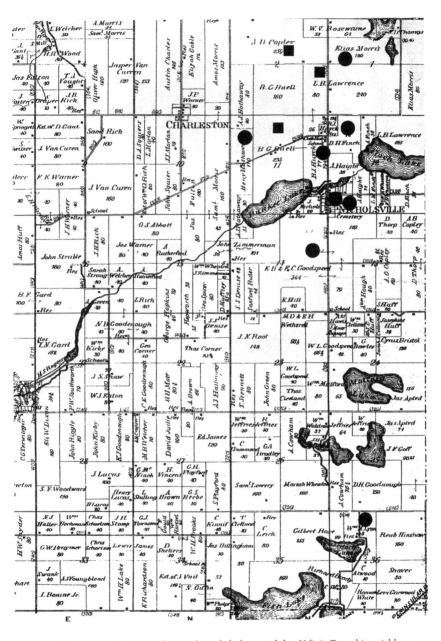

Figure 7.3. Esther Lawrence frequently traded plants with her Volinia Township neighbors. On this map, squares indicate Lawrence, Buell, and Copley family homes, and circles mark the homes of neighbors involved in horticultural trades. From D. J. Lake, *Atlas of Cass County, Michigan* (Philadelphia: C.O. Titus, 1872).

owned nearly five hundred acres, but other families involved in the plant exchanges held much more modest holdings. The Hathaway family, for instance, came to Michigan because of family losses sustained in the 1837 financial panic, and although Benjamin Hathaway had emerged from debt, the family owned only 120 acres in 1870. Other neighbors involved in plant exchanges, like the Finches and Huycks, or the Morris family, cultivated farms ranging in size from 150 to 280 acres.[33] Of those commonly involved in the plant exchange network, William Lyon had the most limited financial resources. He, his wife, and seven children lived on an eighty-acre farm nearly six miles south of the Lawrence and Buell holdings. Although Lyon described himself as the proprietor of "Walnut Hill Grapery," an establishment noted for its production of "genuine" Delaware and Catawba grapes, his total economic worth in 1870 was $4,200. The Lawrence and Buell families, in contrast, each had personal and real property worth over $30,000. Although most of her trades were with neighboring farm families, Esther Lawrence also gave plant materials to several young women of modest means who lived in the nearby village of Nicholsville. Amanda Thorp, the wife of a dry goods merchant, and Mrs. Jerome Bitely, the wife of a blacksmith and veterinarian, often visited Esther Lawrence and sometimes came away with "some plant slips" for their own garden.[34] (See Figure 7.3.)

Through all these gifts and exchanges, the residents of the Nicholsville area expanded their plant collections and enjoyed conversations and contact with their neighbors and friends. More importantly, however, their common interest in plants crossed generational and occupational lines, bringing together older community members, farm families, and village residents. The exchange also ignored some very real economic disparities among local families, enabling those of more modest means, like the Lyon family, to cultivate their own special plants and make meaningful exchanges with their far wealthier neighbors. While commercial nurserymen tended to promote ornamental plants as an expression of individual taste and fashion, these Volinia Township families transformed them into a way to strengthen relationships and to transcend differences. Ornamental plants, flourishing in farm and village yards, offered very visible reminders of community ties, long-standing friendships, and shared values.[35]

When local residents undertook homeground embellishments, their interest in ornamental plants often sparked curiosity, and they sought to learn more about plants and their culture. The acquisition of horticultural knowledge was closely linked to the processes of obtaining ornamental plants, and involved the Lawrence and Buell families in a similar intermingling of local exchanges and a vast new national marketplace of ideas and innovation. Both Esther Lawrence and B. G. Buell satisfied at least some of their horticultural enthusiasm and curiosity through reading, but in addition, their pursuit of horticultural knowledge stimulated social and intellectual interactions among neighbors as they "visited" to examine horticultural innovations.[36]

Among local families, Benjamin Hathaway was an exceptionally good resource for extending horticultural knowledge. A well-known horticultural experimenter, Hathaway kept abreast of new information and often contributed articles on horticulture to journals like the *Prairie Farmer* or the *Michigan Farmer*. In addition, he acted as a kind of cultural mediator between professional horticulturists, large commercial nurseries, and local residents. On a number of occasions, Hathaway ordered plants from Ellwanger and Barry, the nation's largest nursery at mid-century, and his acquisitions enabled local residents to see the best the nation had to offer. For those who hesitated to enter the national market, Hathaway's experiences demystified the process, and made it seem both familiar and accessible to other community members. Hathaway was also involved in statewide horticultural societies from the earliest attempts at organization in the 1850s until the formation of the Michigan State Pomological Society in the 1870s.[37]

A casual visit to the Hathaway home opened many horticultural vistas, for the beauty and diversity of Benjamin Hathaway's plantings were well known across the region. In 1866, the Michigan State Agricultural Society acknowledged his exceptional efforts and awarded him a premium as "best [small] farm in the state." In explaining their choice, the Visiting Committee described Hathaway's holdings and his various experimental efforts, particularly noting his work with hedge plants, his preference for thorn hedges, and his planting of more than two thousand trees including a vast maple grove, Austrian and white pine, and more than ninety Norway spruce. Hathaway's home, the Committee reported, "was a neat building, with very elegant grounds around it, neatly planted with evergreens, and the flower-gardens ornamenting its front, surrounded with hedges of cedar and arbor vitae."[38] Both B. G. Buell and Esther Lawrence recognized Hathaway's achievements, and sometimes recorded their inspection of Benjamin Hathaway's nursery and ornamental grounds in their diaries. In one instance, Esther described a special trip to look over

a newly constructed "plant house" at the Hathaway homestead. B. G. Buell, in turn, noted when he made a trip to inspect Hathaway's raspberry patch, and that he took a visitor from Kalamazoo over to see Hathaway's "new [plant] house and nursery."[39]

On occasion, visitors dropped by to discuss horticultural issues or examine innovations on the Lawrence and Buell farms. In 1872, for instance, Buell noted the visit of well-known Galesburg, Michigan, nurseryman H. Dale Adams, and expressed pleasure at his interest in the Buells' orchard and homegrounds. The two men conversed while walking over the farm, and then inspected neighbor Benjamin Hathaway's nursery before Adams continued on his travels. The exchange not only offered Buell an opportunity to display his farm and horticultural advancements to an appreciative and knowledgeable colleague, but also strengthened his connections to like-minded farmers beyond the immediate area. Potentially, conversations like that with Adams offered new perspectives and helped Buell evaluate his progress and practices in statewide or even national terms.[40]

More formal exchanges of horticultural information took place under the auspices of local organizations and associations. In August 1875, for example, members of the Home Culture Society focused their attention on floriculture as they met at the home of Esther Lawrence. Regular members and a number of guests attended to hear Esther Lawrence along with two of her neighbors present essays on the "cultivation of flowers." The group then examined a "profusion of choice bouquets," shared cuttings of new plants, and enjoyed "pleasant conversation and a good time generally." The meeting required more than casual conversation about plants, however, for participants had to read, prepare their essays, and make a presentation to their friends and neighbors.[41]

Along with local society meetings, both the Lawrence and Buell families routinely attended town, county, and state fairs, and although many exhibits sparked their interest, horticultural displays were an important draw. Every autumn, the families visited the Volinia Farmers' Club Fair or the Cass County Agricultural Society Fair, and often commented on their own entries, the attendance, and the quality of the display.[42] These local fairs showcased their neighbors' productivity, and facilitated the exchange of ideas among a local, or at best, a regional clientele. Yearly trips to the State Fair, however, tended to offer an even broader array of new ideas and experiences for both the Lawrence and Buell families. Esther Lawrence summed up her experiences after a particularly tiring trip to the 1863 State Fair at nearby Kalamazoo, when she declared she had "quite an interesting time."[43]

In contrast to Esther Lawrence's abbreviated remarks, B. G. Buell often provided a much more detailed account of his experiences at State Fairs. In 1869, Buell and Levi Lawrence attended the State Fair in Jackson. With fair entries in tow, they took the express train to Jackson, secured a boarding house, and after meeting Hattie, who had taken a later train, set out to see the sights. The group visited the fairgrounds regularly, and in the floral hall enjoyed a wide-ranging horticultural display replete with a massive collection of flowers from Rochester seedsman James Vick.[44] The trip to Jackson provided an opportunity to take in other local sights as well. The group regularly strolled along Jackson's busy commercial and residential streets, and toured the state prison, but a visit to Jackson's "Greenwood" Cemetery especially impressed Buell. An example of the mid-nineteenth-century rural cemetery movement, Greenwood featured landscape plantings and a variety of ornamental trees and shrubs. Contemporary accounts described the cemetery as a "park for the living as well as a resting place for the dead," and boasted of its "charming undulations and quiet dells; its winding carriage ways; and shadowy paths; its beauty of sward and flower and tree ... the cypress and willow, the parterre and terrace." Buell noted that the cemetery was a "very pleasant spot and well kept; it is thickly set with evergreens." Like Esther's visit to the State Fair in Kalamazoo, Buell's experiences in Jackson opened up new practices, trends, and ideas that extended far beyond the resources or examples of Volinia Township.[45]

Perhaps the most significant opportunity to both extend horticultural horizons and enjoy a national forum for new ideas and practices occurred when Esther Lawrence attended the 1876 Centennial Exhibition in Philadelphia. The Centennial Exhibition had been a focus of interest in Volinia Township for some time, and the Volinia Farmers' Club had made considerable efforts to send off appropriate examples of Michigan agriculture. B. G. Buell, Levi Lawrence, and Benjamin Hathaway, along with many neighboring farmers, displayed marrowfat beans, dent corn, several wheat varieties, clover, and a collection of fruit at the Exhibition. On September 1, 1876, Esther, her youngest son, Austin, and neighbors Mr. and Mrs. Morris began a sixteen-day trip that included an extensive visit to the Exhibition and even a tour of New York City.[46]

Esther recorded the trip in detail, noting how they traveled, their accommodations, and what they enjoyed for tea. She also described the

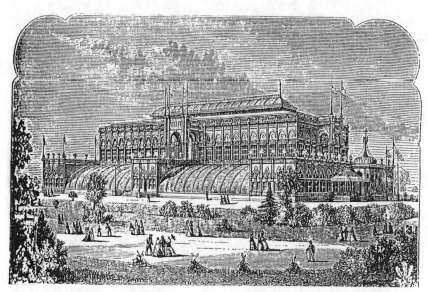

Figure 7.4. Horticultural Hall at the Centennial Exhibition. From James D. McCabe, *The Illustrated History of the Centennial Exhibition* (Philadelphia: The National Publishing Co., 1877), 565.

buildings and crowds of the Exhibition, declaring after the first day that they came away with "an immense amount of mental pictures and few facts, with the knowledge that in months of sightseeing we could neither see nor comprehend one half of what could be learned at the exhibition." Among her many experiences, Esther and her friends enjoyed a visit to Horticultural Hall and the Women's Pavilion, where Esther described her pleasure in the Canadian exhibit that went beyond "fancy needle and worsted work" to display "some beautiful model houses and yards."[47]

Horticultural Hall offered a diverse array of exhibits to spark Esther's interest and broaden her horticultural knowledge. The central building housed a large conservatory, bordered on each end by "forcing-houses" for the propagation of plants. The conservatory was filled with a "superb collection of rare and luxuriant trees and shrubs" including fan palms, orange and lemon trees, eucalyptus, guava, India rubber trees, and many others. The forcing house featured azaleas from Belgium, Japanese maples, and "South Sea Island" pitcher plants. Other areas in the horticultural complex housed exhibits of greenhouse equipment, garden furniture and tools, yard ornaments, a display of the latest in lawn mowers, and even an exhibit of fertilizer from the Pacific Guano Company. Surrounding the horticultural buildings, Esther and her friends enjoyed a twenty-five-acre ornamental garden beautifully designed with both "native and foreign flowers," and enhanced by a number of examples of rustic-work furniture and lawn ornaments. Seedsmen like Henry Dreer also created elaborate displays of their products to remind distant visitors where to place their orders for plants like those they had seen at the Exhibition.[48] The Centennial Exhibition was a grand extravaganza of the latest commercial advances and progressive possibilities. The visit offered Esther and her friends a myriad of fashionable horticultural examples, and a perspective on commercial enterprise that could not be replicated at home. (See Figure 7.4.)

After having their fill of the Centennial Exhibition, Esther and her neighbors proceeded on to New York City. She found the steamboat trip to New York very interesting, but on arrival in the city, could not contain her dismay. "Oh Babel of Babels!" she wrote, and remarked that the travelers spent their first day in New York safely resting in their lodging rooms. Eventually the group ventured out and made a special effort to visit Central Park, a fine example of contemporary landscape architecture frequently discussed in the horticultural and agricultural press. Rain cut the trip short, but despite the interruption, Esther declared the trip a success, for the tour provided an opportunity to see one of the most famed examples of ornamental plant use in the country. Glad to be back on Michigan soil, Esther returned home on September 26, and resumed the daily routine of farm chores, visits with neighbors, and gardening, her mind filled with new horticultural ideas and examples.[49]

Their interest in ornamental plants had social and intellectual implications for the Lawrence and Buell families, but the care of their own trees and flowers impacted both households even more. No matter how practical or simple the embellishment, establishing and maintaining the prescribed lawns and gardens demanded a commitment in time and labor, directing both resources and space away from more traditionally productive farm functions and adding yet another chore to the busy farm day. The diaries of Esther Lawrence and B. G. Buell underscored the level of effort required, and suggested that they met the demands by dividing responsibility for ornamental plant culture among family members. Levi Lawrence and his sons, for example, plowed, laid sod and planted grass seed to establish the lawn at the family's new home, while B. G. Buell or one of his hired men took care of the weekly lawn mowing chores around the Buell farmhouse. In addition, male family members had primary re-

sponsibility for the planting, pruning, mulching, and fertilizing of ornamental trees and shrubs. Esther Lawrence and Hattie Buell, in turn, cultivated flower gardens and indoor plants.[50]

Ornamental plant chores not only required a commitment from all family members but, as Esther's diary revealed, demanded attention throughout the year. Esther began her yearly cycle of plant care in early spring by starting some flower seeds and tubers indoors, and then turning her attention to cleaning and raking the dooryard, tasks that often extended over several days. Levi or the boys normally "fixed" or spaded the garden beds for their mother, so that she could plant some of her annual flower seeds outside. As the spring weather moderated, Esther routinely set out gladiolus corms, and dahlia, tuberose, and Madeira vine tubers, and her frost-sensitive plants like coleus. In addition, she spent many hours trimming her rose bushes and cultivating their beds. Like many other mid-nineteenth-century horticulturists, Esther also placed many of her potted house plants in the garden during the warmer months. As summer approached, Esther usually turned her attention to hoeing and weeding her flower gardens.[51]

When the danger of frost approached in the fall, garden tasks again increased. Esther protected her roses by covering them with straw, "took up" tender plants like verbenas or geraniums for indoor storage, dug tender corms and tubers and stored them in the cellar, and potted some flowering plants to admire on the windowsill during the winter months. She again cleaned flower beds, trimmed back hardy perennials, and with an eye to early spring blossoms, often planted tulip and hyacinth bulbs. Gardening chores continued through the winter as Esther tended her house plants, moving some initially stored in the cellar up to the dining room, and repotting others.[52]

Esther Lawrence juggled her gardening tasks around a myriad of other responsibilities, but never indicated that she found the demands of ornamental plant care burdensome. Several of her actions suggest that, in fact, she especially enjoyed her gardening chores, and altered her other domestic responsibilities to accommodate her love of plants. Esther cultivated many types of flowers including several fashionable varieties like verbenas or geraniums that required extra care. In addition, she purposefully extended the gardening season by nurturing seedlings and potted plants indoors. Esther also appeared to seek out gardening chores in a way or at a time that underscored the pleasure she found in the work. On November 21, 1858, for example, she wrote that she had "trimmed the flowers and dressed the boys, feeling quite able to do so." The comment came just four days after the birth of her youngest son, Austin. Esther had spent the previous three days resting, knitting, and tending the baby, and getting her garden in order for the winter was the first active task she chose to pursue after her confinement. On another occasion, Esther noted that she spent the morning potting plants and the afternoon "admiring them," an obvious indication of pleasure and a willingness to put her plants before other responsibilities.[53]

In the Lawrence and Buell households, both male and female family members appeared to find ornamental plants important enough to devote time and labor to their care, and that shared interest and cooperative spirit may have enhanced Esther's enjoyment of plant culture. Family members routinely took care of their prescribed jobs in a timely fashion and without conflict, but in addition often moved beyond their regular tasks to help others complete ornamental plant chores. Levi, for instance, established a lawn at the new family home, but Esther assisted by cleaning up the yard, raking, and filling in cracks in the newly placed sod with additional soil. Esther often noted that she was "helping Pa" in the vegetable garden or with small fruit culture, even though she did not have primary responsibility for those areas. Levi or the boys, in turn, helped Esther establish new garden beds or spaded already cultivated areas in preparation for spring planting. Their actions relieved Esther of heavy work, but also offered an opportunity to acknowledge their mother's special interests, to show their support, and to express their love.[54]

The Lawrence and Buell families also recognized that ornamental plants could play an important symbolic role in their lives, and very purposefully turned to trees and flowers as a way to cement and signify family bonds. Over the years, family members commonly exchanged plants as a part of daily interactions or as a way to express special sentiments. As sisters and neighbors, Hattie Buell and Esther Lawrence saw each other very regularly, yet often brought along seeds or flowers as special gifts when they visited.[55] This routine pattern of family exchange found its most eloquent expression in the actions of their older sister Charlotte. In the year before the Copley family moved from Ohio to Michigan, seventeen-year-old Charlotte Copley recorded her interest in and knowledge of both cultivated and wild plants in a fragmentary diary that contained little additional information. Her activities varied, but included helping her mother with gardening chores, and the cultivation of her own half-circle garden bed, bordered with "little violets." Charlotte also potted na-

tive bloodroots and enjoyed their blossoms, started some convolvulus and nasturtiums from seed, and "set a hyacinth in a bowl" to force it into bloom.[56]

Eventually Charlotte shared this love of flowers with one of her younger sisters. Charlotte was the oldest of the Copley children. In 1832, her sister Esther was eight, Olivia, another sister, was four, the youngest girl, Flora, was two years old, and Mrs. Copley was about to give birth to another child.[57] On March 14, 1832, Charlotte noted that she "set out some little flower bulbs in a tin cup for Olivia." The sisters then watched their progress and Charlotte recorded the results, noting when the first flower appeared, that they "grew quite fast," and that on March 20, four additional flowers had opened.[58] Charlotte never stated her reasons for giving flowering bulbs to Olivia. Perhaps it was a simple act of kindness for a beloved younger sister, perhaps Olivia had expressed an interest in plants that Charlotte wanted to encourage, or perhaps Charlotte thought Olivia was of an age to be most affected by her mother's impending confinement and the birth of a new sibling. Whatever the reason, Charlotte Copley shared one of her most valued possessions with a family member, and had, in the process, used ornamental plants to express her affection.

Ornamental plants also helped to bridge the distance between scattered family members. Almon Copley, one of Esther Lawrence's older brothers, initially established a home in Missouri, but during the Civil War his family made a forced move into Kansas and remained there after the war ended. Esther's youngest brother, Napoleon, attempted to establish a homestead at Emporia, Kansas, and her sister Euphemia or "Phim" eventually moved to Kansas as a teacher. Several of Esther's aunts and uncles settled in Illinois and other family members lived near Mexico, Indiana. On several occasions, Esther exchanged plants and seeds with these distant family members. In 1861, for example, she carefully packaged seeds and sent them off to Kansas, perhaps hoping to make Almon's family feel at home after their move, or to remind them that they had a loving family in Michigan. Whenever Esther Lawrence visited relatives, she and other family members often worked in the garden or planted special seeds or flowering roots to mark the visit. On her extended stay with western family members in 1878, Esther "helped Eliza plant some flower seeds," and as she visited other relatives in Indiana on her way home, planted some "cypress vine seeds" with them to commemorate her stay. These planting rituals not only enlivened the visit, but created a visible and beautiful reminder of family ties long after family members had parted. The exchange of plant materials moved in both directions, and Esther sometimes returned from her visits with prized specimens. She valued those plants a good deal and treated them with special care, in one case making an additional trip to Hattie's home the day after her return from Kansas "to get my geranium slips that we had brought from Phim's and the specimens of plants we had pressed."⁵⁹

On some occasions, male family members also maintained links with distant relatives through plants. In 1870, B. G. Buell sent a large shipment of trees to his sister-in-law Euphemia. After establishing herself as a teacher in Emporia, Kansas, in the mid-1860s, Phim apparently asked her brother-in-law about cultivating evergreen trees at her new western home. Buell thought they would grow, and enthusiastically took on the task of supplying her request. He eventually assembled a diverse collection of conifers including spruce, balsam fir, Austrian and Scotch pine, and hemlock, prepared them for shipment by "puddling" or coating the roots in mud and wrapping them in moss. He then boxed them up and sent the parcel, weighing 340 pounds, off by express. Typical of his nononsense style, Buell noted in his diary that "I wrote a line to Phim saying I donate the evergreen trees to her Evergreen Place, Emporia." Although he had articulated no special affection for his sister-in-law or concern over her move to Kansas, Buell's actions suggested another reality. He had gone to considerable trouble and expense to fulfill her request and to provide her with living signs of her family in Michigan.⁶⁰

Both B. G. Buell and Esther Lawrence used ornamental plants to solidify family ties, yet there were some subtle distinctions in their actions that underscored how even those gestures varied in meanings. When B. G. Buell sent off his packet of trees to Euphemia, he took particular pride in the diversity of species and the size of the shipment. The large selection of plants highlighted his horticultural expertise, his generosity, and the fact that he had the means to provide such a sizable gift. His efforts reflected his love and concern for Euphemia, but also emphasized his gendered role as successful provider or benefactor. Esther normally sent and received much smaller horticultural offerings, but in addition seemed to derive her meaning from the source of the plants, not necessarily because they were new or unusual varieties. For Esther, geranium cuttings, like those from Euphemia, had significance because her sister had actually nurtured and loved the plant. Once rooted and flourishing on the windowsill, the geraniums offered a concrete and very real connection to Euphemia, bringing her presence into Esther's familiar domestic land-

scape, no matter how great the physical distance between the two sisters.

The need to forge and solidify family ties extended through time as well as space for some family members. As she faithfully recorded her life, Esther often noted her sorrow as family members died and the life around her changed. Over the years, she associated those losses with her parents' home, and lamented its desolate appearance after her parents' death and after the last of her siblings struck out on their own. Usually accompanied by Levi or one of her sons, Esther often walked to the "old homestead" or the "old place," looked about, remembered her childhood or absent family members, and sometimes recorded the sense of loss she experienced. In a typical expression of this darker theme, Esther described her family as "deserting the home of our youth," and then exclaimed, "That sacred spot. The scene of so many innocent pleasures and some heartbreaking woes." Her comments, written in 1856, came as her brother Napoleon left for a trip to Kansas and as her younger sisters packed for their travels. After Napoleon's death, Esther returned to the homestead many times, noting on one occasion that she had come to "see how the *four* graves in the garden appeared." When the Lawrence and Buell families planted evergreens to mark the graves, Esther often walked over to the old place to see the trees "set by the graves" and monitor their condition and growth.[61] Hattie had a similar bond with her parents' home and, like Esther, encouraged her family to develop a shared sense of place and family continuity. In 1871, B. G. Buell noted that "With Hattie and the children went over to the old place as they call it where Hattie used to live and where her father, mother, a sister and a brother are buried."[62] In both families, visits to the old place not only prodded memory, but extended very concrete family ties into yet another generation.

Eventually, Esther Lawrence focused her sentimental attachment on the ornamental plants that continued to grow around the old homestead. On one visit with her four-year-old son Linneaus, Esther described going through the house, and her feeling that the "Spirit of Home was gone." When she visited the garden, however, her mood changed. Flowers, loved and nurtured by her mother, still flourished there, and for Esther "seemed like an offering of affection from the departed."[63] To assuage her loneliness, Esther gradually gathered those meaningful flowers into her own garden, again creating a living presence of absent family members by shaping her domestic space with plants. Sometimes she simply recorded that she had transplanted "flower roots" from the homestead, but the depth and complexity of her feelings were particularly evident when she received a special gift of "bergamot balm" from her neighbor Sarah Goodspeed. When they first arrived in Michigan, Sarah Goodspeed and her husband lived for a time with Alexander and Esther Copley, and the two women developed a long-lasting friendship. The Goodspeeds eventually settled on a farm very to close to the site of the Copleys' first home, and as they parted, Mrs. Copley gave her friend a perennial flower called bergamot balm to brighten their new homestead. Sarah Goodspeed gardened all her life, and long after Mrs. Copley's death in 1852, cherished and nurtured those early gifts from her friend. On June 30, 1876, Esther Lawrence visited Mrs. Goodspeed and reported, "Spent the afternoon at Mrs. Goodspeeds, had a very cordial welcome from her and Eugenie. Brought home a memorial of my visit, some of the bergamot balm that my mother so much prized and brought to Michigan with her forty-three years ago, 1833."[64]

Esther took obvious delight in her gift and its close links to her mother, but the transaction revealed additional layers of meaning. Seventy-four years old in 1876, Sarah Goodspeed had known Esther as a child, and had followed her development as a young wife, mother, and responsible community member. Well aware that Esther loved flowers, Mrs. Goodspeed also understood that Esther and her mother had a strong and very loving relationship. Over the decades, Mrs. Goodspeed and Esther enjoyed many neighborly exchanges and conversations. Why, then, did Mrs. Goodspeed wait so long to share the bergamot balm with Esther? The answer is far from clear, but perhaps had to do with Mrs. Goodspeed's role in the community, and the meaning she attached to the bergamot. An offhand remark from B. G. Buell suggested that her Volinia Township neighbors may have viewed Mrs. Goodspeed as an "elder" resident who sustained a body of traditional, yet very valuable medicinal knowledge. In January 1876, Buell's young daughter Flora was quite ill, and after the family had nursed her for some time, B. G. Buell, a long-time temperance man, turned to Mrs. Goodspeed for assistance. Apparently hoping that Mrs. Goodspeed's advice and her spirits would revitalize Flora when other methods had failed, Buell wrote that he went to "Old Lady Goodspeed's and got some wine for Flora."[65] Mrs. Goodspeed's knowledge and interest in home remedies may have extended to her prized bergamot. Bergamot was an attractive perennial flower, but like many other cultivated herbs, also had a long-standing reputation for

curative powers. Mrs. Goodspeed had protected and nurtured the bergamot for years, and perhaps occasionally used it to brew a medicinal tea. As the end of her life approached, she may have felt that the time had come to give up her traditional role and the plant. For Mrs. Goodspeed, the bergamot represented a tie to her friend Mrs. Copley, but also symbolized her own life, experiences, and accumulated knowledge. Unwilling to abandon all her plants and their associations at one time, Sarah Goodspeed told Esther to come back another day, so that they might together dig up some phlox plants she also treasured.[66]

Esther, in turn, found immediate significance in the bergamot as a gift from her neighbor, and may have sensed the importance that Mrs. Goodspeed attached to passing on the plant. Like other flowers her family had cherished, the bergamot also represented a direct tie to her mother, offering Esther a living reminder of her life, sensibilities, and love that extended well beyond the Michigan years. Mrs. Goodspeed died in 1877, and after her death, Esther wove even more meaning and memories into the bergamot balm in her garden. Several years later, Esther continued to think about the phlox that Mrs. Goodspeed had promised. Perhaps because she hoped to complete the transferal of memory and tradition, Esther visited Mrs. Goodspeed's daughter-in-law Mary in 1879, and reported, "She dug up those phlox for me that Mrs. Goodspeed had promised to do."[67]

The diaries of Esther Lawrence and Charlotte Copley revealed one final dimension of ornamental plants and their significance. As the Copley, Buell, and Lawrence families planted and tended their flowers, the plants themselves took on a powerful emotional import, often providing a welcome source of beauty, hope, or spiritual comfort at particularly difficult times. One of the first hints of their importance and their ability to ameliorate tragedy and sorrow surfaced as Esther confronted the untimely death of her youngest brother, Napoleon. Hoping to establish a homestead and to prosper, Napoleon moved to Kansas in 1855. His dreams were cut short by tuberculosis, and Napoleon came home to Michigan to die. Several days after the death of their beloved brother, Esther and her sisters Hattie and Euphemia gathered at the Lawrence home. Esther carried on the daily routine, noting that she was "choring but [did] not feel at all well." After several days, Levi, knowing the emotional power of plants and perhaps hoping to cheer the sisters, drove them all to a nearby pottery to purchase flower pots. The routine of plant care and the pleasure of flowers appeared to offer relief or at least a diversion from their

sorrow, and Esther reported that they "got two flower crocks and took up our chrysanthemums." When both Hattie and Phim left several weeks after Napoleon's death, Esther once more turned to plants for solace, recording that she planted "Kansas moss seed" on the day Levi took Phim to the train depot. Many years later, Esther again used ornamental plants as a way to relieve a sense of loss. When her youngest son, Austin, entered the "Commercial College" in Kalamazoo on November 21, 1877, Esther's diary entry reported his departure and said simply, "Sad day." Several days later, in a similarly brief entry, Esther noted that she had purchased some new plants from a local horticulturist on the day Austin left home.[68]

Two instances, many years apart and both involving the loss of treasured plants, confirmed the importance of flowers for both Esther Lawrence and her sister Charlotte. Charlotte Copley related the first event as she recorded her family's move from Ohio to Michigan in 1833. On June 9, 1833, Mr. and Mrs. Copley, Charlotte, her six younger siblings, and three hired men loaded the family possessions on two ox-drawn wagons, and started out for a new life in Michigan Territory. Charlotte reported the optimism of the group and the excitement of the journey, as she wrote on the first night out, "All well and in high spirits ... found roads quite good for a few miles." For several days she continued her upbeat assessment, commenting on where they camped and took their meals, on the beauty of the countryside, and sometimes even noting new flowers that she discovered along the way.[69]

On June 12, after a day of mud, bad roads and "snakes aplenty," the "tired, wet, and hungry group" stopped for the night, and the tone of Charlotte's commentary changed. She described the terrible day and then wrote, "still I did not regret leaving Delightful Hill, not half so much as I did in losing most all of my flowers out of the box." Charlotte had packed some of her beloved flowers to bring to the new home, and as teams struggled to pull the wagons out of the mud, her flowers disappeared. The day after her loss, Charlotte clung to her remaining plants, and noted, "I kept close behind the wagon [*sic*] so as not to lose any more of my flowers." Charlotte made no further comments about her flowers, but as the trip unfolded, Mr. Copley appeared to recognize the seriousness of her loss and attempted to comfort her by pointing out particularly beautiful flowers along the way. Charlotte acknowledged her father's concern, and responded to the uniqueness of the plants he found. In one instance, Mr. Copley brought a showy lady's slipper to her, and Charlotte

found its beauty stunning, recording a description of its floral structure and a sketch in her diary. "I wish you could see what a beautiful flower that Pa found this noon," she wrote. Several days later, the group came upon a lake covered with water lilies. "Pa went out in a canoe and got one," Charlotte reported. "They are very beautiful."[70]

Charlotte may well have felt the unmitigated sense of adventure she expressed in her description of the trip, but her comments about the loss of her flowers suggested a more complex reality. In some measure, Charlotte's experience with flowers at Delightful Hill helps to explain what the loss meant to her and other family members. Several years after the family settled in Ohio, Charlotte helped her pregnant mother set out rose bushes, quince, hollyhocks and "lilac sprouts." The selection of plants was both significant and curious, for all but the hollyhocks required some years to establish themselves before coming into their full floral beauty. Either Charlotte and her mother presumed that they would remain at Delightful Hill for an extended period of time and were beginning the long-term improvement of their homegrounds, or their desire for flowers was so strong that they planted available flowers and shrubs without regard to the future. The family's experiences argued for the latter interpretation, for they had settled and re-settled many times in the towns and villages of western New York before the move to Delightful Hill in 1829. No matter where they found themselves, Charlotte and her mother grew accustomed to shaping their domestic space with trees, shrubs, and flowers, and turned to ornamental plants as a portable, yet very meaningful embodiment of the stability, familiarity, and comfort of a well-loved home. When Charlotte lost her "flowers," she lost far more than carefully packed roots, seeds, or cuttings, for she had invested in those plants all the sentiment and security attached to the home she left behind.[71]

Over forty years later, Esther Lawrence experienced a similar loss and treated it with equal seriousness. On January 20, 1877, after an evening of entertaining Hattie and her children, Esther tidied the house and prepared for bed. "Put everything to rights as I supposed," she wrote, "but left the kitchen door unlocked. The wind eddied around the corner, blew it open, and opened the dining room door." The temperature fell below zero, and when the family awoke in the morning, she found that the storm and the open door had "disastrous consequences" for her "pet house plants." After the disaster, Esther took her frozen plants to the cellar, in hopes that perhaps they would revive by a gradual reacclimation to warmth and light.[72]

Bergamot Balm and Verbenas

While she took care of the plants, her response to the disaster was revealing. Linneaus attempted to cheer his mother by hanging a portrait of Timothy Shay Arthur, the editor of her favorite *Arthur's Home Magazine*, in the dining room where she had displayed her favorite house plants. Esther enjoyed Arthur's writings and noted her son's attempt to "help counteract its desolate appearance," but despite his efforts, Esther took her loss as a personal punishment. "[I] felt that I must still be punished," she wrote. "Will it always be so?" Her comment was a cryptic reference to a conflict with her middle son, Archie, that had erupted in December 1874, and continued, although somewhat abated, through the winter of 1877.[73]

Esther Lawrence enjoyed a particularly close relationship to Archie, noting when he was young, how she hated to leave him, and their happiness at being reunited.[74] When he was nineteen years old, Archie abruptly left home after an apparent family disagreement. The day after his departure, Esther eloquently described her sorrow and concern. "Roamed about the house and wept all day," she wrote, "spent the night listening."[75] The household routine gradually resumed although Esther apparently did not hear from Archie or know of his whereabouts until he returned on February 11, 1876, his twenty-first birthday. Although delighted to see Archie when he returned, Esther felt he was not the "same Archie," and when she learned he came home only to visit, described the day as a "strange sad glad day." Archie returned home for good on April 16, 1877, and Esther in a very uncharacteristic show of emotion said, "I felt all day like the enthusiastic little lady we met at the Centennial who wanted to shout, sing, cry all in the same breath."[76]

Despite years of family losses, the deaths of friends and neighbors, and other hardships, nothing appeared to cause Esther as much pain as Archie's departure, and she often commented on her sorrow and emptiness. When she connected the loss of her house plants with Archie's absence and her own guilt and sorrow, Esther quietly revealed what those plants really meant to her. Perhaps in their beauty or perhaps through the complex layers of memory and love they represented, her plants offered Esther encouragement, hope, and comfort. The loss of those beloved "window pets" was a punishment that, in her eyes, matched her guilt and depth of feeling in Archie's departure. Her sense of bereavement lingered on, and even the beautiful mid-January weather seemed to Esther a "continual reminder of my late chastisement."[77]

Frozen plants on a windowsill or flowering roots jostled from a wagon, and the sense of loss they inspired, are fitting reminders of just how im-

portant ornamental plants had become to these rural Midwesterners. By the mid-nineteenth century, domestic reformers, horticultural advisers, and commercial nurserymen had indeed promoted and popularized their culture, and through their advocacy had turned trees and flowers into widely recognized signs of respectability and refinement. The diaries of the Lawrence, Buell, and Copley families revealed, however, that ornamental plants were not just passively accepted and unthinkingly displayed social symbols. As these rural Midwesterners integrated ornamental plants into their daily lives, lovingly nurtured trees and flowers acted as an important mechanism for negotiating, and in some measure ameliorating, the changing circumstances of mid-nineteenth-century American society. On the one hand, ornamental plants offered the Lawrences, Buells, and Copleys an opportunity to enhance community ties, to strengthen family bonds, or to satisfy a simple love of beauty. From another perspective, their interest in trees and flowers drew them into new concepts of refinement and appearances, brought additional contact with a national marketplace, and opened a realm of new ideas and practices. As they selected plants, arranged their yards, attended fairs, and traded flowering roots with their neighbors, the Lawrences, Buells, and Copleys used ornamental plants in varying ways and for diverse purposes, but simultaneously created a satisfying personal amalgamation of traditional practices and modern ideas. In the process, these rural families transformed bergamot balm, verbenas, and many other plants from simple public signs into rich, densely layered, and very meaningful symbols of their own circumstances and interests.

The Lawrence, Buell, and Copley families were not unique in their horticultural interests. They certainly did not construct their meanings in a vacuum, but in the context of family, friends, and neighbors who shared their interest and could read the signs. The letters to Illinois nurseryman John Kennicott also hinted at a range of personal meanings associated with ornamental plant culture. Although far less eloquent than the diaries of Esther Lawrence, B. G. Buell, or Charlotte Copley, the images of Jackson, Lenawee, and Washtenaw County homes, published in combination atlas maps, also speak of a personal significance. The diversity in type, quantity, and arrangement of ornamental plants documented in the many lithographic images suggested that individual agency, personal initiative, and private meaning were at work in dooryards and homegrounds throughout the region. For those interpreting the cultural signposts in mid-nineteenth-century Midwestern domestic landscapes, the widespread presence of ornamental plants does suggest that reformers' improving spirit or commercial nurserymen's urgings had taken hold. A closer look, however, reveals that for some Midwesterners, the meaning of all those trees and flowers was deeper, more personal, and far more complex than adherence to fashion or the desire for public display.

Eight

Conclusion

The mid-nineteenth-century democratization of horticulture, whether encouraged through prescriptive literature, commercial promotion, or local example, left obvious marks on the landscape and traces in the historical record. Midwesterners of varying ranks and circumstances had indeed taken to planting and pruning, hoeing and weeding with an unprecedented enthusiasm, and in the process quite literally transformed the landscapes of their daily lives. While their existence is apparent, the significance of all those trees and flowers is not so easily deciphered. In their efforts to further horticultural interest, proponents attached a set of very clear public meanings to ornamental plant embellishments, insisting that their presence in domestic landscapes represented intelligent home management, an improving spirit, or simply the good taste, refinement, or respectability of family members. The private writings and actions of Midwesterners who planted trees or nurtured flowers suggest that they understood those explicit meanings, but sometimes added layers of very personal significance to the plants they loved and tended. Both these public attributions and the less articulate, but clearly important, private meanings of plant cultivation have their roots in

the broader context of mid-nineteenth-century American society, and it is there that we must look to finally understand the significance of roadside trees or well-tended lawns.

In many ways, the ornamental plant boom blossomed out of paradox, for the escalating pace of change that seemed to pervade American society at mid-century offered horticultural proponents both reason for concern and opportunity for action. Many seized the moment and looked to a flourishing publishing industry, advances in print technology, and expanded communication and transportation networks to broadcast their domestic ideology or practical horticultural information far and wide. These resources were all hallmarks of a modernizing society, and while they spread the horticultural word, they also facilitated movement and linked once isolated communities or regions with an increasingly urbanized national culture. Commercial nurserymen and seedsmen felt the effects of this stirring interest with increased demand for their products, and using the same new transportation networks, could dispense their wares widely and quickly. Communication and publishing advances also enhanced advertising and other promotional devices, creating new business opportunities, defining new needs, encouraging consumerism, and at the same time promoting horticultural interest even further.

Competitive pressures emerged along with economic opportunity and both added another dimension to ornamental plant advocacy. For those who profited from horticultural advice literature or the sale of plants and seeds, growing horticultural interest influenced their own economic well-being. It was, in other words, important to sell books and journal subscriptions or to expand advertising columns if they were to stay in business. This same underlying reality gave the words and actions of commercial nurserymen and seedsmen a more worldly and decidedly modern cast. Whether struggling local nurserymen from Washtenaw, Lenawee, or Jackson Counties or prosperous seed dealers with a national reputation and clientele, these individuals depended on ornamental plants for their livelihood. Each new expression of horticultural enthusiasm contributed to their financial future and, in an increasingly competitive marketplace, might mark the line between financial failure and economic survival or even prosperity. Although most mid-nineteenth-century publishers, advisers, and commercial nurserymen voiced a very sincere interest in advancing horticultural enthusiasm, each new journal subscription or cut-in garden in a Midwestern dooryard translated, at some level, into dollars and cents.

Conclusion

The acceptance of horticultural embellishments on the part of so many Midwestern rural families, villagers, and urban dwellers also grew out of changes in American society. By 1850, residents of southern Michigan and adjoining states had moved beyond the early stages of frontier settlement, and enjoyed the amenities of thriving farms, established towns, and bustling cities. Over the next thirty years, these Midwesterners experienced an agricultural boom promoted, in part, by the rapid growth of regional and eastern cities and by the demands of a war-time economy. Joined to these economic currents by transportation and communication networks that increasingly linked the nation, many farm families found themselves with a market for their goods and at least some discretionary income. At mid-century, technological advances ranging from reapers to sewing machines poured from newly established American factories, and for those who could afford them, eased the daily grind of farm or domestic chores. For some, decreased labor translated into leisure time, and offered an opportunity to at least think about household refinements. As the century wore on, families from varied circumstances, whether well-to-do, progressive farmers like the Lawrences, Buells, and Copleys, or the village and city dwellers of Lenawee, Jackson, and Washtenaw Counties, had the time and financial resources to set aside at least some productive land for ornamental purposes or to purchase, plant, and tend the occasional evergreen tree, flowering roots, or dahlia tubers. For at least some of these families, dooryard trees and flowers served as a daily reminder of their good fortune and participation in a new, less onerous way of life.

While the advancing tide of modernization brought welcome signs of progress, it also had a darker side. To some observers, American society at mid-century offered a spectacle of seemingly constant turmoil and new circumstances. The same cities that created markets for agricultural goods drew rural youth and others into a fast-paced life far from traditional communities and sources of social restraint. The transportation and communication networks that enhanced prosperity and spread the horticultural reform message into Midwestern households brought other news and conflicting values in a never-ending stream of information. The same commercial and industrial capacity that produced labor-saving devices or the resources to enjoy beauty and refinement also brought visions of unrestrained individualism, rampant materialism, and the decline of traditional community values. With the specter of social chaos fueling their reforming fires, ornamental plant advocates turned to trees and flowers as a

way to resolve the multiple paradoxes and perplexing problems that seemed to loom on every front.

For many proponents, interest in horticulture represented a particular and very desirable attitude toward change that balanced openness to innovation with respect for tradition. In 1859, wearing his hat as agricultural reformer, minister Henry Ward Beecher articulated the concept quite clearly as he described the best kind of rural improvement. "We believe in small farms and thorough cultivation," Beecher wrote in his *Plain and Pleasant Talk about Fruits, Flowers and Farming*, in "good fences, good barns, good farmhouses, good stock [and] good orchards," and most of all, in a "spirit of industry, enterprise and intelligence." Beecher went on to advise his readers to take a reasoned, thoughtful course, carefully balanced between the adoption of every new "novelty," and the rejection out-of-hand of all improvement in favor of traditional agricultural ideas or social practices. Evaluate the best of the old and new, Beecher urged, but "hold fast [to] that which is good."[1]

As cultural symbols, ornamental plants blended just the right mix of new ideas and traditional values. According to proponents, they offered beauty and refinement without the threat of ostentatious display or materialistic striving. They shaped character and established ties to the traditional values of home, and yet by their very nature were portable and adaptable. A packet of seeds or flowering roots might follow individuals through challenging or changing circumstances, and even though they finally flourished far from home communities, provided a consistent and very visible reminder of stability, order, and family responsibilities.

Many agricultural reformers understood ornamental plants as a mechanism for counteracting threats to rural life, arguing that they brought a reasonable refinement and new interest that matched whatever excitement the city had to offer, but did so without undue labor or expense or without compromising other traditional rural values. For urbanites, caught up in the fast pace of city life, ornamental plants offered a ubiquitous link to the natural world and an important reminder of the quieter pleasures and virtues of the more traditional rural experience. In addition, well-placed shade trees or neatly pruned flowering shrubs enabled Midwestern families to express their awareness of, and participation in, translocal values and a developing national domestic landscape form without leaving their own rural areas, small towns, or bustling cities. Parochialism had no place in a well-tended dooryard, no matter how isolated or distant from urban centers. Most significantly, ornamental plants were symbols in the landscape and in their concreteness and apparent longevity could simultaneously represent traditional values and new ideas.

While ornamental plants reflected an appropriate middle ground between old and new, many proponents also understood horticulture as a way to direct change and to manifest their own cultural values in a seemingly permanent and inevitable way. Some advocates, like Dr. Jared P. Kirtland, a widely respected Ohio horticulturist and frequent contributor to horticultural and agricultural publications, expressed their concerns and intentions quite openly. In 1855, Dr. Kirtland noted the rising prosperity of the Midwest and its transition from a newly settled frontier to a network of more established and refined communities. But Kirtland warned that as Midwestern society and landscape evolved, it should be "induced in the right direction." Without reasoned horticultural guidance, "fashion, folly, extravagance and demagogism" would shape domestic scenes, and create a "wrong bias" that even the "experience of ages [could] not correct."[2] In a word, the Midwest and the nation needed an appropriate everyday landscape that shaped and reflected desirable social values, beauty, and order. Ornamental plant embellishments were critical to that guiding framework, and despite their claims of democratizing horticulture, proponents wanted to be sure that it was their vision and their values that emerged when residents had completed all their planting and pruning.

A long-running and particularly virulent controversy about plant peddlers underscores what was really at stake for horticultural advocates. Often hired by eastern nurserymen to promote their wares among distant customers, plant peddlers reaped a flurry of complaints so widespread and so frequently repeated that one horticultural editor labeled the never-ending diatribe a "tedious tirade," and voiced his hope that at some point, peddler critics might change their tune.[3] While a few commentators acknowledged the role of plant peddlers in extending horticultural interest to relatively untouched markets, most adversaries decried plant peddler depredations, and accused them of a variety of infractions ranging from unsavory sales practices to outright consumer fraud. Part of this litany of complaint derived from a very real concern about competition, for in many cases, plant peddlers were much more successful in selling trees and flowers than nurserymen and seed dealers who employed less personal sales strategies.

From the perspective of many critics, however, the real problem came as peddlers challenged their attempts to guide and control the meaning

of ornamental plants as cultural symbols. Plant peddlers, their adversaries claimed, sold plants simply to make money, and in their greed, dispensed trees and flowers to people who did not understand or care about their social significance. In 1863, *Moore's Rural New Yorker* offered a very typical perspective on the problem as it pointed out, "persons who know nothing about trees and care but little for them often yield to the solicitations of these agents." Once these unregenerate individuals purchased plants, they failed to care for them properly, and soon, critics claimed, their yards overflowed with poorly planted, neglected, and often dying trees and shrubs. Little more could be expected, the *Rural New Yorker* argued, from people who "never once thought of planting a tree until the agent came along."[4]

By implication, the individuals who "knew nothing about" ornamental plants were also uninformed or cared little about many other issues important to reformers' vision of an improving society. An interest in horticulture, advocates claimed, represented a particular world view that transcended typical boundaries of economic rank, occupation, or even gender. This newly contrived "class" had limits, however, and those who belonged shared a belief in what Henry Ward Beecher termed "industry, enterprise, and intelligence."[5] The lackadaisical, the immoral, the backward, and the improvident had no place in this progressive vision of a new society. When trees and flowers appeared in the dooryards of those who failed to measure up, their presence scrambled the carefully defined social messages, and threatened to make a mockery of horticultural amenities as important conveyors of cultural meaning. At bottom, the controversy surrounding plant peddlers and their practices had its roots in a struggle to control and guide landscape symbols. The virulence of the controversy and its longevity emphasized the significance and meaning that many reformers attached to ornamental plants as they tried to establish and maintain a measure of cultural hegemony in a society fraught with change.

Although perhaps far less consciously than vocal horticultural advocates, families like the Lawrences, Buells, and Copleys also used ornamental plants as a way to cope with a volatile world, and to integrate change and tradition in their own lives. Interest in trees and flowers joined these Midwestern families to national markets and new ideas, and in some measure, they accepted horticultural plantings as signs of personal position and shared progressive values. At the same time, however, ornamental plants enhanced community ties and bound families together

Conclusion 221

through both time and space, creating a means to insure traditional connections despite new circumstances and broadened opportunities. As both Esther Lawrence and her sister Charlotte Copley so eloquently expressed in their words and actions, flowering roots jostled from a wagon or window pets inadvertently frozen on a cold winter night had a meaning far beyond that of immediate loss. In the lives of these Midwestern families, ornamental plants represented a very traditional center of beauty, family affection, and hope. They were quite literally a grounding and an anchor, and their quiet persistence offered comfort, stability, and the courage to meet the challenges of cultural and personal change, uncertainty, and loss.

Sugar maples lining Midwestern roadsides or shading city streets, Norway spruce embellishing rural homegrounds, and lilac and spireas blossoming without fail each spring were indeed important components of Midwestern material culture. As they graced farm and village dooryards or city homegrounds in ever-growing numbers, shade trees and flower gardens became eloquent symbols for a mid-nineteenth-century society that both embraced and feared change. For many reformers, cultural tastemakers, commercial horticulturists, and Midwestern families, this unprecedented interest in ornamental plants represented an attempt to shape the landscapes of daily life in response to both public concerns and individual circumstances. Horticultural proponents and their adherents had no intention of turning away from the spectacle of change and progress that swirled around them, but hoped that in the quiet shade of a dooryard tree or the colorful display of a well-tended garden, they had found a solid, seemingly permanent way to blend the traditional with the new, and to embrace a measured, directed change. Nineteenth-century embellished homegrounds, their presence, their reasons for being, and their varied meanings are eloquent reminders that for many, the march toward modernization was neither a straight path or an uncontested route. The journey brought unprecedented opportunity, a struggle to shape the emerging society, and the literal flowering of domestic landscapes across the Midwest and the nation.

Appendix A

Demographic Diversity of County Residents with Illustrated Homegrounds

TABLE A-1
County Residents with Illustrated Homegrounds— Household Composition in 1870

Household Composition	Number of Families
One or More Domestic Servants	243
No Domestic Servants	422
One or More Farm Laborers	421
No Farm Laborers	244
No Children	91
One to Three Children	408
Four to Six Children	136
More than Six Children	31
Other Household Members (not servants or laborers)	462
No Additional Household Members	203

TABLE A-2
County Residents with Illustrated Homegrounds— Occupations (Head of Household)

Occupations (Head of Household)	Number of Residents
Clergyman, Librarian, Banker	6
Farmer	618
Keeping House	2
Lawyer	9
Livestock Dealer	2
Manufacturer	9
Merchant	27
Miller (grist and sawmill)	16

TABLE A-2 (CONT.)
County Residents with Illustrated Homegrounds—Occupations (Head of Household)

Occupations (Head of Household)	Number of Residents
Newspaper Editor or Publisher	3
Nurseryman	2
Physician, Dentist, or Druggist	15
Real Estate or Insurance Agent	5
Skilled Craftsman (confectioner, blacksmith, etc.)	14
Surveyor, Railroad Employee, Hotel Keeper	3

TABLE A-3
County Residents with Illustrated Homegrounds—Country of Birth (Head of Household)

Country of Birth (Head of Household)	Number of Residents
Canada	9
England	41
Germany	30
Holland	1
Ireland	12
Scotland	3
United States	619
Wales	1

TABLE A-4
County Residents with Illustrated Homegrounds—State of Birth (Head of Household)

State of Birth (Head of Household)	Number of Residents
Connecticut	26
Maine	4
Maryland	1
Massachusetts	22
Michigan	71
New Hampshire	11
New Jersey	16

TABLE A-4 (CONT.)
County Residents with Illustrated Homegrounds—State of Birth (Head of Household)

State of Birth (Head of Household)	Number of Residents
New York	405
Ohio	14
Pennsylvania	14
Rhode Island	2
Vermont	30
Virginia	3

TABLE A-5
County Residents with Illustrated Homegrounds—Value of Personal and Real Property in 1870

Value of Personal and Real Property in Dollars	Number of Residents
<$1,000	2
$1,000–$5,000	67
$5,001–$10,000	204
$10,001–$15,000	183
$15,000–$20,000	96
$20,001–$25,000	46
$25,001–$30,000	25
$30,001–$35,000	12
$35,001–$40,000	8
$40,001–$45,000	5
$45,001–$50,000	4
>$50,000	11

Appendix B

Frequency of Horticultural Elements by Demographic Categories

TABLE B-1
Percentage of Images Exhibiting Specific Tree Types by Demographic Categories

Demographic Category (Number of Images)	Conifers	Deciduous	Original	Weeping	Shrubs
All Images (731)	78%	86%	41%	6%	37%
Men (714)	78%	86%	41%	5%	37%
Women (16)	69%	88%	44%	13%	31%
<$5,000 (61)	79%	79%	21%	3%	20%
$5,000–$15,000 (395)	80%	87%	40%	6%	41%
$15,001–$25,000 (142)	78%	79%	48%	6%	37%
>$25,000 (65)	77%	88%	46%	8%	39%
United States (619)	78%	86%	42%	5%	37%
England (41)	90%	88%	37%	10%	46%
Germany (30)	70%	83%	17%	7%	33%
Farm Home (634)	78%	88%	42%	6%	38%
Town, Village, City (97)	81%	75%	29%	6%	27%
Ann Arbor, Adrian, Jackson Home (27)	78%	78%	37%	7%	30%

TABLE B-2
Percentage of Images Exhibiting Specific Planting Patterns by Demographic Category

Demographic Category (Number of Images)	Random	Linear	Symmetrical	Framing	Clustered	Border
All Images (731)	76%	15%	34%	25%	9%	43%
Men (714)	76%	15%	34%	25%	9%	43%
Women (16)	88%	6%	31%	19%	19%	44%
<$5,000 (61)	77%	13%	38%	15%	7%	25%
$5,000–$15,000 (395)	75%	15%	37%	29%	7%	45%
$15,001–$25,000 (142)	77%	17%	31%	23%	16%	42%
>$25,000 (65)	83%	9%	23%	23%	15%	43%
United States (619)	77%	15%	34%	25%	9%	42%
England (41)	76%	15%	37%	37%	15%	51%
Germany (30)	63%	13%	33%	23%	0%	43%
Farm Home (634)	76%	15%	37%	26%	8%	46%
Town, Village, City (97)	81%	9%	20%	19%	18%	22%
Ann Arbor, Adrian, Jackson Home (27)	89%	7%	7%	19%	22%	19%

Appendix B

TABLE B-3
Percentage of Images Exhibiting Additional Horticultural Elements by Demographic Category

Demographic Category (Number of Images)	Vines	Hedges	Potted Plants	Vases	Hanging Baskets
Images (731)	17%	12%	3%	2%	2%
Men (714)	17%	12%	3%	2%	2%
Women (16)	0%	6%	6%	6%	13%
<$5,000 (61)	9%	10%	2%	2%	0%
$5,000–$15,000 (395)	18%	10%	3%	1%	2%
$15,001–$25,000 (142)	20%	14%	2%	2%	4%
>$25,000 (65)	15%	17%	2%	8%	5%
United States (619)	18%	12%	3%	2%	3%
England (41)	17%	15%	7%	5%	2%
Germany (30)	3%	7%	0%	0%	0%
Farm Home (634)	17%	10%	3%	1%	2%
Town, Village, City (97)	19%	21%	6%	7%	4%
Ann Arbor, Adrian, Jackson Home (27)	26%	26%	7%	19%	7%

TABLE B-4
Percentage of Images Exhibiting Garden Bed Types by Demographic Category

Demographic Category (Number of Images)	Cut-in	Border	Large Bed
All Images (731)	13%	6%	20%
Men (714)	13%	6%	20%
Women (16)	25%	19%	19%
<$5,000 (61)	12%	8%	16%
$5,000–$15,000 (395)	14%	6%	20%
$15,001–$25,000 (142)	16%	6%	23%
>$25,000 (65)	12%	8%	26%
United States (619)	14%	7%	19%
England (41)	20%	5%	29%
Germany (30)	3%	10%	23%
Farm Home (634)	14%	7%	22%
Town, Village, City (97)	9%	3%	8%
Ann Arbor, Adrian, Jackson Home (27)	22%	7%	7%

NOTES

Introduction

1. For typical commentary, see "Gardens," *New England Farmer* 6 (9 November 1827): 126; and "Gardening," *New England Farmer* 5 (6 April 1827): 294.

2. Dr. S. P. Hildreth, "Remarks upon the State of Gardening in Ohio; with Practical Observations on the Cultivation of Many of the Best Varieties of Fruits, Flowers and Vegetables in that Climate," *Magazine of Horticulture* 7, no. 9 (September 1841): 333.

3. Alexander Walsh, "The New York State Report on Horticulture and the Household Arts, as intimately connected with the improvement of Agriculture," *Horticultural Register and Gardener's Magazine* 4 (1839): 94.

4. A number of accounts describe seventeenth and eighteenth-century gardening practices. For accounts of English gardening traditions and influential practitioners, see: Tom Williamson, *Polite Landscapes, Gardens and Society in Eighteenth-Century England* (Baltimore: Johns Hopkins University Press, 1995); Rudy J. Favretti and Joy Pitman Favretti, *Landscapes and Gardens for Historic Buildings, A Handbook for Reproducing and Creating Authentic Landscape Settings* (Nashville, TN: American Association for State and Local History, 1978); G. B. Tobey, *A History of Landscape Architecture, The Relationship of People to Environment* (New York: American Elsevier Publishing Co., 1973); Norman T. Newton, *Design on the Land, The Development of Landscape Architecture* (Cambridge: Harvard University Press, Belknap Press, 1971), especially Chapters XIV and XV; Derek Clifford, *A History of Garden Design*, rev. ed. (New York: Frederick A. Praeger, 1966), especially chapters 5–8. For comments on American public gardens in the period, see: Carlton B. Lees, "The Golden Age of Horticulture," *Historic Preservation* 24–25 (Oct.–Dec. 1972): 33; Barbara W. Sarudy, "Genteel and Necessary Amusements: Public Pleasure Gardens in Eighteenth-Century Maryland," *Journal of Garden History* 9, no. 3 (1989): 118–124; and Therese O'Malley, "Landscape Gardening in the Early National Period," in *Views and Visions, American Landscapes before 1830*, ed. Edward J. Nygren (Washington, DC: Corcoran Gallery of Art, 1986), 145–6. For comment on the history of horticultural organizations, see: U. P. Hedrick, *A History of Horticulture in America to 1860* (New York: Oxford Univer-

sity Press, 1950), 505–6. See also Ernesta D. Ballard, "The Organizations of Horticulture," in *America's Garden Legacy, A Taste for Pleasure*, ed. George H. M. Lawrence (Philadelphia: Pennsylvania Horticultural Society, 1978), 53–64.

5. Jesse Buel, "On the Horticulture of the United States of America," *Gardener's Monthly* 4 (August 1828): 193–4; and C. M. Hovey, "A Retrospective View of the Progress of Horticulture in the United States, during the year 1844," *Magazine of Horticulture* 11, no. 1 (January 1845): 1–9.

6. See "A Talk about 'Horticulture'," *American Agriculturist* 19 (October 1860): 305; and "American Taste in Gardening," *Moore's Rural New Yorker* 14 (6 June 1863): 183.

7. "The Meeting of the State Horticultural Society," *Michigan Farmer* 16 (January 1858): 47; and Edgar Sanders, "Who Buys the Seed?" *Prairie Farmer* 7 (11 April 1861): 239. As an important source for this study, *Michigan Farmer* volume numbers in succeeding citations warrant some clarification. From its founding in 1843 until 1858, volume numbers follow consecutively 1 through 16. A new series began in January 1859 with volume 1 and continued until June 1862 (volume 4). A third series began in July 1862 with volume 1. In 1870, numbering again began with volume 1.

8. As a number of historians have pointed out, the term "Midwest" came into common usage as a regional designation in the early twentieth century. The term itself has referred to various combinations of states including Indiana, Illinois, Michigan, Ohio, Wisconsin, North and South Dakota, Iowa, Nebraska, Kansas, and south to Missouri and Kentucky. In this study, the term Midwest is used to designate those states carved out of the Old Northwest, including Illinois, Indiana, Michigan, Ohio, and Wisconsin. Historian Susan E. Gray has pointed out that although the geographic, economic, and social conditions varied to some degree, the preponderance of regional similarities made these states a "discrete entity in time and place." See Susan E. Gray, *The Yankee West, Community Life on the Michigan Frontier* (Chapel Hill: University of North Carolina Press, 1996), 2–3. For a discussion of the emergence of the Midwest as a regional designation, see James R. Shortridge, *The Middle West: Its Meaning in American Culture* (Lawrence: University Press of Kansas, 1989); and Andrew R. L. Cayton and Susan E. Gray, ed., *The American Midwest, Essays on Regional History* (Bloomington: Indiana University Press, 2001). Although nineteenth-century residents of the region did not refer to themselves as "Midwesterners," the term is useful in both designating a locale and a shared set of environmental, economic, and cultural circumstances.

9. This study draws upon the methodologies and theoretical underpinnings of material culture studies. An investigative model that includes description of ornamental plant use as a part of material culture, exploration of social and cultural context, and interpretation of the broader social and cultural significance structures and guides analysis. For further discussion, see E. McClung Fleming, "Artifact Study: A Proposed Model," in *Material Culture Studies in America*, ed. Thomas J. Schlereth (Nashville, TN: American Association for State and Local History, 1986), 162–173; and Kenneth L. Ames, "Material Culture as Non-Verbal Communication, A Historical Case Study," *Journal of American Culture* 3, no. 4 (Winter 1980): 619.

Notes to Introduction   233

10. Charles Van Ravenswaay, *A Nineteenth-Century Garden* (New York: Universe Books, Main Street Press, 1977), 18. Historian and botanist George H. M. Lawrence has also described the second half of the nineteenth century as a period of unprecedented horticultural activity, unrivaled "before or since." See George H. M. Lawrence, "The Development of American Horticulture," in *America's Garden Legacy, A Taste for Pleasure*, ed. George H. M. Lawrence (Philadelphia: Pennsylvania Horticultural Society, 1978), 93. For comment on the decline of horticultural interest, see M. Christine Doell, *Gardens of the Gilded Age, Nineteenth-Century Gardens and Homegrounds of New York State* (Syracuse, NY: Syracuse University Press, 1986), 113. Other garden historians have proposed different chronological divisions. See Lees, "The Golden Age of Horticulture," 35–6. Lees labels the entire nineteenth century as the "golden age of horticulture." Landscape historian Rudy J. Favretti delimits 1860–1900 as a consistent period in garden interest and design. See Favretti, *Landscapes and Gardens for Historic Buildings*, 45–9. Patricia M. Tice, *Gardening in America, 1830–1910* (Rochester, NY: Strong Museum, 1984), 71–73, points out the emergence of new gardening styles at the end of the nineteenth century and continuing into the early twentieth century.

11. Although probing historians' traditional portrayal of modernization, Sally McMurry provides a concise discussion of the characteristics of a modern society. See McMurry, *Transforming Rural Life, Dairying Families and Agricultural Change, 1820–1885* (Baltimore: The Johns Hopkins University Press, 1995), 230. For additional commentary on the characteristics of a modernizing society, see Richard D. Brown, "Modernization and the Modern Personality in Early America, 1600–1865: A Sketch of a Synthesis," *Journal of Interdisciplinary History* 2 (Winter 1972): 201–3; Richard D. Brown, "Modernization: A Victorian Climax," *American Quarterly* 27 (December 1975): 533–548; and Richard D. Brown, *Modernization: The Transformation of American Life, 1600–1865*.(New York: Hill and Wang, 1976). A number of scholars have examined the stresses accompanying these cultural changes from political, social, and cultural perspectives. See especially Ronald G. Walters, *American Reformers 1815–1860* (New York: Hill and Wang, 1978), 3–19; Daniel Walker Howe, *The Political Culture of the American Whigs* (Chicago: University of Chicago Press, 1979); Peter Dobkin Hall, *The Organization of American Culture, 1700–1900* (New York: New York University Press, 1982); and Thomas Bender, *Community and Social Change in America* (New Brunswick: Rutgers University Press, 1978). For commentary on modernization and its significance in rural society, see Christopher Clark, *The Roots of Rural Capitalism, Western Massachusetts, 1760–1860* (Ithaca: Cornell University Press, 1990).

12. For particularly useful discussions of the developing region and its unique features, see R. Douglas Hurt, "Midwestern Distinctiveness," in *The American Midwest*, 165–6; and Andrew R. L. Cayton, "The Anti-region: Place and Identity in the History of the American Midwest," in *The American Midwest*, 157.

13. As European settlement began, the Potawatomis were a significant presence in southeastern Michigan, but in a series of treaties with the federal government beginning in 1807 and ending in 1833, they gradually ceded their southern

Michigan holdings, further hastening European settlement in the region. The federal government actually "removed" many of the remaining Native Americans in southern Michigan between 1838 and 1840. For a concise discussion of the treaties and treaty negotiations, see Helen Hornbeck Tanner, ed., *Atlas of Great Lakes Indian History* (Norman, OK: Published for the Newberry Library by University of Oklahoma Press, 1986), 155–161.

14. Gray, *The Yankee West*, 10–13.

15. Jon Gjerde, *The Minds of the West. Ethnocultural Evolution in the Rural Midwest, 1830–1917* (Chapel Hill: University of North Carolina Press, 1997), 5.

16. R. Douglas Hurt, "Midwestern Distinctiveness," 169.

17. For commentary on the commercial advancement of the Old Northwest, see Andrew R. L. Cayton and Peter S. Onuf, *The Midwest and the Nation, Rethinking the History of an American Region* (Bloomington: Indiana University Press, 1990), 34–5.

18. Andrew R. L. Cayton and Susan E. Gray, "The Story of the Midwest: An Introduction," in *The American Midwest*, 10 and 16–17.

19. Sally McMurry, *Transforming Rural Life, Dairying Families and Agricultural Change, 1820–1885*, 230–1.

20. See Cayton and Gray, "The Story of the Midwest: An Introduction," in *The American Midwest*, 15; and Cayton, "The Anti-region," in *The American Midwest*, 146.

21. For commentary on the use of regions as a "setting" for understanding broader phenomenon, see Cayton, "The Anti-region," in *The American Midwest*, 147.

22. Several historic geographers and historians have documented the development of the atlas map and county history industry. See especially Walter W. Ristow, *American Maps and Mapmakers, Commercial Cartography in the Nineteenth Century* (Detroit: Wayne State University Press, 1985); Michael P. Conzen, ed., *Chicago Mapmakers, Essays on the Rise of the City's Map Trade*, ed. Michael P. Conzen (Chicago: Chicago Historical Society for the Chicago Map Society, 1984), 4–10 and 47–63; Michael P. Conzen, "The County Landownership Map in America, Its Commercial Development and Social Transformation 1814–1939," *Imago Mundi* 36 (1984): 9–31; Richard W. Stephenson, compiler, *A Checklist of Nineteenth Century United States County Maps in the Library of Congress* (Washington, DC: Library of Congress, 1967), vii–xxv; Walter W. Ristow, "Nineteenth-Century Cadastral Maps in Ohio," *Papers of the Bibliographic Society of America* 59, no. 3 (1965): 306–15; and Norman J. W. Thrower, "The County Atlas of the United States," *Surveying and Mapping* 21 (1961): 365–373. See also John W. Reps, *Views and Viewmakers of Urban America, Lithographs of Towns and Cities in the United States and Canada, Notes on the Artists and Publishers, and a Union Catalog of Their Work, 1825–1925* (Columbia: University of Missouri Press, 1984).

23. See *The National Union Catalog Pre-1956 Imprints*, vol. 164 (London: Mansell Publishing, 1971), 279–280 and 283–285, for listing of Everts and Stewart's, or Everts's *Combination Atlas Maps* in other states.

24. See Bates Harrington, *How 'Tis Done, A Thorough Ventilation on the Numerous Schemes Conducted by Wandering Canvassers Together with the Various Advertising Dodges for the Swindling of the Public* (Chicago: Fidelity Publishing Co., 1879), 28–9, 61–67, for a critical description of how canvassers worked. See also Conzen, "Maps for the Masses," in *Chicago Mapmakers*, 50.

25. See Ristow, *American Maps and Mapmakers*, 414–5, for commentary on the elements normally pictured in the illustrations.

26. See especially Michael P. Conzen, "Landownership Maps and County Atlases," *Agricultural History* 58 (1984): 120–1; and Russell Swenson, "Illustrations of Material Culture in Nineteenth-Century County and State Atlases," *Pioneer Society Transactions* 5 (1982), 63–4. Also see Peter C. Marzio, *The Democratic Art, Chromolithography 1840–1900, Pictures for a Nineteenth-Century America* (Boston: David R. Godine, 1979), 5, 47 and 109. John Reps also argues that the earliest lithographic artists took their work and its accuracy quite seriously. See Reps, *Views and Viewmakers of Urban America*, 4.

27. Local newspapers often announced the arrival of the firm, reported its progress through the county, and admonished citizens to support such a worthy undertaking. For local perspectives on the introduction of *Combination Atlas Maps* in Lenawee, Jackson, and Washtenaw Counties, see "A County Atlas," *Dexter (Michigan) Leader*, 25 April 1873; "The County Atlas," *Jackson (Michigan) Daily Citizen*, 15 January 1874; "The New Atlas," *Jackson Daily Citizen*, 21 February 1874; and "A Historical Atlas," *Ypsilanti (Michigan) Commercial*, 11 March 1874.

28. In his critique of the county atlas industry, Bates Harrington, a former employee of an atlas map firm, provided an estimate of the cost of atlases and charges for engravings, portraits, and biographies. Harrington noted that the price of lithographic views varied according to their size. See Harrington, *How 'Tis Done*, 61–70. For a more contemporary analysis of costs, see Gerald Carson, "Get the Prospect Seated ... And Keep Talking," *American Heritage* 9 (August 1958): 38–41 and 77–80; and Betty and Raymond Spahn, "Wesley Raymond Brink, History Huckster," *Journal of the Illinois State Historical Society* 58 (Summer 1965): 117–138. In Lenawee County, 2,328 citizens or just over 5 percent of the county's population bought the *Combination Atlas Map of Lenawee County, Michigan*. Slightly over 4 percent or 1,550 Washtenaw County citizens purchased the new book for their county, while in Jackson County, 1,837 residents or 4.8 percent paid for the *Combination Atlas Map of Jackson County*. See "Business Directory of Lenawee County, Michigan," *Combination Atlas Map of Lenawee County, Michigan* (Chicago: Everts and Stewart, 1874), 132–143; "Business Directory of Washtenaw County, Michigan," *Combination Atlas Map of Washtenaw County, Michigan* (Chicago: Everts and Stewart, 1874), 113–124; "Business Directory of Jackson County, Michigan," *Combination Atlas Map of Jackson County, Michigan* (Chicago: Everts and Stewart, 1874), 136–144; and *Census of the State of Michigan 1874* (Lansing: W. S. George and Co., 1875), 20, 26, and 44. See also Thrower, "The County Atlas of the United States," 369, for an industry-wide comparison of sales.

29. For a concise discussion of content analysis of visual images, see Michael S. Ball and Gregory W. H. Smith, *Analyzing Visual Data, Qualitative Research Methods*, vol. 24 (London: Sage Publications, 1992), 20–31.

30. As historian Mary Ryan so eloquently argued in her study of middle class formation in Oneida County, New York, the "raw details of local history" can reveal "historical structures that have existed in many other times and locations." These focused accounts based on local circumstances and experiences serve to "illuminate, enliven, and give a human dimension" to our understanding of general social patterns, developments, and practices. See Mary P. Ryan, *Cradle of the Middle Class, The Family in Oneida County, New York, 1790–1865* (New York: Cambridge University Press, 1981), 17.

31. See Sally McMurry, *Families and Farmhouses in Nineteenth-Century America, Vernacular Design and Social Change* (New York: Oxford University Press, 1988), 4. McMurry uses the term cultural mediator to describe the role of progressive farm families in initiating new practices in local communities.

32. A number of anthropologists and historic geographers have provided helpful discussions of the power and meaning of symbols with particular emphasis on cultural symbols in the landscape. See especially Raymond Firth, *Symbols, Public and Private* (Ithaca: Cornell University Press, 1973), 54; and Lester B. Rowntree and Margaret W. Conkey, "Symbolism and the Cultural Landscape," *Annals of the Association of American Geographers* 70, no. 4 (December 1980): 460.

33. Family members included B. G. Buell, Esther Copley Lawrence, and Charlotte Copley. Their papers and diaries are available at the Bentley Historical Library, University of Michigan, Ann Arbor, Michigan.

34. In his study of parlor organs, historian Kenneth Ames provides an invaluable model for understanding the multiple meanings of material objects. See Ames, "Material Culture as Nonverbal Communication: A Historical Case Study," *Journal of American Culture* 3, no. 4 (Winter 1980): 619–641, especially 620–1 and 625. Historian Bernard Herman also argues that material culture is often imbued with multiple and frequently shifting meanings. Herman then suggests that one of the primary roles of material culture analysis should be to determine how "things" gain significance, and how that significance changes over time. See Bernard L. Herman, *The Stolen House* (Charlottesville: University Press of Virginia, 1992), 3.

Chapter One: Embellished Landscapes

1. S. Q. Lent, "Ornamenting Farms," *Ninth Annual Report of the Secretary of the State Pomological Society of Michigan, 1879* (Lansing: W. S. George and Co., 1880), 377–8.

2. See William L. Jenks, "History in County Names," in *Historic Michigan*, vol. 1, ed. George N. Fuller (Dayton, OH: National Historical Association, 1924), 463–4, and 466; and Willis F. Dunbar, rev. ed by George S. May, *Michigan, A History of the Wolverine State* (Grand Rapids: William B. Eerdmans Publishing Co., 1980), 197.

3. For a good general discussion of Michigan's glacial geology and its resultant land forms, see John A. Dorr, Jr. and Donald F. Eschman, *Geology of Michigan* (Ann Arbor: University of Michigan Press, 1970), 158–60; and Lawrence M. Sommers, ed., *Atlas of Michigan* (East Lansing: Michigan State University Press, 1978), 16–17, 20–24, and 32. For discussions of southern Michigan plant associations and presettlement vegetation, see Burton V. Barnes and Warren H. Wagner, Jr., *Michigan Trees, A Guide to the Trees of Michigan and the Great Lakes Region* (Ann Arbor: University of Michigan Press, 1981), 33–46; Sheridan L. Dodge, "Presettlement Forest of South-Central Michigan," *Michigan Botanist* 26 (1987): 139–152; and L. G. Brewer, T. W. Hodler, and H. A. Rapp, "The Presettlement Vegetation of Southwestern Michigan," *Michigan Botanist* 23 (1984): 153–156.

4. Dunbar, *Michigan, A History of the Wolverine State*, 190–1, 198–9, and 272; and George N. Fuller, "Michigan Territory 1805–1837," in *Historic Michigan*, vol. 1, ed. George N. Fuller (Dayton, OH: National Historical Association, 1924), 244–5.

5. Dunbar, *Michigan, A History of the Wolverine State*, 268–279. See also Willis F. Dunbar, *All Aboard! A History of Railroads in Michigan* (Grand Rapids: William B. Eerdmans Publishing Co., 1969), for a comprehensive discussion of the development of Michigan's railroads. Also see Brian Deming, *Jackson, An Illustrated History* (Woodland Hills, CA: Windsor Publications, 1984), 42.

6. Charles N. Lindquist, *Lenawee County, A Harvest of Pride and Promise* (Chatsworth, CA: Windsor Publications, 1990), 16–17, 31 and 34; Lula A. Reed, compiler, *The Early History, Settlement and Growth of Jackson, Michigan* (Jackson, MI: Jackson County Historical Society, 1965), 16–17; Jonathan L. Marwil, *A History of Ann Arbor* (Ann Arbor: Ann Arbor Observer Co., 1987), 2 and 13; and Deming, *Jackson, An Illustrated History*, 35.

7. "Michigan as It Was and Is," *Michigan Farmer* 11 (January 1853): 28–29.

8. William S. Maynard, Washtenaw County, Michigan, to Thomas Ewbank, Commissioner of Patents, November 25, 1850, in *Report of the Commissioner of Patents for the Year 1850, Part II Agriculture* (Washington, DC: Office of Printers to House of Reps., 1851), 330–31.

9. "Statistics of the State of Michigan," *Michigan Farmer* 4 (22 February 1862): 52; Lindquist, *Lenawee County, A Harvest of Pride and Promise*, 25–6; Deming, *Jackson, An Illustrated History*, 36 and 39–42; Dunbar, *Michigan, A History of the Wolverine State*, 471; Marwil, *A History of Ann Arbor*, 49; and *Manufactures of the United States in 1860 Compiled from the Original Returns of the Eighth Census, 1860* (Washington, DC: Government Printing Office, 1865), 263–272.

10. S. B. McCracken, ed., *Michigan and the Centennial, Being a Memorial Record Appropriate to the Centennial Year* (Detroit: Printed for the Publisher S. B. McCracken at the Office of the Detroit Free Press, 1876), 668–670; and Francis A. Walker, compiler, *The Statistics of the Population of the United States, Ninth Census*, vol. 1 (Washington, DC: Government Printing Office, 1872), 543. See also *History of Jackson County, Michigan*, vol. 1 (Chicago: Inter-State Publishing Co., 1881), 523–540; *History of Washtenaw County, Michigan*, vol. 1 (Chicago: Chas C. Chapman and Co., 1881), 951–958; Richard I. Bonner, ed., *Memoirs of Lenawee County, Michigan*, vol. 1 (Madison, WI: Western Historical Association,

1909), 673; and William A. Whitney and R. I. Bonner, *History and Biographical Record of Lenawee County, Michigan*, vol. 1 (Adrian, MI: W. Stearns and Co., 1879), 60–63, for discussions of organizations and other social and cultural activities.

11. Within the state, only the populations of Kent and Wayne Counties exceeded Washtenaw and Lenawee, but Saginaw, Oakland, St. Clair, and Calhoun Counties had more citizens than Jackson County. See Walker, *The Statistics of the Population of the United States, Ninth Census*, 38 and 168–176.

12. Demographic information for the families with homes pictured in the atlas maps of the three-county area was extracted from the 1870 Federal Population Census for Lenawee, Washtenaw, and Jackson Counties. See 1870 Federal Population Census, Jackson County, Michigan, Microfilm Copy, M-593, roll 678 and 679; 1870 Federal Population Census, Lenawee County, Michigan, Microfilm Copy, M-593, roll 685 and 686; and 1870 Federal Population Census, Washtenaw County, Michigan, Microfilm Copy, M-593, roll 707 and 708. Occasionally, the 1870 Agricultural Census provided economic information if those figures were not available in the Population Census. See also 1870 Products of Agriculture, Lenawee County, Michigan, Microfilm Copy, T-1164, roll 20; 1870 Products of Agriculture, Jackson County, Michigan, Microfilm Copy, T-1164, roll 19; and 1870 Products of Agriculture, Washtenaw County, Michigan, Microfilm Copy, T-1164, roll 23. Complete census information was available for 663 of the 731 families with homes pictured in the atlas maps. Percentages reported here are based on the number of families for which information on that particular demographic variable was available rather than the total number of images.

13. Town and village population figures were derived from Walker, *The Statistics of the Population of the United States, Ninth Census*, 171, 172, and 176. The place of residence, whether farm, town, village, or city, was derived from designations used in the atlas maps. Information on occupations was derived from both the Federal Population Census and the atlas maps from the three counties.

14. The 1870 Federal Census included the value of both real and personal estate. Personal estate represented the value of all "property, possessions, or wealth of each individual" which was not reported in the value of property. See Lee Soltow, *Patterns of Wealthholding In Wisconsin Since 1850* (Madison: University of Wisconsin Press, 1971), 141–145, for information on instructions to census takers and what they were to include in the real and personal property information. Soltow also provides some context for these figures by calculating the arithmetic mean of total worth for males in the northern United States in 1870 as $2,921, and for adult males in Wisconsin in 1870 as $2,369. Soltow based his work on a spin sampling technique used on the manuscript census and including both landholders and non-landholders. See Lee Soltow, *Men and Wealth in the United States 1850–1870* (New Haven: Yale University Press, 1975), 65; and Soltow, *Patterns of Wealthholding in Wisconsin Since 1850*, 80. In the illustrations of the three counties, eighteen families recorded total wealth below the $2,921 that Soltow has calculated as the mean, even though real estate values were included in the total figure. Economic information was calculated for the 663 residents for which information was complete. In some cases, the value was derived from the 1870

Agricultural Census which listed value of property and of farm equipment rather than total personal property, and subsequently may provide a slightly lower total figure.

15. Of the families with children, 61 percent had one to three children, 20 percent had four to six, and only 5 percent had more than six children. Gender of the head-of-household was available for 730 families, and age information for 665 families. Data for family size was available for 666 families. Data regarding domestic servants and farm laborers was available for 665 families.

16. Place of birth was available for 717 families. The information was drawn from either the Federal Population Census or the "Business Directory" provided in the atlas maps for the three counties. The dominance of settlers from New York State was quite typical for other regions in southern Michigan. For additional information on the ethnic background of Michigan settlers, see Gregory S. Rose, "The County Origins of Southern Michigan's Settlers: 1800–1850," *East Lakes Geographer* 22 (1987): 74–87; and Hemalata C. Dandekar and Mary Bockstahler, "The Changing Farmscape, A Case Study of German Farmers in Southeast Michigan," *Michigan History* 74, no. 1 (1990): 42–7.

17. All data on ornamental plant use was based on content analysis of lithographs in the *Combination Atlas Map of Jackson County*; *Combination Atlas Map of Lenawee County*; and the *Combination Atlas Map of Washtenaw County, Michigan*.

18. For example, 76 percent of the families with a clearly defined front yard also had an additional fenced space, as did 36 percent of the residents with a front/side yard.

19. Numbers and types of ornamental plants, along with their arrangement, were tabulated only in front yards, front/side yards, or in the extended yard areas clearly visible to the front or to the sides of the homes.

20. Among pictured homes owned by women, 69 percent featured coniferous trees, while 70 percent of German homes featured coniferous trees, compared to the overall rate of 78 percent. Ninety percent of families from England displayed conifers in their yards.

21. See *Combination Atlas Map of Washtenaw County*, 29. Thirteen percent of women heads-of-household, 10 percent of English immigrants, and 8 percent of residents in the highest economic rank selected weeping trees.

22. Although the actual origin of these trees was impossible to determine, they clearly varied from the much younger trees often present in the yards. Trees in this study were categorized as "original" if they extended over the top of the house, and if they had a large, well-developed trunk. These trees were all deciduous, suggesting that they did represent original growth or local trees that were readily available for transplanting.

23. In sixteen illustrations, original trees were the only type of tree present, and in twelve instances were the only type of ornamentation visible in the home-grounds.

24. Twenty-one percent of families in the lowest economic rank had original trees in their dooryards, and only 17 percent of German immigrants.

25. Among all the categories of ornamental plant types, shrubs were the most difficult to distinguish and to establish an accurate and consistent definition. Size

in relation to other plantings and the appearance of multiple woody stems emerging from the ground were the primary criteria for inclusion in this category.

26. See *Combination Atlas Map of Lenawee County*, 92.

27. Of farm residents, 38 percent pictured shrubs in their homegrounds. In contrast, 27 percent of the town, village, and city homes pictured shrubs.

28. See *Combination Atlas Map of Washtenaw County*, 46.

29. These structures may have sometimes supported grape vines, but since they were included in the ornamental space and the vines themselves could not be identified, they were numbered among the ornamental vines.

30. In contrast to 17 percent for all images, vines appeared in only 3 percent of the German households, and in 9 percent of those families in the lowest economic rank.

31. For the purposes of this study, ornamental hedges were defined as a very closely spaced and highly pruned plantings.

32. See *Combination Atlas Map of Washtenaw County*, 57 and 76.

33. Twenty-five percent or 184 residences of those pictured in the three-county area planted trees and shrubs against or very near their homes. Urban dwellers and families in the lowest economic rank were less likely to select a framing design, while families from England selected that planting pattern more frequently than average.

34. Of women, 13 percent selected this garden type, as did 10 percent of those heads of the household born in England.

35. Lent, "Ornamenting Farms," 377–8.

36. See *Combination Atlas Map of Jackson County*, 35 and 57.

37. See *Combination Atlas Map of Washtenaw County*, 67.

38. Bedding-out or ribbon design plans used a combination of plants with varying blossom color to create often complex geometric patterns. Horticultural advice literature often recommended suitable plants for this kind of display and included ornamental design plans. See, for example, "Design for a Flower Garden, with a selection of Plants adapted to the same," *Magazine of Horticulture* 18, no. 5 (May 1852): 207–209.

39. Among the illustrated images, 22 percent of the farm families cultivated this garden type compared to only 8 percent of town, village, and city residents. Among English immigrants, 29 percent of the illustrated homes showing that type of garden design, compared to 19 percent of the families born in the United States and 23 percent of German immigrants.

40. It is impossible to distinguish plant varieties grown in cut-in gardens from atlas map illustrations. The low growth form and nonwoody character of the plants, along with the prominent position of the gardens in front or just to the side of the house, suggests that residents emphasized ornamentals and planted perennial, biennial, or annual flowers in these beds.

41. See *Combination Atlas Map of Lenawee County*, 61.

42. See *Combination Atlas Map of Jackson County*, 94; *Combination Atlas Map of Lenawee County*, 104; and *Combination Atlas Map of Washtenaw County*, 62 and 81.

43. See *Combination Atlas Map of Washtenaw County*, 102 and 65; and *Combination Atlas Map of Lenawee County*, 62.

44. Twenty-five percent of female households, 22 percent of city homes, and 20 percent of homes of English immigrants featured cut-in gardens.

45. See *Combination Atlas Map of Lenawee County*, 49 and 75. A number of horticultural commentators described rockeries and warned their readers about the danger of a tasteless or inappropriate use of rocks to embellish the garden space. See especially A. J. Downing, *A Treatise on the Theory and Practice of Landscape Gardening* (New York: Orange Judd and Co., 1865; reprint, New York: Dover Publications, 1991), 399–404 (page references are to reprint edition).

46. For examples of gazebos, see the residence of William A. Wilcox, *Combination Atlas Map of Lenawee County*, 76; and the residence of Henry Krause, *Combination Atlas Map of Washtenaw County*, 61. An example of a large arbor appeared on the homegrounds of Sydney T. Smith, *Combination Atlas Map of Jackson County*, 93.

47. For examples of benches surrounding trees, see residence of Sydney Harwood, C. H. Millen, and Andrew Tenbrook, *Combination Atlas Map of Washtenaw County*, 73, 61 and 58. For an example of a free-standing bench, see the residence of T. W. Thompson, *Combination Atlas Map of Jackson County*, 69. Of town, village, and city families, 8 percent had yard furniture compared to 5 percent of farm families. Of those families with an economic worth more than $25,000, 11 percent displayed yard furniture.

48. Eighteen residences (2 percent) displayed hanging baskets, and while most appeared conspicuously on a front or side porch, several residents selected other options. See *Combination Atlas Map of Lenawee County*, 114. Decorative vases, used as the focal point of a cut-in garden or placed on a pedestal directly on the lawn, embellished fourteen illustrated home.

49. Women (6 percent), English immigrants (7 percent), and residents of Ann Arbor, Adrian, and Jackson (7 percent) showed evidence of potted plant culture. These percentages include only those illustrations in which potted plants were actually visible. For comments on bay windows, see Henry T. Williams, ed., *Window Gardening Devoted Specially to the Culture of Flowers and Ornamental Plants for In Door Use and Parlor Decoration* (New York: Henry T. Williams, 1872), 22.

50. See *Combination Atlas Map of Lenawee County*, 75. Dr. Owen had eleven distinctive ornamental features on his highly cultivated homegrounds.

51. See *Combination Atlas Map of Jackson County*, 31, 37, 49, and 75; *Combination Atlas Map of Lenawee County*, 77 and 87; and *Combination Atlas Map of Washtenaw County*, 59 and 87.

52. For an analysis of native forest composition in southern Michigan, see Barnes and Wagner, *Michigan Trees*, 38–49. Barnes and Wagner note that conifers, except for eastern red cedar, wetland plants like black spruce and white cedar, and occasional pockets of white pine, were uncommon in southern Michigan. For contemporary comments on the mid-nineteenth-century preference for Norway spruce, see "Evergreens for the Farm," *Tenth Annual Report of the Secretary of the State Horticultural Society of Michigan, 1880* (Lansing: W. S. George and Co., 1881), 12; and George Taylor, "When and How To Plant Evergreens," *Fifth Annual Report of the Secretary of the State Horticultural Society of Michigan, 1875* (Lansing: W. S. George and Co., 1876), 54.

53. Many nurserymen's catalogues and advertisements featured weeping trees. For a widely disseminated example, see "Abridged Descriptive Catalogue of Fruit Trees, Ornamental Trees, Shrubs, and Greenhouse Plants Cultivated and For Sale by Hubbard and Davis, Detroit, Michigan," *Michigan Farmer* 2 (10 March 1860): 76.

54. According to the horticultural literature, most plants commonly trained and cultivated in hedgerows were non-native species, and southern Michigan homeowners probably purchased them in large quantities from commercial nurserymen. See, for example, "Live Fences, Planting and Management of Quickset Hedges," *Report of the Commissioner of Patents, for the Year 1854, Agriculture* (Washington, DC: A.O.P. Nicholson, 1855), 393–418. Nurserymen often advertised hedge plants for sale. See, for example, "Trees! Flowers! Bulbs! Seeds! Hedge Plants! [advertisement] *Adrian Times and Expositor*, 17 September 1872.

55. Typical prices for conifers and deciduous trees were drawn from both a well-known Midwestern firm, the F. K Phoenix Nursery, and the prominent Mt. Hope Nursery operated by George Ellwanger and Patrick Barry in Rochester, New York. See *General Descriptive Catalogue of Fruit and Ornamental Trees, Shrubs, Roses, Bulbs, Green-House and Garden Plants Cultivated and For Sale at the Bloomington Nursery, 1859* (Chicago: Daily Democrat Steam Printing House, 1859), 24–5, and 22–23, National Agricultural Library, United States Department of Agriculture, Beltsville, Maryland; and *1870 Abridged Catalogue of Select Fruit and Ornamental Trees, Cultivated and Sold by Ellwanger and Barry* (Rochester, NY: Mount Hope Nurseries, 1870), 27–32, and 16–24, Ellwanger and Barry Company Papers, Department of Rare Books and Special Collections, Rush Rhees Library, University of Rochester, Rochester, New York.

56. See *1870 Abridged Catalogue of Select Fruit and Ornamental Trees, Cultivated and Sold by Ellwanger and Barry*, 40–41, 33–40, and 24–26; and *Illustrated Descriptive Catalogue of the Monroe Nursery, Monroe, Michigan* (Monroe, MI: Commercial Steam Printing House, 1874), 41–48, University Archives and Historical Collection, Michigan State University, East Lansing, Michigan.

57. *Wm. Adair and Company Descriptive Catalogue of Vegetable, Flower and Agricultural Seeds, Vines, Trees, Plants, Etc.* (1872), 56–61; and *Catalogue of New and Rare Green House Plants, Evergreen Shrubs, Bedding Plants, Dahlias, Verbenas, Roses, Petunias, Fuchsias, Chrysanthemums, Phlox, Geraniums, Bulbs, Etc., Grown and For Sale by Hubbard and Davis* (Detroit: William H. Thompson, 1871), 6–10, Clarke Historical Library, Central Michigan University, Mount Pleasant, Michigan.

58. See *Illustrated Descriptive Catalogue of the Monroe Nursery, Monroe, Michigan* (1874), 64–65; *1870 Abridged Catalogue of Select Fruit and Ornamental Trees, Cultivated and Sold by Ellwanger and Barry*, 44–5; and *A Descriptive Catalogue of a Choice Collection of Flower, Vegetable and Agricultural Seeds Cultivated and For Sale by Benjamin K. Bliss* (Springfield, MA: Samuel Bowles and Co., 1872), 96–100, Special Collections, Michigan State University, East Lansing, Michigan.

59. See *Wm. Adair and Company Descriptive Catalogue of Vegetable, Flower and Agricultural Seeds, Vines, Trees, Plants, Etc.* (1872), 60–66; and *A Descriptive Cat-*

alogue of a Choice Collection of Flower, Vegetable and Agricultural Seeds Cultivated and For Sale by Benjamin K. Bliss (1872), 116–120, 121, and 142.

60. Orders from Lenawee, Jackson and Washtenaw County residents were listed in Ellwanger and Barry Cash Books. See Cash Book, vol. 55, 13 April 1852–18 October 1853; Cash Book, vol. 56, 1 March 1854–19 July 1856; Cash Book, vol. 57, 1 January 1859–30 November 1861; Cash Book, vol. 58, 2 December 1861–30 November 1864; Cash Book, vol. 59, 18 November 1864–29 October 1867, Ellwanger and Barry Company Papers, Department of Rare Books and Special Collections, Rush Rhees Library, University of Rochester, Rochester, New York. Local nurserymen sometimes ordered plants from Ellwanger and Barry for propagation or resale, and once their names were removed from the list, the average order size dropped to $17.30.

61. Mr. Dean's detailed orders appeared in Day Books, vol. 90, April 1854–April 1856, Ellwanger and Barry Company Papers, Department of Rare Books and Special Collections, Rush Rhees Library, University of Rochester, Rochester, New York. Information on A. J. Dean was drawn from the 1850 Federal Population Census, Lenawee County, Michigan, Microfilm Copy, M-432, roll 355, 494; and 1860 Federal Population Census, Lenawee County, Michigan, Microfilm Copy, M-653, roll 551. The Dean family had modest holdings, with real estate valued at $6,000 in both the 1850 and 1860 censuses, and an additional $1,000 in personal property reported in 1860. Mr. Dean's Ellwanger and Barry order represented 3.6 percent of his 1860 personal estate.

Chapter Two: A Reason for Planting

1. See Sally McMurry, *Transforming Rural Life*, 230, for a contemporary discussion of the characteristics and complexities of a modernizing society. Also see Brown, "Modernization and the Modern Personality in Early America," 201–3, for a working definition of modernization and the social characteristics that accompany it.

2. Historian John Higham has suggested that the primary reasons for this conservative turn resided in the "confluence of modernizing forces like urbanization and industrialization." Concern over these issues led many reformers to look for "new possibilities for organizing and disciplining a culture of rampant individualism." See John Higham, *From Boundlessness to Consolidation, The Transformation of American Culture 1848–1860* (Ann Arbor: William Clements Library, 1969), 24 and 26.

3. Paul S. Boyer, *Urban Masses and the Moral Order in American, 1820–1920* (Cambridge: Harvard University Press, 1978), ix.

4. For an account of the transmission of social values in the context of family life, see Ryan, *Cradle of the Middle Class*, especially 237–8. See also Stuart M. Blumin, *The Emergence of the Middle Class, Social Experience in the American City, 1760–1900* (New York: Cambridge University Press, 1989). Both David P. Handlin, *The American Home: Architecture and Society, 1815–1915* (Boston: Little,

Brown and Co., 1979); and Clifford E. Clark, Jr., *The American Family Home, 1800–1960* (Chapel Hill: University of North Carolina Press, 1986), discuss the home as a source of social stability.

5. A number of recent studies have considered these social aspects of home, with particular emphasis on women's roles and the concept of domesticity. See, for example, Glenna Mathews, *"Just a Housewife," The Rise and Fall of Domesticity in America* (New York: Oxford University Press, 1987); Kathryn Kish Sklar, *Catharine Beecher, A Study of American Domesticity* (New Haven: Yale University Press, 1973); Linda Kerber, *Women of the Republic, Intellect and Ideology in Revolutionary America* (New York: W. W. Norton, 1986); and Nancy Cott, *The Bonds of Womanhood, Woman's Sphere in New England, 1780–1835* (New Haven: Yale University Press, 1977). Christopher Lasch, *Haven in a Heartless World, The Family Besieged* (New York: Basic Books, 1977), especially 4–12, presents a broad overview of the social history of the family touching upon nineteenth-century developments.

6. Bushnell republished and expanded on some of his early ideas in a volume entitled *Christian Nurture*, and that work remained immensely influential for decades. See Horace Bushnell, *Christian Nurture* (New York: C. Scribners, 1863). For comments on the declining influence of organized religion in the mid-nineteenth-century, see Colleen McDannell, *The Christian Home in Victorian America, 1840–1900* (Bloomington: Indiana University Press, 1986), xv; and Maxine Van-DeWetering, "The Popular Concept of 'Home' in Nineteenth-Century America," *Journal of American Studies* 18 (April 1984), 7–10. See also Clifford E. Clark, Jr., "Domestic Architecture as an Index to Social History: The Romantic Revival and the Cult of Domesticity in America, 1840–1870," *Journal of Interdisciplinary History* 7, no. 1 (Summer, 1976): 33–56, for an intriguing analysis of home religion manifested in material culture.

7. A number of commentators have discussed concepts of home as a physical place, and the nineteenth-century interest in environment and architecture as a means to influence human behavior. See, for example, D. Geoffrey Hayward, "Home as an Environmental and Psychological Concept," *Landscape* 20, no. 1 (October 1975): 2–9; Delores Hayden, *Seven American Utopias, The Architecture of American Socialism, 1790–1975* (Cambridge: MIT Press, 1976); David J. Rothman, *The Discovery of the Asylum, Social Order and Disorder in the New Republic*, rev. ed. (Boston: Little, Brown and Co., 1990); Gwendolyn Wright, *Moralism and the Model Home* (Chicago: University of Chicago Press, 1980); Gwendolyn Wright, *Building the Dream: A Social History of Housing in America* (Cambridge: MIT Press, 1983); Handlin, *The American Home*; Clark, *The American Family Home*; and Dell Upton, "Pattern Books and Professionalism, Aspects of the Transformation of Domestic Architecture in America, 1800–1860," *Winterthur Portfolio* 19 (1984): 107–150.

8. *Round Table* VI (23 November 1867), 337, quoted in Frank Luther Mott, *A History of American Magazines, 1865–1885*, vol. 3 (Cambridge: Harvard University Press, 1938), 3–9. For additional commentary on the increase in journals and newspapers, see Frank Luther Mott, *A History of American Magazines, 1850–1865*, vol. 2 (Cambridge: Harvard University Press, 1938), 3–9; and Wal-

ters, *American Reformers, 1815–1860*, 6. For numerical analysis of newspapers and journals, see Frances A. Walker, compiler, The *Statistics of the Population of the United States Embracing the Tables of Race, Nationality, Sex, Selected Ages and Occupations to which are added The Statistics of School Attendance and Illiteracy, of Schools, Libraries, Newspapers and Periodicals, Churches, Pauperism, and Crime and of Areas, Families, and Dwellings* (Washington, DC: Government Printing Office, 1872), 482 and 497.

9. See Alan Nourie and Barbara Nourie, eds., *American Mass-Market Magazines* (Westport, CT: Greenwood Press, 1990), especially 33–39. See also Mott, *A History of American Magazines, 1850–1865*, 103–107, 218–261, and 494–515; and *A History of American Magazines, 1865–1885*, 25–39. Mott notes that over five hundred general interest magazines were available in 1870. For additional commentary on the growth of western journals, see Mott, *A History of American Magazines, 1850–1865*, 113–116; Mott, *A History of American Magazines, 1865–1885*, 50–55. For commentary on periodicals directed to a female audience, see Kathleen L. Endres and Therese L. Lueck, eds., *Women's Periodicals in the United States, Consumer Magazines* (Westport, CT: Greenwood Press, 1995), xii; Mott, *A History of American Magazines, 1850–1865*, 416–418; and *A History of American Magazines, 1865–1885*, 90–103.

10. For commentary on nineteenth-century religious journals, see P. Mark Fackler and Charles H. Lippy, eds., *Popular Religious Magazines of the United States* (Westport, CT: Greenwood Press, 1995), 123, and 520–525; and Mott, *A History of American Magazines, 1850–1865*, 113–116; Mott, *A History of American Magazines, 1865–1885*, 66–89.

11. Walker, *The Statistics of the Population of the United States*, 497.

12. Albert L. Demaree, *The American Agricultural Press, 1819–1860* (New York: Columbia University Press, 1941), xi and 12–19. For additional analysis of the mid-nineteenth-century agricultural press, and its influence, see A. L. Demaree, "The Farm Journals, Their Editors, and Their Public, 1830–1860," *Agricultural History* 15 (October 1941): 182–188; Donald B. Marti, "Agricultural Journalism and the Diffusion of Knowledge: The First Half-Century in America," *Agricultural History* 54 (January 1980): 28–37; Donald Marti, *To Improve the Soil and the Mind. Agricultural Societies, Journals and Schools in the Northeastern States, 1791–1865* (Ann Arbor: University Microfilms International, 1979), 124–162; and Clarence H. Danhof, *Change in Agriculture: The Northern United States, 1820–1870* (Cambridge: Harvard University Press, 1969), 54–64.

13. Walker, *The Statistics of the Population of the United States*, 487 and 497. Although this number includes both agricultural and horticultural journals, agricultural journals were far more common. See also Mott, *A History of American Magazines, 1865–1885*, 151.

14. United State Post Office Records, South Lyon, Michigan, 1853–1865, Bentley Historical Library, University of Michigan, Ann Arbor, Michigan. The postmaster listed the names of journals and the individuals in the area who subscribed as a method of recording postage payments.

15. "The Atlantic Monthly," *Tecumseh (Michigan) Herald*, 1 October 1863. In the middle decades of the nineteenth century, newspapers or journals "sent by way

of exchange between publishers" did not have to pay postage, thus encouraging the free exchange of information. See *United States Domestic Postage Rates 1789–1956* (Washington, DC: Post Office Department, 1956): 54. For an insightful study of the effects of these free exchange policies, and an analysis of the relationship between the newspaper press and the postal service, see Richard B. Kielbowicz, *News in the Mail, The Press, Post Office, and Public Information, 1700–1860s* (Westport, CT: Greenwood Press, 1989).

16. "Home Adornment," *Transactions of the Illinois State Horticultural Society for 1868* (Chicago: Prairie Farmer Co., Published by the Society, 1869), 291.

17. Lewis Allen, *Rural Architecture, Being A Complete Description of Farm Houses, Cottages and Out Buildings* (New York: C. M. Saxton, 1863), xii.

18. Rev. H. L. Baugher, "A Plea for Home Education and Family Religion," *Evangelical Review* 3 (January 1852): 314.

19. See Andrew Jackson Downing, *The Architecture of Country Houses Including Designs for Cottages, and Farm-Houses, and Villas, with Remarks on Interiors, Furniture, and the Best Modes of Warming and Ventilating* (New York: D. Appleton, 1850; reprint, New York: Dover Publications, 1969), xx (all references to reprint edition).

20. "Get a Home and Keep It," *American Agriculturist* 23 (January 1864): 21.

21. Rev. E. C. Guild, "Home Comforts and Amusements at Small Costs," *Unitarian Review and Religious Magazine* 8 (1880): 526.

22. Horace W. Cleveland, William Backus, and Samuel D. Backus, *The Requirements of American Village Homes Considered and Suggested; with Designs for Such Houses of Moderate Cost* (New York: Appleton, 1856), 4–5.

23. Mrs. Jeremiah Brown, "The Home, and Ornamentation of Homes," *Fourth Annual Report of the Secretary of the State Pomological Society of Michigan, 1874* (Lansing: W. S. George and Co., 1875), 111.

24. A number of contemporary historians have discussed the nineteenth-century concept of association. For a particularly interesting discussion, see James A. Schmiechen, "The Victorians, the Historians, and the Idea of Modernism," *American Historical Review* 93 (1988): 287–316.

25. H. Harbaugh, "Home Feeling," *Hours at Home* 1 (September 1865): 410 and 409.

26. H. Harbaugh, "Home Feeling," *Hours at Home* 2 (November 1865): 57 and 60.

27. "The Moral Uses of Plants," *Western Journal of Agriculture, Manufactures, Mechanic Arts, Internal Improvement, Commerce, and General Literature* 1 (1848): 46.

28. "Get a Home and Keep It," 21.

29. H. Harbaugh, "Home Feeling," *Hours at Home* 2 (November 1865): 57.

30. "Rooms and Their Ornaments," *Mother's Magazine and Family Monitor* 21 (1853): 209.

31. Marian H. Ford, "Attractive Homes," *Potter's American Monthly* 16 (1881): 53.

32. Fredrika Bremer, *The Homes of the New World*, vol. 2, trans. Mary Howitt (New York: Harper and Brothers, 1853), 141.

33. Charles W. Garfield, "Beautiful Homes. Their Influence and How to Make Them," *Second Annual Report of the Secretary of the State Pomological Society of Michigan, 1872* (Lansing: W. S. George and Co., 1873), 432.

34. Mrs. H. M. Barker, "Home-Keeping and House-Keeping," *Ladies' Floral Cabinet and Pictorial Home Companion* 4 (December 1875): 186. In succeeding references this journal is cited as *Ladies' Floral Cabinet*.

35. Dr. William Darlington, "The Influence of Horticulture upon the Human Character," *Michigan Farmer* 4 (March 1847): 197; "Gardens," *Michigan Farmer* 1 (15 March 1843): 26; and George E. Blakelee, "An Attractive and Happy Home," *Michigan Farmer* 4 (18 January 1862): 26.

36. John F. W. Ware, "The New England Home," *Monthly Religious Magazine* 26 (December 1861): 349–50 and 354; and "Rooms and Their Ornaments," *Mother's Magazine and Family Monitor* 21 (1853): 211. Contemporary historians have labeled this measured acquisition of objects for moral purposes "pious consumption" or "moral materialism" and have suggested that during the nineteenth century, it transformed the home into a socially acceptable object for material indulgence. See Lori Merish, "'The Hand of Refined Taste' in the Frontier Landscape: Caroline Kirkland's *A New Home, Who'll Follow?* and the Feminization of American Consumerism," *American Quarterly* 45, no. 4 (December 1993): 487. Tamara Plakins Thornton's *Cultivating Gentlemen, The Meaning of Country Life Among the Boston Elite, 1785–1860* (New Haven: Yale University Press, 1979) explores horticulture as a way to reconcile the problems inherent in the simultaneous pursuit and transcendence of wealth and material display. Daniel Horowitz also traces American concern for materialism and consumption patterns and their influence on the moral health of American society. See Daniel Horowitz, *The Morality of Spending, Attitudes toward Consumer Society in America 1875–1940* (Baltimore: Johns Hopkins University Press, 1985), xvii. A number of other scholars have also examined material goods as a way to enhance the home. See, for example, Sally McMurry, "City Parlor, Country Sitting Room, Rural Vernacular Design and the American Parlor, 1840–1900," *Winterthur Portfolio* 20, no. 4 (1985): 261–280. Although mid-nineteenth-century commentators liked to boast about widely available ornamentation and its appearance in homes of all ranks, some historians have suggested their comments require judicious appraisal. See, for example, Blumin, *The Emergence of the Middle Class*, 139–140, and 157–8.

37. "Landscape Gardening," *Methodist Quarterly Review* 28 (1846): 376; and James Vick, "Rural Homes," *Second Annual Report of the Secretary of the State Pomological Society of Michigan, 1872* (Lansing: W. S. George and Co., 1873), 437.

38. "Floriculture," *Michigan Farmer* 6 (15 August 1848): 243.

39. J. J. Thomas, "Horticulture for the People," *Third Annual Report of the Secretary of the State Pomological Society of Michigan, 1873* (Lansing: W. S. George and Co., 1874), 487.

40. See Cleveland, "Landscape Gardening," 391. A number of authors have discussed concepts of nature and moral order in the mid-nineteenth century. See, for example, Catharine L. Albanese, *Nature Religion in America From the Algonkian Indians to the New Age* (Chicago: University of Chicago Press, 1990); Lawrence Buell, *New England Literary Culture from Revolution through Renais-*

sance (New York: Cambridge University Press, 1986); and Barbara Novack, *Nature and Culture, American Landscape and Painting, 1825–1875* (New York: Oxford University Press, 1980). See also Neil Evernden, *The Social Creation of Nature* (Baltimore: Johns Hopkins University Press, 1992), for a discussion of the process of transmuting social concepts onto the natural world.

41. "The Moral Uses of Plants," 41; and Mrs. Jeremiah Brown, "The Home, and Ornamentation of Homes," 115.

42. Ware, "The New England Home," 354; Bremer, *The Homes of the New World*, 140; and Downing, *The Architecture of Country Houses*, xix.

43. Cleveland and Backus, *The Requirements of American Village Homes*, 14–15.

44. Ware, "The New England Home," 354; and H. W. S. Cleveland, "Landscape Gardening," *Christian Examiner* 58(1855): 385. A number of historians have discussed the nineteenth-century desire to unite the best in rural and urban settings in a kind of middles landscape. See, for example, David Schuyler, *The New Urban Landscape, The Redefinition of City Form in Nineteenth-Century America* (Baltimore: Johns Hopkins University Press, 1986); James L. Machor, *Pastoral Cities, Urban Ideals and the Symbolic Landscape of America* (Madison: University of Wisconsin Press, 1987); Thomas Bender, *Toward an Urban Vision, Ideas and Institutions in Nineteenth-Century America* (Lexington: University Press of Kentucky, 1975); John Stilgoe, *Borderland, Origins of the American Suburb, 1820–1939* (New Haven: Yale University Press, 1988); and Kenneth Jackson, *Crabgrass Frontier, The Suburbanization of the United States* (New York: Oxford University Press, 1985).

45. "The Age and the Home," *Monthly Religious Magazine* 23 (January 1860): 2; and "Home Life," *Hearth and Home* 7 (10 October 1874): 307. Several scholars have commented on the concept of home as haven. See, for example, Clark, *The American Family Home*, 29; and Kirk Jeffrey, "The Family as Utopian Retreat from the City: The Nineteenth-Century Contribution," *Soundings* 55 (1972): 21–41.

46. Andrew Jackson Downing, *Cottage Residences or, A Series of Designs for Rural Cottages and Cottage Villas, and their Gardens and Grounds. Adapted to North America*, ed. George E. Harney (New York: John Wiley and Son, 1873; reprint, New York: Dover Publications, 1981), ix (page references to reprint edition). See also Jackson T. Lears, *Fables of Abundance, A Cultural History of Advertising in America* (New York: Basic Books, 1994), 103.

47. G. H. Kern, *Practical Landscape Gardening* (Cincinnati: Moore, Wilstach, Keys and Co., 1855), 248.

48. Walter Elder, "Rural Notes," *Gardeners' Monthly* 5, no. 9 (September 1863): 265.

49. "Farmer, 'Fix-Up' Your Homes," *American Agriculturist* 15 (February 1856): 100.

50. "Auction Sale," [advertisement] *Adrian Times and Expositor*, 23 October 1872. "Farm for Sale or Exchange," [advertisement] *Adrian Times and Expositor*, 29 May 1873.

51. For commentary on this process, see Rowntree and Conkey, "Symbolism and the Cultural Landscape," 467. Rowntree and Conkey stress that as symbols

are tied to key cultural values, their meaning becomes "unquestioned and sanctified," and that eventually the "interests and needs of the society are presented to the individual as his own ultimate interests and needs."

52. Several scholars have explored the symbolic power of goods or objects as a means of expressing cultural categories, maintaining order, or protecting cultural ideals. For an insightful analysis of material objects, their symbolic meaning and their use in maintaining social order, see Ames, "Material Culture as Non-Verbal Communication: A Historical Case Study," 620. Grant McCracken also provides a particularly useful discussion of the diffusion of cultural meaning from cultural tastemakers to goods or objects themselves, and on to the consumer. See Grant McCracken, "Culture and Consumption: A Theoretical Account of the Structure and Movement of the Cultural Meaning of Consumer Goods," in *Perspectives in Consumer Behavior*, 4th ed., ed. Harold H. Kassarjian and Thomas S. Robertson (Englewood Cliffs, NJ: Prentice-Hall, 1991), 581–599. See also Mary Douglas and Baron Isherwood, *The World of Goods* (New York: Basic Books, 1979), 59 and 65–67, for a discussion of goods as critical for "making visible and stable the categories of culture." Like McCracken, Douglas argues that goods, in their concreteness, help to "pin down" social meaning and give a physical reality to a particular set of values.

53. Rowntree and Conkey, "Symbolism and the Cultural Landscape," 462. See also Mark P. Leone, "Interpreting Ideology in Historical Archaeology: Using the Rules of Perspective in the William Paca Garden, Annapolis, Maryland," in *Ideology, Power, and Prehistory*, ed. Daniel Miller and Christopher Tilley (New York: Cambridge University Press, 1984), 25–35.

54. Mrs. L. B. Adams, "Two Ways of Taking Things Easy," *Michigan Farmer* 1 (25 June 1859): 206.

55. Joseph Breck, *The Flower Garden* (Boston: John P. Jewett and Co., 1851), 15; and "Surroundings Indicate Character," *Prairie Farmer* 5 (16 February 1860): 106.

56. Hon. Simon Brown, "Farmer's Gardens," *Report of the Commissioner of Agriculture for 1863* (Washington, DC: Government Printing Office, 1863), 363.

57. "Attractions of Home," *Michigan Farmer* 6 (1 November 1848): 329.

58. Ware, "Home Life," 15.

59. See "Suburban Residences," *Horticulturist* 9 (1 July 1854): 297; and "Taste versus Fashion," *Horticulturist* 19 (July 1864): 202.

60. John Ware, for example, noted that the ideal home should convey "immediately and only ... the idea of home." See Ware, "The New England Home," 354; and John Ware, "Home, The Residence," *Monthly Religious Magazine* 27 (1862): 96.

61. A number of scholars have commented on the seeming conflict between the promotion of cultural ideals and the need to express social differences. See, for example, Daniel Walker Howe, *The Political Culture of the American Whigs* (Chicago: University of Chicago Press, 1979), 20. Howe has argued that this discrepancy was a clear component of Whig ideology which willingly accepted differences in economic rank, but emphasized the importance of social cohesion based on shared values. See also Lawrence W. Levine, *Highbrow/Lowbrow. The*

Emergence of Cultural Hierarchy in America (Cambridge: Harvard University Press, 1988), especially 227 and 229. Levine notes the ambivalence of cultural arbiters as they promoted culture as a way to elevate the masses, and also as a way to distinguish social groups from the general mass culture. Richard L. Bushman, *The Refinement of America, Persons, Houses, Cities* (New York: Alfred A. Knopf, 1992) also suggests that the idea of gentility enabled anyone to be a part of the middle class if they would display certain marks of refinement.

62. Edward Sprague Rand, *Flowers for the Parlor and Garden* (Boston: J. E. Tilton and Co., 1869), 9; and "Flowers as an Influence," *Prairie Farmer* 19 (2 April 1867): 286.

63. A Lady, "A Word for Flowers," *Tilton's Journal of Horticulture* 8 (1870): 153.

64. Mr. Jason Woodman, "Beautiful Homes for Farmers," *Nineteenth Annual Report of the Secretary of the State Board of Agriculture of the State of Michigan, 1880* (Lansing: W. S. George and Co., 1880), 157. A number of scholars have discussed goods as a means of social transformation. See, for example, David Jaffee, "Peddlers of Progress and the Transformation of the Rural North, 1760–1860," *Journal of American History* 78 (September 1991): 511–535; and David Jaffee, "One of the Primitive Sort, Portrait Makers of the Rural North, 1760–1860," in *The Countryside in the Age of Capitalist Transformation, Essays in the Social History of Rural America*, ed. Steven Hahn and Jonathan Prude (Chapel Hill: University of North Carolina Press, 1985), 104. Lears, *Fables of Abundance*, 42–3, also offers an insightful discussion on the role of advertising in promoting goods as a way to achieve self-transformation.

65. Garfield, "Beautiful Homes, Their Influence and How to Make Them," 426 and 429.

66. For a rural perspective, see, for example, McMurry, *Families and Farmhouses in Nineteenth-Century America*, especially 56–60. John Mack Faragher, *Sugar Creek, Life on the Illinois Prairie* (New Haven: Yale University Press, 1986), also provides an insightful analysis of gender roles and responsibilities among the nineteenth-century Midwestern rural population.

67. Several historians have pointed out that farm improvement in the nineteenth century often involved a new use of physical space. See, for example, Thomas Hubka, *Big House, Little House, Back House, Barn , The Connected Farm Buildings of New England* (Hanover: University Press of New England, 1985), 70 and 73–4; J. Ritchie Garrison, *Landscape and Material Life in Franklin County, Massachusetts, 1770–1860* (Knoxville: University of Tennessee Press, 1991), 3–4 and 117; and John Brinkerhoff Jackson, "A New Kind of Space," in *Changing Rural Landscapes*, ed. Ervin H. Zube and Margaret J. Zube (Amherst: University of Massachusetts Press, 1977), 69–71.

68. "Front Yards," *New England Farmer* 16 (1 July 1837): 3. Also see Bushman, *The Refinement of America*, 262, for a discussion of the extension of the home's domestic space into the front yard during the nineteenth century.

69. Several scholars have explored the relationship between domestic consumption and gender roles in the nineteenth century. See especially Lori Merish, "'The Hand of Refined Taste' in the Frontier Landscape," 485–523; and Mc-

Murry, "City Parlor, Country Sitting Room, Rural Vernacular Design and the American Parlor, 1840–1900,"485–523. Others have emphasized the expanding role of women in domestic consumption in the last decades of the nineteenth century and in the early twentieth century. See, for example, Thomas J. Schlereth, *Victorian America, Transformations in Everyday Life, 1876–1915* (New York: Harper Collins, 1991), 141; and William R. Leach, "Transformation in a Culture of Consumption: Women and Department Stores, 1890–1925," *Journal of American History* 71, no. 2 (September 1984): 319–342.

70. M. B. Bateham, "Floriculture for the Million," *Second Annual Report of the Secretary of the State Pomological Society of Michigan, 1872* (Lansing: W. S. George and Co., 1873), 418; Jane Grey Swisshelm, *Letters to Country Girls* (New York: J. C. Riker, 1853), 57, 27, and 23–5; and "Home and the Household," *Ladies' Floral Cabinet* 3 (September 1874): 138.

71. Mrs. M. P. A. Crozier, "Farm Adornment," *The Seventh Annual Report of the Secretary of the State Pomological Society of Michigan, 1877* (Lansing: W. S. George and Co., 1878), 295. A number of scholars have discussed these conceptual links between women and the natural world in nineteenth-century thought. See especially Vera Norwood, *Made From This Earth, American Women and Nature* (Chapel Hill: University of North Carolina Press, 1993); Carolyn Merchant, *The Death of Nature, Women, Ecology and the Scientific Revolution* (San Francisco: Harper Collins, 1983); Carolyn Merchant, *Ecological Revolutions, Nature, Gender, and Science in New England* (Chapel Hill: University of North Carolina Press, 1989); and Sherry B. Ortner, "Is Female to Male as Nature Is to Culture?" in *Women, Culture, and Society*, ed. Michelle Zimbalist Rosaldo and Louise Lamphere (Stanford, CA: Stanford University Press, 1974), 67–87.

72. Downing, *A Treatise on the Theory and Practice of Landscape Gardening*, 10; and "Neighborhood Improvement," *Horticulturist* 23 (April 1868): 124.

73. O. Ordway, "Treatise on the Advantages to be Derived from the Cultivation of Flowers," *Transactions of the Illinois State Agricultural Society, 1856–7*, vol. 2 (Springfield: Lanphier and Walker, 1857), 405; and T. T. Lyon, "A Plea for the Flowers," *Michigan Farmer* 3 (21 December 1861): 498.

Chapter Three: Hands-on Horticulture

1. "Laws of Association in Ornamental Gardening," *North American Review* 87 (July 1858): 159.

2. For analysis of some of these early sources of horticultural information, see Brenda Bullion, "Early American Farming and Gardening Literature: 'Adapted to the Climates and Seasons of the United States'," *Journal of Garden History* 12, no. 1 (1992): 29–34; Therese O'Malley, "Appropriation and Adaptation: Early Gardening Literature in America," *Huntington Library Quarterly* 55 (1992): 401–431; Hedrick, *A History of Horticulture in America*, 467–497; and Liberty Hyde Bailey, *The Standard Cyclopedia of Horticulture* (New York: Macmillan Co., 1937), 1521–1523.

3. For a description of Jefferson's interpretation of ornamented farm concepts, see Rudy J. Favretti, "Thomas Jefferson's 'Ferme Ornee' at Monticello," *Proceedings of the American Antiquarian Society* 103, no. 1 (1993): 17–29. For a discussion of seventeenth and eighteenth-century English gardening traditions and influential practitioners, see Williamson, *Polite Landscapes, Gardens and Society in Eighteenth-Century England*; Favretti, *Landscapes and Gardens for Historic Buildings*; Tobey, *A History of Landscape Architecture;* Newton, *Design on the Land*, especially chapters 14 and 15; Clifford, *A History of Garden Design*, especially chapters 5–8; and Christopher Thacker, *The History of Gardens* (Berkeley: University of California Press, 1992), especially 209–212 and 228–230. See George B. Tatum, "The Gardens They Made," in *America's Garden Legacy, A Taste for Pleasure*, ed. George H. B. Lawrence (Philadelphia: Pennsylvania Horticultural Society, 1978), 65–88, for a concise history of European influence on American gardens. Also see O'Malley, "Landscape Gardening in the Early National Period," 137–142.

4. For commentary on Loudon's contributions, see Thacker, *The History of Gardens*, 229–230.

5. O'Malley, "Appropriation and Adaptation: Early Gardening Literature in America," 425–26; and Hedrick, *A History of Horticulture in America*, 197.

6. See Demaree, *The American Agricultural Press 1819–1860*, xi and 12–19. For additional analysis of the mid-nineteenth-century agricultural press, and its development, issues, and influences, see Demaree, "The Farm Journals, Their Editors, and Their Public, 1830–1860," 182–188; Marti, "Agricultural Journalism and the Diffusion of Knowledge: The First Half-Century in America," 28–37; Marti, *To Improve the Soil and the Mind*, 124–162; and Danhof, *Change in Agriculture*, 54–64.

7. Hedrick, *A History of Horticulture in America*, 482.

8. "The Agricultural Press," *Gardener's Monthly* 3, no. 1 (January 1861): 23–4.

9. J. L. Tappan, "History of the Agricultural Press of Michigan," *Michigan Farmer* 1 (1 January 1859): 5.

10. The *Michigan Farmer* frequently reported on who was heading the horticultural department and their goals and interests. See especially J. C. Holmes, "Valedictory," *Michigan Farmer* 7 (1 November 1849): 326; S. B. Noble, "To Our Horticultural Readers," *Michigan Farmer* 11 (January 1853): 24; J. C. Holmes, "Salutatory," *Michigan Farmer* 13 (January 1855): 12; and "Our Horticultural Department," *Michigan Farmer* 14 (January 1856): 23. For comments on Johnstone, see Demaree, *The American Agricultural Press, 1819–1860*, 383–385. Also see T. T. Lyon, *History of Michigan Horticulture* (Lansing: Thorp and Godfrey, 1887), 200.

11. Erik A. Ernst, "John A. Kennicott of The Grove, Physician, Horticulturist, and Journalist in Nineteenth-Century Illinois," *Illinois State Historical Society Journal* 74 (Summer 1981): 114; "Our Twenty-Six Years," *Prairie Farmer* 19 (5 January 1867): 1; and "Editorial Notes," *Horticulturist* 27 (February 1872): 53.

12. "The Agricultural Press," *Gardener's Monthly* 3, no. 1 (January 1861): 23; and "Agriculture in Cities," *Gardener's Monthly* 10, no. 4 (April 1868): 122.

13. See, for example, "Farm and Garden—Forest Trees," *Adrian Times and Expositor*, 15 March 1869; "Fruiting Vines," *Tecumseh Herald*, 21 September 1865; and "Landscape Architecture," *Ypsilanti Commercial*, 19 July 1873.

14. Bailey, *The Standard Cyclopedia of Horticulture*, 1561–2.

15. Bailey, *The Standard Cyclopedia of Horticulture*, 1559; Hedrick, *A History of Horticulture in America*, 490 and 496; "The Close of the Magazine," *Magazine of Horticulture* 34, no. 12 (December 1868): 354–356; "Suspension of Hovey's Magazine," *Gardener's Monthly* 11, no. 1 (January 1869): 19–20; and "A Quarter of a Century," *Magazine of Horticulture* 25, no. 12 (December 1859): 530–1.

16. Bailey, *The Standard Cyclopedia of Horticulture*, 1559; and "Commercial Nurseries," *Horticulturist* 19 (October 1864): 323.

17. "Our Magazine for 1864," *Gardener's Monthly* 5, no. 11 (November 1863): 336; Bailey, *The Standard Cyclopedia of Horticulture*, 1559 and 1587–8; and "Extension of Gardener's Monthly," *Gardener's Monthly* 7, no. 11 (November 1865): 336.

18. The Ladies' Floral Cabinet," *Horticulturist* 29 (March 1874): 64; "Articles Wanted," *Ladies' Floral Cabinet* 3 (June 1874): 83; and Bailey, *The Standard Cyclopedia of Horticulture*, 561–2.

19. Bailey, *The Standard Cyclopedia of Horticulture*, 1561–2. The *Western Horticultural Review* was published for several years in the 1850s, while the *Western Pomologist* lasted for a few years in the 1870s.

20. Subscription rates varied somewhat, both over the span of this study and among journals. The subscription rate for the *Michigan Farmer* was $.50 a year in 1845, but had risen to $2.00 yearly by 1860. The *Magazine of Horticulture* charged $2.00 for a yearly subscription in 1868. In 1874, a subscription to the *American Agriculturist* cost $4.00 yearly. See *Michigan Farmer* 3 (June 1845): 33; *Michigan Farmer* 2 (15 December 1860): 393; *Magazine of Horticulture* 34 (December 1868): 353; and *Hearth and Home* 6 (31 January 1874): 80.

21. See "Plant Trees," *Dexter Leader*, 30 May 1873; "Progress in Horticulture," *Adrian Times and Expositor*, 18 January 1869; and "The Michigan Farmer," *Tecumseh Herald*, 1 October 1863.

22. "Seventy-five Excellent Premiums," *American Agriculturist* 25 (October 1866): 349; and Walker, *The Statistics of the Population of the United States*, 497.

23. Patrick Barry, "To the Readers of the Horticulturist," *Horticulturist* 10 (July 1855): 298; "The Gardener's Monthly and the Horticulturist," *Horticulturist* 20 (November 1865): 369; and "To New Readers," *Horticulturist* 29 (March 1874): 64.

24. "Advertising Farms," *Michigan Farmer* 12 (January 1854): 16; and William Bacon, "The Horticulturist," *Horticulturist* 18 (March 1863): 91.

25. Bailey, *The Standard Cyclopedia of Horticulture*, 1523–1533.

26. Frederika Bremer, *The Homes of the New World: Impressions of America*, 43.

27. "Book Notices," *American Agriculturist* 14 (3 May 1855): 121; and "Scott's Suburban Home Grounds," *Horticulturist* 26 (September 1871): 238–9.

28. Hedrick, *A History of Horticulture in America*, 487; Bailey, *The Standard Cyclopedia of Horticulture*, 1567 and 1578–9; and Liberty Hyde Bailey, ed., *Cyclopedia of American Agriculture*, vol. 4 (New York: Macmillan Co., 1909), 584.

29. Bailey, *The Standard Cyclopedia of Horticulture*, 1592, 1551, and 1602.

30. "Gardening," *Report of the Commissioner of Patents for the Year 1854* (Washington, DC: A. O. P. Nicholson, 1855), 322. For accounts of the early history of the United States Department of Agriculture and its reports, see Gladys L. Baker, Wayne D. Rasmussen, Vivian Wiser, and Jane M. Porter, *Century of Service, The First 100 Years of the United States Department of Agriculture* (Washington, DC: Centennial Committee, U.S.D.A., 1963), 2–19; Wayne D. Rasmussen and Gladys L. Baker, *The Department of Agriculture* (New York: Praeger Publishers, 1972), 3–10; and B. P. Poore, "History of the Agriculture of the United States," *Report of the Commissioner of Agriculture for the Year 1866* (Washington, DC: Government Printing Office, 1867), 524–7.

31. *Report of the Commissioner of Agriculture for the Year 1862* (Washington, DC: Government Printing Office, 1863), 20.

32. See "The Report of the Department of Agriculture, 1869," *Tilton's Journal of Horticulture* 9 (1871): 87. For typical comments and criticism of the annual report, see "The Commissioner of Agriculture," *Prairie Farmer* 44 (3 May 1873): 137; and "The Agricultural Bureau—Who Shall Be at the Head of It?" *American Agriculturist* 21 (May 1862): 136. For an insightful evaluation of the value and the defects of these reports, see Paul W. Gates, *Agriculture and the Civil War* (New York: Alfred W. Knopf, 1965), 310–323.

33. See, for example, *Report of the Commissioner of Patents for the Year 1851, Part II Agriculture* (Washington, DC: Robert Armstrong, 1852), 409–413; and *Report of the Commissioner of Patents for the Year 1850, Part II Agriculture* (Washington, DC: Office of Printers to House of Reps., 1851), 410–411.

34. "Michigan State Agricultural Society," *Report of the Commissioner of Patents for the Year 1858, Agriculture* (Washington, DC: James B. Steedman, 1859), 147.

35. Bailey, *The Standard Cyclopedia of Horticulture*, 1553–1559; and "Laws Affecting Horticulture," *Transactions of the Illinois State Horticultural Society for 1869*, vol. 3 (St. Louis: R. P. Studley and Co., 1870), xiv. State funding for horticultural society reports varied a good deal. The states of Indiana and Wisconsin published their respective horticultural society reports, while the Ohio Horticultural Society published its own report beginning in 1848. The Columbus (Ohio) Horticultural Society published its proceedings in local newspapers from 1845 until 1886.

36. Frank W. King, "Annual Statement of Librarian," *Tenth Annual Report of the Secretary of the State Pomological Society of Michigan, 1880* (Lansing: W. S. George and Co., 1881), 530–1. In 1887, T. T. Lyon reported that 8,400 copies of the report were printed, and the Pomological Society distributed 6,000 copies, with the idea that those who do "something in aid of horticulture in Michigan shall be first served." See Lyon, *History of Michigan Horticulture*, 188.

37. King, "Annual Statement of Librarian," 530–1.

38. "List of Books belonging to the Cincinnati Horticultural Society," *Western Farmer and Gardener and Horticultural Magazine* 4 (June 1844): 277; "Catalogue," *Transactions of the Michigan State Agricultural Society with Reports of the County Agricultural Societies for 1851*, vol. 3 (Lansing: Ingals, Hedges and Co.,

1852), 285–6; and "Reading Room of the Y.M.C.A.," *Adrian Times and Expositor*, 22 February 1871.

39. "The New System of Premiums," *Second Annual Report of the Secretary of the State Pomological Society of Michigan, 1872* (Lansing: W. S. George and Co., 1873), 184–6.

40. "To Agricultural and Horticultural Societies," [advertisement] *Michigan Farmer* 14 (January 1856): 26.

41. Wilson Flagg, "On the Embellishment of Dwelling Houses and their Enclosures," *Magazine of Horticulture* 19, no. 10 (October 1853): 439; and William Webster, "Cottage Gardening," *Horticulturist* 8 (1 February 1853): 87.

42. "Mr. Downing and the Horticulturist," *Horticulturist* 7 (September 1852): 397.

43. "Design for a Farm with Grounds and Farm Buildings," *Michigan Farmer* 17 (January 1855): 16; and "A Mechanic's Garden or What Has Been Grown on Half an Acre," *American Agriculturist* 13 (22 November 1854): 166.

44. Downing, *A Treatise on the Theory and Practice of Landscape Gardening*, 18, 53, and 22, 66, 77, 80, and 112.

45. See William Webster, "Cottage Gardening," 87; Frank J. Scott, *The Art of Beautifying Suburban Home Grounds of Small Extent* (New York: D. Appleton, 1870; facsimile edition, Watkins Glen, NY: Library of Victorian Culture, American Life Foundation, 1982), 14; B. S. Olmstead, "Home Scenes," *Horticulturist* 24 (March 1869): 86; and P. B. M., "Suburban Gardening," *Horticulturist* 7 (October 1852): 447–8.

46. Downing, *A Treatise on the Theory and Practice of Landscape Gardening*, 75, 17, and 20; and "Parks and Pleasure Grounds for Farmers," *Michigan Farmer* 13 (March 1855): 69–70.

47. A. D. G., "Hints to Ornamental Planters," *Horticulturist* 20 (July 1865): 201; "The Embellishment of City Door-Yards," *American Agriculturist* 21 (December 1862): 370–1; and Scott, *The Art of Beautifying Suburban Home Grounds*, 30–31.

48. Downing, *A Treatise on the Theory and Practice of Landscape Gardening*, 63.

49. P. Barry, "Ornamental Gardening—The Lawn," *Annual Report of the Indiana State Horticultural Society, 1871* (Indianapolis: R. J. Bright, 1871), 102. Many advisers commented on the attractiveness of a lawn. See, for example, "Ornamentation of City and Village Lots," *Eighth Annual Report of the Secretary of the State Pomological Society of Michigan, 1878* (Lansing: W. S. George and Co., 1879), 134; William Webster, "Laying Out and Planting Lawns," *Horticulturist* 5 (October 1850): 162; "Lawns and Lawn-Mowers," *American Agriculturist* 32 (April 1873): 129. For a historical perspective on lawns in early nineteenth-century American domestic landscapes, see Virginia Scott Jenkins, *The Lawn, A History of an American Obsession* (Washington, DC: Smithsonian Institution Press, 1994), 12–28.

50. See Isaac C. Ferris, "On the Adornment of Our Homes," *Transactions of the Indiana Horticultural Society with Proceedings of the Fourteenth Annual Session* (Indianapolis: Sentinel Co., 1875), 114; Scott, *The Art of Beautifying Suburban Home Grounds*, 22; "Laying Out a Garden," *American Agriculturist* 15 (April

1856): 160; and "Lawns and Flower Beds," *Prairie Farmer* 12 (24 March 1859): 181.

51. For comments on lawn preparation and care, see Scott, *The Art of Beautifying Suburban Home Grounds*, 108; "The Lawn," *Magazine of Horticulture*, 21, no. 4 (April 1855): 163–7; "What Shall I Do With My Front Yard?" *Michigan Farmer* 16 (April 1858): 111; "The Lawn," *Horticulturist* 20 (April 1865): 98–9; "The Lawn," *Magazine of Horticulture* 21, no. 4 (April 1855): 165–6; Rev. A. D. Gridley, "Farm Embellishments," *Magazine of Horticulture* 33, no. 5 (May 1867): 133; and "Mowing Machines," *Horticulturist* 7 (April 1852): 197.

52. For comments on prices and availability of lawn mowers, see "A Lawn Cutting Machine," *Michigan Farmer* 14 (October 1856): 309; *Horticulturist* quoted in "Lawn Mowing Machines," *Michigan Farmer* 1 (3 September 1859): 283. "Great Improvement in Lawn Mowing Machines," *Gardener's Monthly* 2, no. 8 (August 1860): 243; "Lawn Mowers at Home and Abroad," *American Agriculturist* 29 (September 1870): 339; "Floriculture," *Michigan Farmer* 3 (February 1872): 35; and *Vick's Illustrated Floral Guide for 1873* (Rochester, NY: James Vick, 1873), 10, Department of Rare Books and Special Collections, Rush Rhees Library, University of Rochester, Rochester, New York.

53. Downing, *A Treatise on the Theory and Practice of Landscape Gardening*, 69 and 7; and "The Lawn," *Horticulturist* 21(April 1855): 161.

54. For examples of the wide range of comments on evergreen trees and the particular virtues of the Norway spruce, see "A Chapter on Evergreens," *American Agriculturist* 16 (May 1857): 108; George Jacques, "Evergreens," *Magazine of Horticulture* 31, no. 4 (April 1865): 109; "Popular Evergreen Trees," *Horticulturist* 22 (May 1867): 132–3; and George Taylor, "When and How to Plant Evergreens," *Fifth Annual Report of the Secretary of the State Pomological Society of Michigan, 1875* (Lansing: W. S. George and Co., 1876), 54.

55. Many advisers commented on the benefits of trees. See, for example, Downing, *A Treatise on the Theory and Practice of Landscape Gardening*, 260–1; and "Design for a Small Place of Twenty Acres," *Gardener's Monthly* 2, no. 4 (April 1860): 115.

56. For examples of commentary on shrubbery and its importance in embellishment, see "Planting Shrubbery," *American Agriculturist* 16 (April 1857): 88; "Planting Shrubbery," *American Agriculturist* 19 (May 1860): 145; and Walter Elder, "The Modern Style of Planting Shrubbery," *Gardener's Monthly* 7, no. 9 (September 1865): 265.

57. "Climbing Plants as Helps to Home Adornment," *Horticulturist* 24 (October 1869): 289; and "Vines in the Flower Garden," *American Agriculturist* 22 (June 1863): 180.

58. Thomas G. Fessenden, *The New American Gardener* (Boston: Odiorne and Co., 1833), 109; Hortus, "Notes on Flower Garden Plants," *Magazine of Horticulture* 19, no. 2 (February 1853): 72; Calla, "Flower-Gardening for Cottage Homes," *Ladies' Floral Cabinet* 6 (July 1877): 90; and "How to Lay Out a Good Garden," *American Agriculturist* 13 (13 September 1854): 7.

59. "Landscape Gardening—The Practice of Floriculture," *Godey's Ladies' Book* 45 (August 1852): 128; Charles Hovey, "Some Remarks on the Formation

of the Margins of Flower Beds on Grass Plots or Lawns," *Magazine of Horticulture* 6, no. 3 (March 1840): 84; and "Design for a Flower-Garden," *Horticulturist* 24 (April 1859): 112.

60. "Planning the Flower Garden," *American Agriculturist* 19 (May 1860): 147; "The Flower Border," *Michigan Farmer* 15 (May 1857): 148; Scott, *The Art of Beautifying Suburban Home Grounds*, 247; and "Landscape Gardening—Of Planting the Flower Garden," *Godey's Ladies' Book* 45 (December 1852): 528. See also R. Carmichael, "On Grouping and Arranging Plants in the Flower Garden," *Magazine of Horticulture* 11, no. 3 (March 1845): 105–6.

61. Edward S. Rand, Jr., "Flowers in Masses," *Horticulturist* 20 (June 1865): 165; and "Planning the Garden," *American Agriculturist* 18 (March 1859): 84.

62. W. B. Wicken, "The Parterre," *Gardener's Monthly* 15, no. 2 (February 1873): 40–1; Charles Hovey, "Design for a Flower Garden," *Magazine of Horticulture* 21, no. 1 (January 1855): 22; "Floriculture," *Michigan Farmer* 3 (1 February 1872): 35; and "Garden Plans," *American Agriculturist* 32 (July 1873): 263.

63. See "Ribbon Gardening," *American Agriculturist* 21 (April 1862): 113; Walter Elder, "Ribbon Flower-Beds," *Gardener's Monthly* 5, no. 5 (May 1863): 136; and "Old and New Styles of Gardening," *American Agriculturist* 25 (June 1866): 225.

64. "Carpet Bedding," *Tilton's Journal of Horticulture* 9 (1871): 122–3; and "Carpet System of Bedding Out," *Magazine of Horticulture* 34, no. 6 (June 1868): 186–7.

65. "Flowers and Flower-Beds," *Vick's Illustrated Monthly Magazine* 1, no. 2 (February 1878): 39.

66. "Hardy Perennials," *Horticulturist* 22 (February 1867): 40; and Mrs. L. M. McFarland, "Flowers Easily Grown," *Ladies' Floral Cabinet* 3 (April 1874): 50.

67. Breck, *The Flower Garden*, 31; and Charles Hovey, "Observations on the Formation of Rock-work in Gardens," *Magazine of Horticulture* 6, no. 1 (January 1840): 15–6.

68. "Ornamentation of City and Village Lots," *Eighth Annual Report of the Secretary of the State Pomological Society of Michigan, 1878* (Lansing: W. S. George and Co., 1879), 137.

69. "Vases and Statuary in Ornamental Grounds," *American Agriculturist* 20 (March 1861): 84; and "Design for a Rustic Summer House," *American Agriculturist* 18 (July 1859): 212.

70. See "Vases for Garden Decoration," *Horticulturist* 25 (March 1870): 92; "A Talk About Vases, Gardens, Etc.," *Horticulturist* 19 (April 1864): 124; "Ornamental Stone Works," [advertisement] *Adrian Times and Expositor*, 12 May 1870; and "The Michigan Ornamental Stone Company," *Adrian Times and Expositor*, 12 May 1870.

71. For a definition and examples of rustic work, see "A Few Words on Rustic Arbours," *Horticulturist* 4 (January 1850): 320; "Structures in Rustic Work," *American Agriculturist* 29 (December 1870): 458; and Frank Calvert, "Rustic Work," *Prairie Farmer* 5 (22 March 1860): 182; Also see "Ornaments for the Garden and Lawn," *American Agriculturist* 24 (January 1865): 20; and Miss Frances E. Willard, "On the Embellishment of a Country Home," *Transactions of*

the Illinois State Agricultural Society, vol. 3-1857–8 (Springfield: Bailhache and Baker, 1859), 468.

72. See "Design for a Garden-Seat," *Horticulturist* 24 (December 1869): 356; and "Ornamental Structures for the Garden," *American Agriculturist* 16 (October 1857): 229; and "Garden Furniture," *Horticulturist* 8 (July 1853): 301.

73. Edwin A. Johnson, *Winter Greeneries at Home* (New York: Orange Judd Co., 1878), 13; and "House Plants," *Michigan Farmer* 2 (1 July 1871): 203. For additional comments on the benefits of houseplants, see Ruth Hall, "House Plants," *Report of the Commissioner of Agriculture for 1863* (Washington, DC: Government Printing Office, 1863), 367; and "House Plants," *Sixth Annual Report of the Secretary of the State Pomological Society of Michigan, 1876* (Lansing: W. S. George and Co., 1877), 183.

74. For commentary on the changing conditions of plant culture, see Andrew Bridgman, "Room Plants," *Horticulturist* 17 (April 1862): 169; "Growing Plants in Rooms," *Horticulturist* 19 (March 1864): 73; "Window Flowers," *Gardener's Monthly* 12, no. 3 (March 1870): 82; and "Window Gardening," *Gardener's Monthly* 4, no. 5 (May 1862): 129; and "House Plants," *Michigan Farmer* 2 (1 July 1871): 203.

75. "To Keep House Plants from Freezing," *Jackson Daily Citizen*, 15 November 1873; and "Window Plants," *Adrian Times and Expositor*, 24 September 1864. For additional comments on the dangers of frost and how to prevent it, see "Window Plants and Frost," *American Agriculturist* 29 (February 1870): 63; Hortense Share, "Window Plants," *Ladies' Floral Cabinet* 3 (January 1874): 2; and Mrs. Fannie E. Briggs, "Window Gardening in the Wilds of Iowa," *Ladies' Floral Cabinet* 4 (November 1875): 178.

76. For an example of general advice on indoor plant culture, see "House Plants," *Michigan Farmer* 2 (1 July 1871): 203; Room Plants," *Horticulturist* 16 (February 1861): 91; and "Window Green-House," *Horticulturist* 4 (January 1850): 302.

77. "Growing Plants in Rooms—II," *Horticulturist* 19 (April 1864): 105; and Edgar Sanders, "Window Gardening," *Prairie Farmer* 5 (5 January 1860): 6. Also see Mary F. Williams, "Experiences in Growing House Plants," *Ladies' Floral Cabinet* 3 (February 1874): 18; and S. E. Jones, "Out-door Gardening," *Ladies' Floral Cabinet* 3 (June 1874): 81, for comments on placing houseplants outside.

78. For examples of commentary on the best plants to select, see "Preparing for Window Gardening," *American Agriculturist* 32 (October 1873): 383; "Twelve Best Plants for Windows," *Gardener's Monthly*, 4, no. 3 (February 1863: 50; "Window Plants," *Michigan Farmer* 3 (18 April 1872): 123; "Growing Plants in Rooms—II," *Horticulturist* 19 (April 1864): 106–7; "Parlor Plants," *American Agriculturist* 15 (July 1856): 232–3; Williams, *Window Gardening*, 257 and 247–8; and "Plants for Room Culture," *Michigan Farmer* 2 (May 1864): 513.

79. Mrs. Polyanthus Periwinkle, "My Bay-Window," *Ladies' Floral Cabinet* 5 (March 1856): 34; and "Rustic Window-Boxes," *American Agriculturist* 32 (November 1873): 425.

80. Peter Henderson, *Practical Floriculture, A Guide to the Successful Cultivation of Florists' Plants, for the Amateur and Professional Florist* (New York: Or-

ange Judd and Co., 1869), 163. For examples of the variety of comments on hanging baskets and indoor plant culture, see "Hanging Baskets for Flowers," *Prairie Farmer* 6 (13 December 1860): 375; and "Ornamental Hanging Baskets for Plants," *American Agriculturist* 20 (July 1861): 212.

81. O. H. Peck, "Wardian Cases," *Horticulturist* 22 (March 1867): 80. For comments on the history of Wardian Cases, see Williams, *Window Gardening*, 160; and Rand, *Flowers for the Parlor Garden*, 233. For commentary on how Wardian Cases worked, see "Household Ornaments," *American Agriculturist* 26 (October 1867): 371.

82. See Rand, *Flowers for the Parlor Garden*, 238; O. H. Peck, "Wardian Cases," *Horticulturist* 22 (March 1867): 80; and Charles Hovey, "Wardian Cases or Parlor Conservatories," *Magazine of Horticulture* 24, no. 3 (March 1858): 136.

Chapter Four: Commercial Realities

1. See William J. Green, Green Bay, Wisconsin, to John Kennicott, The Grove, Illinois, 27 March 1862, Kennicott Business Papers, 1194, The Grove National Historic Landmark, Glenview, Illinois. In succeeding citations, the abbreviation KBP, followed by the entry number, will be used to identify the Kennicott Business Papers.

2. Hortense Share, "My Flower Garden," *Ladies' Floral Cabinet* 3 (April 1874): 54.

3. "Shade Trees," *Tecumseh Herald* (21 September 1865); and Lily of the Valley, "Flower Gardening," *Ladies' Floral Cabinet* 4 (April 1875): 53.

4. "Flower Seeds for Distribution," *American Agriculturist* 17 (January 1858): 20; and "Flower Seeds," *American Agriculturist* 17 (April 1858): 118.

5. For commentary on growth of the program, see "Notes on the Seeds for Distribution in 1859," *American Agriculturist* 17 (December 1858): 358; "The Seed Distribution," *American Agriculturist* 18 (November 1859): 352. The *Agriculturist* ended its massive general seed distribution program in 1864 citing a new tariff on imported seeds and plants, the rising cost of paper, and a drought that cut local yields.

6. James Vick, "A Retrospect," *Vick's Illustrated Monthly Magazine.* 1, no. 4 (April 1878): 117; and Note to Rural Floriculturists," *Moore's Rural New Yorker* 33 (15 January 1876): 37. In contrast to the *Rural New Yorker's* assertion, these seed distribution programs were almost without exception limited to journal subscribers or agents, individuals with enough resources and interest to acquire a subscription.

7. See "Flower Seeds for Distribution," 20; and "Flower Seeds," 118.

8. "Further Offer of Seeds-Free," *American Agriculturist* 16 (March 1857): 56; "About Seed Envelopes," *American Agriculturist* 16 (March 1857): 68; and "Catalogue of Seeds for Free Distribution in 1859," *American Agriculturist* 17 (December 1858): 377.

9. For accounts of these early efforts at plant introductions and the history of the Department of Agriculture, see Nelson Klose, *America's Crop Heritage, The*

History of Foreign Plant Introduction by the Federal Government (Ames: Iowa State College Press, 1950), 24–65, and 141–145 (the appendix includes text of the "Treasury Department Circular Regarding Plant Introduction, September 6, 1827"); Gladys L. Baker, Wayne D. Rasmussen, Vivian Wiser, and Jane M. Porter, *Century of Service, The First 100 years of the United States Department of Agriculture* (Washington, DC: Centennial Committee, USDA, 1963), 1–20; Wayne D. Rasmussen and Gladys L. Baker, *The Department of Agriculture* (New York: Praeger Publishers, 1972), 3–13; and Jack Shephard, "Seeds of the Presidency, the Capitol Schemes of John Quincy Adams," *Horticulture* 61 (January 1983): 38–47.

10. Klose, *America's Crop Heritage*, 39–44 and 54–8.

11. A number of commissioners from both the Patent Office and the Department of Agriculture commented on the purpose of the seed distribution program. See, for example, Isaac Newton, *Report of the Commissioner of Agriculture for 1862* (Washington, DC: Government Printing Office, 1863), 20–21.

12. *Report of the Commissioner of Patents for the Year 1861* (Washington, DC: Government Printing Office, 1862), 3; *Report of the Commissioner of Agriculture for the Year 1863* (Washington, DC: Government Printing Office, 1863), 10–11; Frederick Watts, *Report of the Commissioner of Agriculture for the Year 1873* (Washington, DC: Government Printing Office, 1874), 7–8; Frederick Watts, *Report of the Commissioner of Agriculture for the Year 1874* (Washington, DC: Government Printing Office, 1875), 13–14.

13. For comments on its beginnings, see Isaac Newton, *Report of the Commissioner of Agriculture for 1863* (Washington, DC: Government Printing Office, 1863), 11. Also see Klose, *America's Crop Heritage*, 56; and Baker, Rasmussen, Wiser, and Porter, *Century of Service*, 14–15.

14. Klose, *America's Crop Heritage*, 39; and Isaac Newton, *Report of the Commissioner of Agriculture for 1864* (Washington, DC: Government Printing Office, 1865), 11. For the proportion of seeds distributed to various parties, see, for example, *Report of the Commissioner of Agriculture for 1866* (Washington, DC: Government Printing Office, 1867), 8; *Report of the Commissioner of Agriculture for 1867* (Washington, DC: Government Printing Office, 1868), xv; and *Report for the Commissioner of Agriculture for 1869* (Washington, DC: Government Printing Office, 1870), 17.

15. See Charles Mason, "Agricultural Report," *Report of the Commissioner of Patents for the Year 1856, Agriculture* (Washington, DC: Cornelius Wendell, 1857), vi–viii; and "The Seed Distribution," *Report of the Commissioner of Agriculture for the Year 1867* (Washington, DC: Government Printing Office, 1868), xv. For detailed comments on these various failings, see "Report of the Superintendent of Seed Division," *Report of the Commissioner of Agriculture for 1868* (Washington, DC: Government Printing Office, 1869), 125; and "Report of the Commissioner of Agriculture for 1868," *Gardener's Monthly* 12, no. 3 (March 1870): 87.

16. L., "Our Washington Letter—The Storehouse of the Agricultural Department," *Detroit Advertiser and Tribune*, 1 February 1864. Eventually, Congress responded to requests for a new building and in 1867 allocated $100,000 for that

Notes to Chapter 4   261

purpose. The new Department of Agriculture facility was completed in 1868. See Baker, Rasmussen, Wiser and Porter, *Century of Service, The First 100 Years of the United States Department of Agriculture*, 18.

17. George W. Bowlsby, "Hudson, Lenawee County, Michigan, 6 January 1851," in *Report of the Commissioner of Patents for 1850, Agriculture* (Washington, DC: Office of Printers to the House of Rep., 1851), 411; O. H. Kelley, "Wright County, Minnesota," in *Report of the Commissioner of Patents for the Year 1859, Agriculture* (Washington, DC: George W. Bowman, 1860), 564; and Barber G. Buell, "Diary," 10 April 1870, Barber G. Buell Papers, 1763–1934, Bentley Historical Library, University of Michigan, Ann Arbor, Michigan.

18. "Agricultural Humbug at Washington," *American Agriculturist* 17 (February 1858): 40. For examples of the complaints, see "Patent Office Seeds," *Prairie Farmer* 2 (30 September 1850): 216; "The Patent Office and Its Agricultural Department," *Michigan Farmer* 16 (August 1858): 233; "Seeds from the Patent Office," *Magazine of Horticulture* 25, no. 5 (May 1859): 232; and Dr. George Pepper Norris, "The Patent Office Report of 1860," *Horticulturist* 17 (May 1862): 232.

19. For commentary on seed distribution as direct competition for the seed industry and agricultural press promotions, see Paul W. Gates, *Agriculture and the Civil War* (New York: Alfred A. Knopf, 1965), 314; and Jack Ralph Kloppenburg, Jr., *First the Seed. The Political Economy of Plant Biotechnology 1492–2000* (New York: Cambridge University Press, 1988), 65. Kloppenburg suggests that the USDA generally offered better quality seed than what was available through private sources.

20. See "Occupations of the Male Inhabitants," *U. S. Census of Population 1850*, 1850.1 through 1850.3, Microfilm Copy, roll 1, lxxi; Joseph C. G. Kennedy, compiler, "Occupations in the United States," in *Population of the United States in 1860; Compiled from the Original Returns of the Eighth Census* (Washington, DC: Government Printing Office, 1864), 662, 664, and 674; Francis A. Walker, compiler, "Persons Engaged in Each Occupation," *The Statistics of the Population of the United States from the Original Returns of the Ninth Census*, vol. 1 (Washington, DC: Government Printing Office, 1872), 674; *Compendium of the Tenth Census 1880*, part 2 (Washington, DC: Government Printing Office, 1885), Microfilm Copy, roll 3, 1382–3. The actual occupational categories included in these total numbers varied over the years, and makes a direct comparison somewhat problematic. In 1850, the total included gardeners, florists, and nurserymen. In 1860, seedsmen and vinegrowers were added to the list. In both 1870 and 1880, the total included florists, gardeners, nurserymen, and vinegrowers.

21. See O. B. Galusha, "Nurseries of Illinois," *Transactions of the Illinois State Agricultural Society*, vol. 3-1857–58 (Springfield: Bailhache and Baker, 1859), 372–377; and "Horticultural Register 1867, of Nurserymen, Fruit-Growers, Agents, Dealers and Publishers," *Transactions of the Illinois State Horticultural Society for 1866* (Chicago: Emery and Co., 1867), 109–122. Mr. Galusha noted that the list was incomplete, but the best that could be assembled in a short time. Many of the nurseries listed appeared to deal solely in fruit trees and were not included among the forty-six ornamental plant nurseries.

22. See "List of the Principal Nurserymen, Florists, and Seedsmen," *American Horticultural Annual 1867* (New York: Orange Judd and Co., 1867), 147; and "Nurserymen, Florists, Seedsmen, and Dealers in Horticultural Stock," *American Horticultural Annual 1871* (New York: Orange Judd and Co., 1871): 140–152. These numbers represent only those firms who appeared to sell ornamental plants, not those selling only fruit trees, small fruits, or vines.

23. See Albert S. Bolles, *Industrial History of the United States* (Norwich, CT: The Henry Bill Publishing Co., 1879), 169, for commentary on how few nurseries were available to supply public demand before the 1830s. See also George H. M. Lawrence, "The Development of American Horticulture," in *America's Garden Legacy, A Taste for Pleasure*, ed. George H. M. Lawrence (Philadelphia: Pennsylvania Horticultural Society, 1978), 92–94, for a concise account of early American nurseries and seed houses. For commentary on the changing geographic requirements of American nurseries, see Henry W. Lawrence, "The Geography of the United States Nursery Industry: Locational Change and Regional Specialization in the Production of Woody Ornamental Plants" (Ph.D. diss., University of Oregon, 1985), 129–138.

24. See Hedrick, *A History of Horticulture in America*, 71–2 and 209; and Charles Van Ravenswaay, "Horticultural Heritage—The Influence of the U.S. Nurserymen, A Commentary," in *Proud Heritage, Future Promise: A Bicentennial Symposium* (Washington, DC: Associates of the National Agricultural Library, 1977), 144. For an account of the eighteenth-century Chesapeake nursery trade, see Barbara W. Sarudy, "Nurserymen and Seed Dealers in the Eighteenth-Century Chesapeake," *Journal of Garden History* 9, no. 3 (1989): 111–117. U. P. Hedrick also provides commentary on early southern nurserymen. See Hedrick, *A History of Horticulture in America*, 283–4.

25. Lawrence, "The Geography of the United States Nursery Industry: Locational Change and Regional Specialization in the Production of Woody Ornamental Plants," 147–8. See also Bailey, *The Standard Cyclopedia of Horticulture*, 1567. For comments on Robert Buist and his contributions, see "The Rosedale Nurseries," *Gardener's Monthly* 17, no. 5 (May 1875): 145. See also Hedrick, *A History of Horticulture in America*, 214 and 245.

26. A. J. Downing, "Notes on the Progress of Gardening in the United States during the Year 1840," *Gardener's Magazine* 16 (1840): 643–4; and "Henry A. Dreer," *Gardener's Monthly* 16, no. 1 (January 1874): 19. See also Hedrick, *A History of Horticulture in America*, 203–204 and 249–251; and Bailey, *The Standard Cyclopedia of Horticulture*, 1518. In addition see Leslie R. Hawthorn and Leonard H. Pollard, *Vegetable and Flower Seed Production* (New York: Blakiston Co., 1954), 23–26, for commentary on increasing acreage devoted to seed production in the United States after 1850.

27. Hedrick, *A History of Horticulture in America*, 205. See also Margaret Fisbee Somer, *The Shaker Garden Seed Industry* (Old Chatham, NY: Shaker Museum, 1972), for a detailed account of the development and extent of the Shaker seed industry.

28. Diane Holahan Grosso, "From the Genesee to the World," *University of Rochester Library Bulletin* 35(1982): 8. Also see Daryl G. Watson, "Shade and

Ornamental Trees in the Nineteenth-Century Northeastern United States" (Ph.D. diss., University of Illinois, 1978), 201–2, for commentary on the importance of this geographical location.

29. See Grosso, "From the Genesee to the World," 7 and 9–11. For commentary on the rise of Rochester as a nursery center, see *Garden of the Genesee* (Rochester, NY: Rochester Historical Society, 1940), 11. Also see Blake McKelvey, "The Flower City: Center of Nurseries and Fruit Orchards," In *The Rochester Historical Society Publication XVIII* (Rochester, NY: Rochester Historical Society, 1923), 127; Blake McKelvey, *Rochester, The Flower City 1855–1890* (Cambridge: Harvard University Press, 1949); and Dan Parks, "The Cultivation of Flower City," *Rochester History* 45, no. 3–4 (1983): 25–45.

30. For a good discussion of the rise of Rochester seedsmen, see Parks, "The Cultivation of Flower City," 25–45. For specific commentary on James Vick, see Karl Sanford Kabelac, "Advice for Gardeners: Vick's Monthly Magazine (The First Series, 1878–1891)," *The University of Rochester Library Bulletin* 39 (1986): 24–35. For a history of the Joseph Harris seed firm, see William A. Aeberli and Margaret Becket, "Joseph Harris—Captain of the Rochester Seed Industry," *University of Rochester Library Bulletin* 35 (1982): 69–83. For mid-nineteenth-century commentary on the growth of Vick's, see "An Immense Seed Establishment," *Horticulturist* 22 (October 1867): 320; and "Seed Business of James Vick," *Horticulturist* 25 (April 1870): 118.

31. For a study of Kennicott and his accomplishments, see Erik A. Ernst, "John A Kennicott of The Grove, Physician, Horticulturist, and Journalist in Nineteenth-Century Illinois," *Illinois State Historical Society Journal* 74 (Summer 1981): 109–118. See Watson, "Shade and Ornamental Trees in the Nineteenth-Century Northeastern United States," 209–11, for commentary on the importance of Kennicott and his customer base.

32. For numbers of Illinois nurseries and their location, see "Nurserymen, Florists, Seedsmen, and Dealers in Horticultural Stock," *American Horticultural Annual 1871*, 147. Also see F. K. Phoenix, "Some Observations on the Climate and Soil, and the State of Horticulture in Wisconsin Territory," *Magazine of Horticulture* 11, no. 2 (February 1845): 56–8; *Wholesale Catalogue of the Bloomington Nursery, Bloomington Illinois, for the Fall of 1860 and Spring of 1861* (Bloomington: Steele and Carpenter, 1860), 1, KBP, 720; *General Descriptive Catalogue of Fruit and Ornamental Trees, Shrubs, Roses, Bulbs, Green-House and Garden Plants Cultivated and For Sale at the Bloomington Nursery* (Chicago: Daily Democrat Steam Printing House, 1859), 3–5, National Agricultural Library, United States Department of Agriculture, Beltsville, Maryland; "Bloomington Nursery," [advertisement], *American Agriculturist* 27 (October 1868): 382; and "Trees, Plants, Seeds," [advertisement], *American Agriculturist* 31 (March 1872): 115.

33. "Horticultural Register 1867, of Nurserymen, Fruit-Growers, Agents, Dealers and Publishers," *Transactions of the Illinois State Horticultural Society for 1866*, 112. See also Bailey, *The Standard Cyclopedia of Horticulture*, 1572; and Hedrick, *A History of Horticulture in America*, 325.

34. For commentary on the Fahnstock Nursery, see "The Fahnstock Nursery," *Michigan Farmer* 2 (30 June 1860): 103. For an overview of the number of Ohio

nurserymen, see "Nurserymen, Florists, Seedsmen, and Dealers in Horticultural Stock," *American Horticultural Annual 1871*, 146. Also see Hedrick, *A History of Horticulture in America*, 311.

35. For information on Hubbard and Davis, see Lyon, *History of Michigan Horticulture*, 251; "Death of A. C. Hubbard, Esq.," *Michigan Farmer* 2 (26 August 1871): 268; "Abridged Descriptive Catalogue of Fruit Trees, Ornamental Trees, Shrubs, and Greenhouse Plants Cultivated and For Sale by Hubbard and Davis," *Michigan Farmer* 2 (10 March 1860): 76–7; *Descriptive Catalogue of Fruits, Ornamental Trees, Flowering Shrubs, and Plants Cultivated and For Sale by Hastings, Hubbard and Davis at the Detroit and Oakland Horticultural Gardens* (Detroit: Bagg and Harmon, 1846), Michigan Library and Historical Center, Lansing, Michigan; *Descriptive Catalogue of Fruits, and Ornamental Trees, Shrubs, and Plants Cultivated and For Sale by Hubbard and Davis at the Detroit Horticultural Garden* (Detroit: Dunckler and Wales, 1853), Clarke Historical Library, Central Michigan University, Mount Pleasant, Michigan. For general information on early Michigan nurseries including Hubbard and Davis, see J. C. Holmes, "The Early History of Horticulture in Michigan," *Collection and Researches Made by the Pioneer Society of the State of Michigan*, 2nd ed., vol. 10 (Lansing: Wynkoop, Hollendeck, Crawford Co., 1908), 74–5.

36. For information on Adair's nurseries, see "Horticultural Notes by the Way—No. 16," *Michigan Farmer* 7 (1 December 1849): 350; and "Adair's Nursery," *Michigan Farmer* 2 (19 May 1860): 155. For an example of an Adair advertisement, see "Fruit and Ornamental Trees," [advertisement], *Michigan Farmer* 14 (May 1856): 96. See also *First Annual Descriptive Catalogue of Choice and Select Flower, Vegetable and Agricultural Seeds for sale by Wm. Adair and Co., 1871* (Detroit: Wm. Graham's Steam Press, 1871), 3, Clarke Historical Library, Central Michigan University, Mount Pleasant, Michigan.

37. See "Nurseries," *Michigan Farmer* 11 (August 1853): 243; and *Descriptive Catalogue of Ornamental Shrubs, Select Fruit Trees, Roses, Dahlias and Other Plants Cultivated and For Sale at the Elmwood Garden and Nursery* (Detroit: Barns, Brodhead and Co., 1852), 3–4, Michigan Library and Historical Center, Lansing, Michigan.

38. See "Monroe Nursery," [advertisement], *Michigan Farmer* 15 (December 1857): 383; and "The Nursery Business in the City of Monroe," *Michigan Farmer* 2 (July 1863): 30. For information on the Ilgenfritz Monroe Nursery, see H. Dale Adams, "Among the Vineyards and Orchards of Michigan," *Fourth Annual Report of the Secretary of the State Pomological Society of Michigan, 1874* (Lansing: W. S. George and Co., 1875), 138–9; "Michigan Nurseries," *Seventh Annual Report of the Secretary of the State Pomological Society of Michigan, 1877* (Lansing: W. S. George and Co., 1878), 270; "Nurseries," *Third Annual Report of the Secretary of the State Pomological Society of Michigan, 1873* (Lansing: W. S. George and Co., 1874), 214–5; Lyon, *History of Michigan Horticulture*, 213; and "The Nursery Business in the City of Monroe, Michigan," *Gardener's Monthly* 5, no. 9 (September 1863): 279.

39. For a brief history of the D.M. Ferry Company, see *Eighty Years of Growing 1856–1936* (Detroit: Ferry-Morse Seed Co., 1936), 6–7; and *The Seeds of To-*

morrow: Ferry-Morse Seed Company 1856–1956 (Detroit: Ferry-Morse Seed Co., 1956), 5–6. Also see Hedrick, *A History of Horticulture in America*, 322. For more detailed accounts of the early years of the company, see H. K. White, "D. M. Ferry and Company," typescript reminiscence of the company, Ferry Family Papers, Bentley Historical Library, University of Michigan, Ann Arbor, Michigan.

40. John Lightfoot, of Mercer County, Kentucky, for example, ran "two or three wagons" filled with fruit trees each spring, urging neighboring settlers to buy until his stock was gone. For accounts of the establishment of local Midwestern nurseries, see "Memoirs of the Pioneer Fruit Growers and Nurserymen of the Ohio Valley," *Annual Report of the Indiana Horticultural Society* (Indianapolis: R. J. Bright, 1877), especially 112–113, 122–3, and 124.

41. R. O. Thompson, Nebraska City, Nebraska Territory, to John Kennicott, The Grove, Illinois, 24 July 1860, KBP, 889; and Andrew L. Siler, Fountain Green, Utah Territory, to John Kennicott, The Grove, Illinois, 22 March 1862, KBP, 1183.

42. See "Fruit Trees-Shrubbery," [advertisement], *Michigan Farmer* 2 (15 April 1844): 40. See also "Ypsilanti Horticultural Garden and Nursery," [advertisement], *Michigan Farmer* 4 (December 1847): 168. For Lay's own account of the early years of his nursery, see Holmes, "The Early History of Horticulture in Michigan," 73.

43. "Pickings by the Way—No. 6," *Michigan Farmer* 10 (December 1852): 376. Also see Lyon, *History of Michigan Horticulture*, 255.

44. For commentary on the Ann Arbor Nursery, see "Greenhouse Plants," *Michigan State Journal* (19 April 1838); "Dahlias," *Michigan State Journal* (30 March 1841); "Ann Arbor Garden and Nursery," [advertisement], *Michigan Farmer* 3 (April 1845): 16.

45. For B. W. Steere's description of his early years as a nurseryman, see Lyon, *History of Michigan Horticulture*, 220–1. For an example of Steere's orders, see "Daybooks April 1854–April 1856," vol. 90, p. 85; and "Cash Books—March 1, 1854–July 19, 1856," vol. 56; and Jan. 1, 1859–Nov. 30, 1861," vol. 57, Ellwanger and Barry Company Papers, Department of Rare Books and Special Collections, Rush Rhees Library, University of Rochester, Rochester, New York.

46. See "Jackson Nursery," [advertisement], *Michigan Farmer* 15 (December 1857): 383; and "The Jackson Nursery," *Michigan Farmer* 15 (November 1857): 341; "Choice Seeds by Mail," [advertisement], *Michigan Farmer* 2 (3 March 1860): 71; "Garden Seeds," *Michigan Farmer* 2 (3 March 1860): 67; and "Trees," [advertisement], *Jackson Weekly Citizen*, 18 February 1873, 8; and "Oak Grove Nursery," [advertisement], *Jackson Weekly Citizen*, 17 February 1874, 3. For a listing of Jackson nurserymen and florists, see *Jackson City Directory 1873–4* (Jackson, MI: Polk, Murphy and Co., 1873), 244; *Jackson City Directory, 1876* (Jackson, MI: R. L. Polk and Co., 1876), 235; and *Jackson City Directory, 1883* (Detroit: R. L. Polk and Co., 1883), 310–311.

47. See "Rose, Rose, Roses," [advertisement], *Adrian Times and Expositor*, 26 March 1870; and "Fruit and Ornamental Trees and Shrubs," [advertisement], *Tecumseh Herald*, 4 February 1864. Also see "Edmiston's Nursery and Fruit Farm," *Adrian Times and Expositor*, 29 February 1872.

48. See, for example, "It Concerns Ladies Most," [advertisement], *Ypsilanti Commercial*, 12 April 1873. In these early decades of horticultural advertising, "gendered" notices like Toms' were relatively rare. For additional commentary on gender and advertising, see Ellen G. Garvey, *The Adman in the Parlor: Magazines and the Gendering of Consumer Culture, 1880s to 1910s* (New York: Oxford University Press, 1996), 8.

49. "Rose, Rose, Roses," [advertisement], *Adrian Times and Expositor*, 26 March 1870; and "Hendrick's Garden," *Ypsilanti Commercial*, 28 May 1870.

50. "Fruit and Ornamental Trees, Shrubs," [advertisement], *Tecumseh Herald*, 4 February 1864; and "Rose, Rose, Roses," [advertisement], *Adrian Times and Expositor*, 26 March 1870.

51. See Sarah P. Stetson, "The Traffic in Seeds and Plants from England's Colonies in North America," *Agricultural History* 23 (1949): 48–9. See also Ray Desmond, "Technical Problems in Transporting Living Plants in the Age of Sail," *Canadian Horticultural History* 1, no. 2 (1986): 74–90, for a discussion of the problems of sea transport, and the various solutions.

52. For an early history of the express industry, see Levi C. Weir, "The Express," in *One Hundred Years of American Commerce 1795–1895*, vol. 1, ed. Chauncey M. DePew (New York: D. O. Haynes and Co., 1895), 137–140; and Franklin W. Ball, "Just Express It," *Railroad Magazine* 47, no. 1 (October 1948): 94–103.

53. H. A. Terry, Crescent City, Iowa, to John Kennicott, The Grove, Illinois, 1 October 1861, KBP, 103; 31 October 1861, KBP, 1050; and 20 March 1862, KBP, 1171.

54. For a broad overview of postal reforms in the mid-nineteenth century, see Carl H. Scheele, *A Short History of the Mail Service* (Washington, DC: Smithsonian Institution Press, 1970), 73–96. See also *United States Domestic Postage Rates, 1789–1956* (Washington, DC: Post Office Department, 1956), 4–6, 10, 56–64.

55. "Annual Report of the Postmaster General 1860," in *Message from the President of the United States to the Two Houses of Congress at the Commencement of the Second Session of the 36th Congress*, vol. 3 (Washington, DC: George W. Bowman, 1860), 449. Serial Set 1080, 36th Congress, 2nd Session, December 3, 1860–March 2, 1861, Senate Executive Document I, part 3, Readex microprint. Congress was not so pleased with the Committee's work. They publicly criticized the group, called for an investigation, and cut appropriations. Holt resigned in response to the Congressional backlash. See Baker, Rasmussen, Wiser and Porter, *Century of Service*, 8–9.

56. See *United States Domestic Postage Rates, 1789–1956*, 33, and 58–9; and Scheele, *A Short History of the Mail Service*, 91.

57. For an insightful commentary on the role of the postal service in setting in motion a "communication revolution" and in creating a national marketplace for "commercial information," see Richard R. John, *Spreading the News: The American Postal System from Franklin to Morse* (Cambridge: Harvard University Press, 1995), 282. See also Wayne E. Fuller, *The American Mail: Enlarger of the Common Life* (Chicago: University of Chicago Press, 1972); and Theda Skocpol, "The Tocqueville Problem, Civic Engagement in American Democracy," *Social Science*

History 21, no. 4 (Winter 1997): 455–479. Skocpol comments on the role of the postal service in spreading ideas and promoting a national vision.

58. "Post-Office Rulings," *Gardener's Monthly* 15, no. 3 (March 1873): 86; "Seeds by Mail—The New Law," *American Agriculturist* 22 (May 1863): 143; and "A Nursery in Every Town," *American Agriculturist* 26 (October 1867): 352. For comments on the democratizing effect of the laws, see "Watch Congress!" *American Agriculturist* 33 (April 1874): 127.

59. The horticultural or agricultural press frequently published advice on methods of packing plants for shipping. See, for example, "Taking Up and Packing Trees and Plants," *American Agriculturist* 17 (April 1858): 117; "A Visit Among the Flowers," *Prairie Farmer* 13 (2 April 1864): 230; "Packing Living Plants," *American Agriculturist* 33 (March 1874): 102; and Henderson, *Practical Floriculture*, 188–9.

60. For examples on commentary on how to treat newly received plants, see "Care of Trees Received from the Nursery," *Horticulturist* 21 (November 1866): 344; "Plants Received by Mail—How to Treat," *American Agriculturist* 33 (December 1874): 459; and Edwin A. Johnson, *Winter Greeneries at Home* (New York: Orange Judd Co., 1878), 100–1.

Chapter Five: Commercial Realities

1. "Catalogues," *Gardener's Monthly* 4, no. 1 (January 1862): 24; and "Notes on Western Travel," *Gardener's Monthly* 9, no. 12 (December 1867): 369.

2. George Ellwanger, "Report on Ornamental Trees and Shrubs," *Third Annual Report of the Secretary of the State Pomological Society of Michigan, 1873* (Lansing: W. S. George and Co., 1874), 419–426; and James Vick, "Flowers at the Michigan State Fair," *Third Annual Report of the Secretary of the State Pomological Society of Michigan, 1873* (Lansing: W. S. George and Co., 1874), 225.

3. "List of Premiums," *Sixth Annual Report of the Secretary of the State Board of Agriculture of the State of Michigan for the Year 1867* (Lansing: W. S. George and Co., 1867), 335–6; and "List of Premiums," *Second Annual Report of the Secretary of the State Pomological Society of Michigan, 1872* (Lansing: W. S. George and Co., 1873), 258–260.

4. "Annual Meeting of the Michigan State Agricultural Society," *Michigan Farmer* 2 (30 September 1871): 308; and "Our Flowers at State Fairs," *Vick's Illustrated Floral Guide for 1873* (Rochester, NY: James Vick, 1873), 131, Department of Rare Books and Special Collections, Rush Rhees Library, University of Rochester, Rochester, New York.

5. "Vick's Special Floral Premiums," *Annual Report of the Indiana Horticultural Society* (Indianapolis: R. J. Bright, 1872), 143–4; and "Vick's Floral Premiums," *Vick's Illustrated Monthly Magazine* 1, no. 6 (June 1878): 192.

6. For examples of the many accounts of nursery visits, see "The Nurseries at Rochester, New York," *American Agriculturist* 23 (November 1864): 307; Dr. J. S. H., "A Trip to the New York Nurseries," *Gardener's Monthly* 2, no. 1 (January 1860): 8; "How Things Are Done in the West," *Prairie Farmer* 20 (16 November

1867): 310. *Moore's Rural New Yorker* also ran a series on "Eminent Horticulturists" that provided detailed biographical information, a portrait, and often a sketch of the nurseryman's homegrounds. See, for example, F. R. Elliott, "Eminent Horticulturists-VII-George Ellwanger," *Moore's Rural New Yorker* 24 (8 July 1871): 9. Also see Mr. J. W. Wood, "One Hour at the Nurseries of Messrs. Hoopes and Company, Westchester, Pennsylvania," *Gardener's Monthly* 12, no. 8 (August 1870): 227.

7. For typical examples of this type of correspondence, see J. Griffen, Prairie City, Illinois, to John Kennicott, The Grove, Illinois, 17 March 1860, KBP, 780; J. B. Hamilton, Neenah, Wisconsin, to John Kennicott, The Grove, Illinois, 14 April 1862, KBP, 1251; Julius Crone, Chicago, Illinois, to John Kennicott, The Grove, Illinois, 21 November 1859, KBP, 706; and F. Fairman, Chicago, Illinois, to John Kennicott, The Grove, Illinois, 14 November 1860, KBP, 927.

8. See *American Agriculturist* 17 (March 1858): 92–94; and *American Agriculturist* 23 (March 1874): 109–120. In both instances, the numbers represent only advertisements for ornamental plants. A number of historians have documented the history of advertising and have emphasized its increasing importance in the last quarter of the nineteenth century, but almost without exception have ignored the early advertising efforts of the horticultural industry and its effect on establishing a national market for horticultural goods. See, for example, Daniel Pope, *The Making of Modern Advertising* (New York: Basic Books, 1983); Charles A. Goodrum and Helen Dalrymple, *Advertising in America, The First 200 Years* (New York: Harry N. Abrams, 1990); Robert Jay, *The Trade Card in Nineteenth-Century America* (Columbia: University of Missouri Press, 1987); James D. Norris, *Advertising and the Transformation of American Society, 1865–1920* (New York: Greenwood Press, 1990); and Frank Presbrey, *The History and Development of Advertising* (New York: Doubleday and Co., 1929, reprint New York: Greenwood Press, 1968).

9. For examples of early regional advertising, see "Prince's Linnean Botanic Garden," [advertisement], *Michigan Farmer* 2 (1 February 1845): 168; "Ypsilanti Horticultural Garden and Nursery,"[advertisement], *Michigan Farmer* 4 (December 1847): 165; "Buffalo Nursery and Horticultural Garden," and "Detroit Nursery and Garden," [advertisements], *Michigan Farmer* 5 (August 1847): 96. See the *Michigan Farmer* 11 (September 1853): 283, for an example of a Genesee Valley Nursery and Syracuse Nursery advertisement; *Michigan Farmer* 14 (September 1856): 283, for an early Ellwanger and Barry advertisement; *Michigan Farmer* 16 (March 1858): 91, for an early Thorburn advertisement; *Michigan Farmer* 1 (1 October 1859): 319, for an Andre Leroy Nursery advertisement; *Michigan Farmer* 4 (8 March 1862): 76, for an early Vick's advertisement; and *Michigan Farmer* 2 (23 December 1871): 404, for an early Briggs and Brothers advertisement.

10. For examples of these advertisements, see *Adrian Times and Expositor*, 3 January 1868; 26 February 1869; 9 September 1869; 30 September 1871; 17 January 1872; 16 March 1872; 27 September 1872; and 12 March 1873.

11. For examples of advertisements in the *Jackson Weekly Citizen*, see 23 March 1869; 14 March 1871; 12 March 1872; 18 March 1873; 14 October 1873; and 28 April 1874.

12. For examples of these advertisements, see *Tecumseh Herald*, 8 February 1866; *Dexter Leader*, 15 April 1869; 21 February 1873; and 19 March 1875. Circulation figures are drawn from "Table VI-Periodicals," *Statistics of the State of Michigan Collected for the Ninth Census of the United States, June 1, 1870* (Lansing: W. S. George and Co., 1873), 670–676.

13. "Advertisements," *Michigan Farmer* 4 (June 1846): 56; and "Our Nursery Advertisements," *Michigan Farmer* 11 (March 1853): 81.

14. "Horticultural Notes," *Michigan Farmer* 16 (March 1858): 80; and "The Toledo Nurseries," *Michigan Farmer* 15 (March 1857): 89.

15. "Packing Plants," *Vick's Illustrated Monthly Magazine* 1, no. 4 (April 1878): 116.

16. John L. Wilson, Chicago, Illinois, to John Kennicott, The Grove, Illinois, 27 March 1862, KBP, 1193.

17. "Detroit Nursery," [advertisement], *Michigan Farmer* 10 (April 1852): 126; and "Hubbard and Davis Abridged Descriptive Catalogue of Fruit Trees, Ornamental Trees, Shrubs, and Greenhouse Plants," *Michigan Farmer* 2 (10 March 1860): 76.

18. "Lovers of Flowers, Attention!," [advertisement], *American Agriculturist* 23 (March 1864): 90; and "Plants by Mail," [advertisement], *American Agriculturist* 33 (March 1874): 111. For commentary on the limitations of early advertisements and the development of display advertising, see Goodrum and Dalrymple, *Advertising in American*, 20; Presbrey, *The History and Development of Advertising*, 232–252; and Pope, *The Making of Modern Advertising*, 235.

19. "Vick's," [advertisement], *Michigan Farmer* 2 (March 1864): 431; and "Fruit and Ornamental Trees for Spring of 1861," [advertisement], *Prairie Farmer* 7 (28 February 1861): 144.

20. "Vick's Illustrated Catalogue of Seeds and Guide to Flower Gardens," *Horticulturist* 20 (February 1865): 60. Few historians have documented the history of nursery catalogues, but several have pointed out the need and their richness as a source of information on marketing and horticultural history. For commentary on the history of nursery catalogues, see Van Ravenswaay, "Horticultural Heritage—The Influence of the U.S. Nurserymen: A Commentary," 143–145; Jane Gates, "Old Nursery Catalogues," *Pacific Horticulture* 42, no. 2 (1981): 9–11; and Jayne T. Maclean, "Nursery and Seed Trade Catalogues," *Journal of the NAL Associates* 5, no. 3–4 (July/December 1980): 88–92. See also Thomas J. Schlereth, "Country Stores, County Fairs, and Mail-Order Catalogues; Consumption in Rural America," in *Consuming Visions, Accumulation and Display of Goods in America, 1880–1920*, ed. Simon J. Bronner (New York: W. W. Norton and Co., 1989), 364–375.

21. "Catalogues," *Michigan Farmer* 5 (August 1847): 95; and "The Combined Catalogue," *Ypsilanti Commercial*, 22 February 1873.

22. "Nursery Catalogues," *Gardener's Monthly* 12, no. 3 (March 1870): 88.

23. *H. A. Dreer's General Plant Catalogue for 1860* (Philadelphia), National Agricultural Library, United States Department of Agriculture, Beltsville, Maryland.

24. For examples of these varying forms, see *Washburn and Co.'s Amateur Cultivator's Guide to the Flower and Kitchen Garden: Containing a Descriptive List of Two Thousand Varieties of Flower and Vegetable Seeds; Also a List of French Hybrid Gladiolus* (Boston: Washburn and Co., 1868); and *D. M. Ferry and Co.'s Illustrated, Descriptive, and Priced Catalogue of Garden, Flower and Agricultural Seeds* (Detroit: Gulley's Print, 1877), Clarke Historical Library, Central Michigan University, Mount Pleasant, Michigan. Also see "D. M. Ferry and Co.'s Illustrated, Descriptive, and Priced Catalogue," [advertisement], *Michigan Farmer* 3 (22 February 1872): 61.

25. "Vick's Floral Guide for 1872," [advertisement], *Michigan Farmer* 3 (11 January 1872): 13. For an example of Vick's presentation and comments on the quarterly publication, see *Vick's Floral Guide for 1875* (Rochester, NY: Charles F. Muntz and Co., 1875), especially 128, Department of Rare Books and Special Collections, Rush Rhees Library, University of Rochester, Rochester, New York.

26. *Vick's Illustrated Floral Guide for 1873* (Rochester, NY: James Vick, 1873), 4, Department of Rare Books and Special Collections, Rush Rhees Library, University of Rochester, Rochester, New York.

27. For several examples of clubs and their terms, see *A Descriptive Catalogue of a Choice Collection of Flower, Vegetable and Agricultural Seeds Cultivated and For Sale by Benjamin K. Bliss* (Springfield, MA: Samuel Bowles and Co., 1860), 8, National Agricultural Library, United States Department of Agriculture, Beltsville, Maryland; and *Vick's Illustrated Floral Guide for 1865* (Rochester, NY: James Vick, 1865), 3, Department of Rare Books and Special Collections, Rush Rhees Library, University of Rochester, Rochester, New York.

28. See *A Descriptive Catalogue of a Choice Collection of Flower, Vegetable and Agricultural Seeds Cultivated and For Sale by Benjamin K. Bliss* (1860), 6; and *A Descriptive Catalogue of a Choice Collection of Flower, Vegetable and Agricultural Seeds Cultivated and For Sale by Benjamin K. Bliss* (1872). For commentary on the use of color prints as promotional devices, see Van Ravenswaay, "Horticultural Heritage—The Influence of the U.S. Nurserymen: A Commentary," 147; and Gates, "Old Nursery Catalogues," 10.

29. See *A Descriptive Catalogue of a Choice Collection of Flower, Vegetable and Agricultural Seeds Cultivated and For Sale by Benjamin K. Bliss* (1872), 104–141.

30. For examples of this familiar, friendly tone, see *Washburn and Co.'s Amateur Cultivator's Guide to the Flower and Kitchen Garden* (1868), 11; and *Vick's Illustrated Floral Guide for 1871* (Rochester, NY: James Vick, 1871), 1, Department of Rare Books and Special Collections, Rush Rhees Library, University of Rochester, Rochester, New York.

31. See "Briggs and Brothers," [advertisement], *Michigan Farmer* 2 (23 December 1871): 404; and *A Descriptive Catalogue of a Choice Collection of Flower, Vegetable and Agricultural Seeds Cultivated and For Sale by Benjamin K. Bliss* (1860), 46; and *A Descriptive Catalogue of a Choice Collection of Flower, Vegetable and Agricultural Seeds Cultivated and For Sale by Benjamin K. Bliss* (1872), 9. Also see Daniel Boorstin, "Welcome to the Consumption Community," *Fortune* 76 (1 September 1967): 118; and Daniel Boorstin, *The Americans: The Democratic Experiences* (New York: Random House, 1973; reprint, New York: Vintage

Books, 1974), 89–90, and 118–119. Boorstin suggests that the consumption of certain items became a way to form links between individuals regardless of other social or cultural characteristics.

32. *Vick's Illustrated Catalogue and Floral Guide for the Spring of 1865*, i–iv; *Catalogue of New and Rare Green House Plants, Evergreen Shrubs, Bedding Plants, Dahlias, Verbenas, Roses, Petunias, Fuchsias, Chrysanthemums, Phlox, Geraniums, Bulbs, Etc., Grown and For Sale by Hubbard and Davis* (Detroit: William H. Thompson, 1871), 8–9, Clarke Historical Library, Central Michigan University, Mount Pleasant, Michigan; and *Dreer's Garden Calendar for 1868* (Philadelphia: Henry A. Dreer, 1868), National Agricultural Library, United States Department of Agriculture, Beltsville, Maryland.

33. *Descriptive Catalogue of Select Fruit Trees, Ornamental Trees, Shrubs, Roses, and Other Plants Cultivated and For Sale at the Mount Hope Nurseries, Ellwanger and Barry, Proprietors* (Rochester, NY: Press of Lee, Mann and Co., 1850–1851), Ellwanger and Barry Company Papers, Department of Rare Books and Special Collections, Rush Rhees Library, University of Rochester, Rochester, New York.

34. See *Vick's Illustrated Catalogue and Floral Guide for 1873*, 21–24; and *D. M. Ferry and Co.'s Illustrated, Descriptive, and Priced Catalogue of Garden, Flower, and Agricultural Seeds* (1877), 1–7; and *D. M. Ferry and Company, 1881 Seed Annual* (Detroit: Calvert Lith. Co., 1881), 2–14, Clarke Historical Library, Central Michigan University, Mount Pleasant, Michigan.

35. James Vick typified the practice of many other commercial horticulturists by offering the catalogue free to customers, and for a nominal fee to others who could then subtract the price of the catalogue from their first order. See, for example, *Vick's Floral Guide for 1875* (Rochester: Charles F. Muntz and Co., 1875), especially 128, Department of Rare Books and Special Collections, Rush Rhees Library, University of Rochester, Rochester, New York.

36. A number of scholars have explored itinerancy as a nineteenth-century marketing device. For a general background on types of peddlers and their techniques, see David Jaffee, "Peddlers of Progress and the Transformation of the Rural North, 1760–1860," *Journal of American History* 78 (September 1991): 511–535; Peter Benes, ed., *Itinerancy in New England and New York, The Dublin Seminar for New England Folklife, Annual Proceedings 1984* (Boston: Boston University Press, 1986); J. R. Dolan, *The Yankee Peddlers of Early America* (New York: Clarkson N. Potter, 1964); Richardson Wright, *Hawkers and Walkers in Early America, Strolling Peddlers, Preachers, Lawyers, Doctors, Players, and Others, from the Beginning to the Civil War* (Philadelphia: J. B. Lippincott Co., 1927); and Gerald Carson, "The Indomitable Peddler," in *Readings in the History of American Marketing, Settlement to Civil War*, compiled by Stanley J. Shapiro and Alton F. Doody (Homewood, IL: Richard D. Irwin, 1986), 321–330. For specific comments on peddlers' roles in expanding markets, see R. Malcolm Keir, "The Tin-Peddler," *Journal of Political Economy* 21 (1913): 255–258; Fred Mitchell Jones, *Middlemen in the Domestic Trade of the United States 1800–1860*, Illinois Studies in the Social Sciences 21, no. 3 (Urbana: University of Illinois, 1937), 12; Lee M. Friedman, "The Drummer in Early American Merchandise Distribution," *Bulletin of the Business Historical Society* 21 (April 1947): 39; Glenn Porter and

Harold C. Livesay, *Merchants and Manufacturers, Studies in the Changing Structure of Nineteenth-Century Marketing* (Baltimore: Johns Hopkins University Press, 1971), 4 and 11. See also Timothy B. Spears, *100 Years on the Road, The Traveling Salesman in American Culture* (New Haven: Yale University Press, 1995), 79, for a similar assessment of traveling salesmen and their role in extending consumer culture. For one of the few assessments of plant peddling in the historical literature, see Earl W. Hayter, "Horticultural Humbuggery Among the Western Farmers, 1850–1890," *Indiana Magazine of History* 43, no. 3 (September 1947): 205–224.

37. See Barbara W. Sarudy, "Nurserymen and Seed Dealers in the Eighteenth-Century Chesapeake," *Journal of Garden History* 9, no. 3 (1989): 111–117; and James Hirn to W. R. Prince, 2 December 1823, folder 132; and Joseph Simmons to W. R. Prince, 28 August 1818, folder 80, The Prince Family Manuscript Collection, National Agricultural Library, United States Department of Agriculture, Beltsville, Maryland.

38. See Grosso, "From the Genesee to the World," 7 and 9–11; and "Purchasing Fruit Trees," *Moore's Rural New Yorker* 14 (21 November 1863): 375.

39. See, for example, *Abridged Catalogue of Select Fruit and Ornamental Trees Cultivated and Sold by Ellwanger and Barry Mt. Hope Nurseries, Rochester, New York for 1860* (Rochester, NY: Steam Press of A. Strong and Co., 1860) Ellwanger and Barry Company Papers, Department of Rare Books and Special Collections, Rush Rhees Library, University of Rochester, Rochester, New York; and "Purchasing Fruit Trees," *Moore's Rural New Yorker* 14 (21 November 1863): 375.

40. "Nurserymen's Association," *Prairie Farmer* 3 (19 May 1859): 311; and "Dishonest Tree Peddlers," *Horticulturist* 26 (March 1871): 45.

41. "Purchasing Fruit Trees," *Moore's Rural New Yorker* 14 (21 November 1863): 375; and "How Shall We Preserve Ourselves from the Wiles of the Fruit Tree Agent?" *Seventh Annual Report of the Secretary of the State Pomological Society of Michigan, 1877* (Lansing: W. S. George and Co., 1878), 232. For examples of varying arrangements, see William S. Green, Green Bay, Wisconsin, to John A. Kennicott, The Grove, Illinois, 8 March 1862; and 10 April 1862, KBP, 1141 and 1240; and Levi Thumb, Irving, Illinois, to John Kennicott, The Grove, Illinois, 26 February 1860, KBP, 754.

42. "Bulbs and Bulb Peddlers," *American Agriculturist* 29 (October 1870): 383.

43. "How Shall We Preserve Ourselves from the Wiles of the Fruit Tree Agent?" *Seventh Annual Report of the Secretary of the State Pomological Society of Michigan, 1877*, 232.

44. D. M. Dewey, *The Tree Agents' Private Guide: A Manual for the Successful Work in Canvassing for the Sale of Nursery Stock* (Rochester, NY: D. M. Dewey, 1875), 12. See also Karl S. Kabelac, "Nineteenth-Century Rochester Fruit and Flower Plates," *The University of Rochester Library Bulletin* 35 (1982): 99 and 93–4; and Charles Van Ravenswaay, *Drawn from Nature. The Botanical Art of Joseph Prestele and His Sons* (Washington, DC: Smithsonian Institution Press, 1984); Charles Van Ravenswaay, "Drawn and Colored from Nature. Painted Nurserymen's Plates," *Magazine Antiques* 123 (January–March 1983): 594–599; and Carl W. Drepperd, "The Tree, Fruit and Flower Prints of D. M. Dewey, Rochester, New York from 1844," *Spinning Wheel* 12, no. 5 (May 1956): 12–15, 46.

45. D. M. Dewey, *The Tree Agents' Private Guide*, 7–9, 21, 29, 22, 15 and 18.

46. Untitled, *Vick's Illustrated Monthly Magazine* 2, no. 10 (October 1879): 289.

47. Harrington, *How 'Tis Done*, 246 and 248; and H. Dale Adams, "Fruit Culture on the Farm," *Sixteenth Annual Report of the Secretary of the State Board of Agriculture of the State of Michigan* (Lansing: W. S. George and Co., 1878), 321.

48. For a variety of examples of accusations against plant peddlers, see "Tree Peddlers," *Gardener's Monthly* 16, no. 5 (May 1874): 133; "Beware of Traveling Tree Peddlers," *American Agriculturist* 22 (May 1863): 186; "Imposition Exposed," *Horticulturist* 10 (May 1855): 235; "'Hard Times' Tree Sellers," *American Agriculturist* 20 (April 1861): 104; and "Imposition," *Horticulturist* 10 (May 1855): 142.

49. Levi Thumb, Irving, Illinois, to John Kennicott, The Grove, Illinois, 26 February 1860, KBP, 754.

50. E. P. Powell, "Concerning Nursery Agents," *Adrian Times and Expositor*, 4 January 1860.

51. See, for example, William S. Green, Green Bay, Wisconsin to John Kennicott, The Grove, 8 March 1862, KBP, 1141; 27 March 1862, KBP, 1194; and 10 April 1862, KBP, 1240.

52. "Tree Agents," *Prairie Farmer* 45 (24 October 1874): 338.

53. Andrew S. Fuller, "Humbugs in Horticulture," *Horticulturist* 24 (June 1869): 169.

54. For an insightful commentary on advertising as an important technique for transferring meaning from cultural tastemakers to an object, see Grant McCracken, "Culture and Consumption: A Theoretical Account of the Structure and Movement of the Cultural Meaning of Consumer Goods," in *Perspectives in Consumer Behavior*, 4th ed., ed. Harold H. Kassarjian and Thomas S. Robertson, 581–599. Historians of advertising have suggested that until the late nineteenth century, most advertisers presumed a demand for their products and used advertisements to supply information about cost and availability of their products. Gradually, however, competitive pressures and an expanding array of consumer goods encouraged a new advertising strategy. Advertisers worked to create new demands and develop consumer wants along with simply supplying information. For commentary on this change in advertising strategy, see Norris, *Advertising and the Transformation of American Society*, xviii; and David M. Potter, *People of Plenty. Economic Abundance and the American Character* (Chicago: University of Chicago Press, 1954), 173–5.

55. See *Vick's Illustrated Catalogue and Floral Guide for 1871*, 9; and *A Descriptive Catalogue of a Choice Collection of Flower, Vegetable and Agricultural Seeds Cultivated and For Sale by Benjamin K. Bliss* (1860), 46.

56. See, for example, "Bloomington Nursery," [advertisement], *Michigan Farmer* 2 (23 September 1871): 301; and "Fine Mixed Tulips for Fall," [advertisement] *Michigan Farmer* 2 (2 September 1871): 277. Also see *D. M. Ferry and Co.'s Illustrated, Descriptive, and Priced Catalogue of Garden, Flower and Agricultural Seeds* (1877), 5; and *Vick's Illustrated Catalogue and Floral Guide for 1865*, 2.

57. See *Washburn and Co.'s Amateur Cultivator's Guide to the Flower and Kitchen Garden* (1868), 42–43.

58. See, for example, "Flower Seeds," [advertisement], *Michigan Farmer* 16 (1858): 90; "Flower and Garden Seeds," [advertisement], *Michigan Farmer* 2 (March 1864): 434; "Toledo Nursery," [advertisement], *Michigan Farmer* 12 (1854): 288; "William Adair," [advertisement], *Michigan Farmer* 1 (5 November 1859): 359; *Washburn and Co.'s Amateur Cultivator's Guide to the Flower and Kitchen Garden* (1868), 11; and "Advertisement," *Descriptive Catalogue of Hardy Ornamental Trees, Shrubs, Roses, Etc., Cultivated and For Sale at the Mount Hope Nurseries, Rochester, New York* (1860), n. p.; and *D. M. Ferry and Co.'s Illustrated, Descriptive, and Priced Catalogue of Garden, Flower and Agricultural Seeds* (1877), 62.

59. Dewey, *The Tree Agents' Private Guide*, 15; and *Washburn and Co.'s Amateur Cultivator's Guide to the Flower and Kitchen Garden* (1868), 50–51.

60. See "Hubbard and Davis Abridged Descriptive Catalogue of Fruit Trees, Ornamental Trees, Shrubs, and Greenhouse Plants," *Michigan Farmer* 2 (10 March 1860): 76; and *Descriptive Catalogue of a Choice Collection of Vegetable, Agricultural, and Flower Seeds, For Sale by B. K. Bliss* (1872), 57 and 27.

61. Bateham, "Floriculture for the Million," 416 and 419–20.

Chapter Six: A Neighborly Nudge

1. "Visit to the Farm of Brother Benjamin Miller by Committee of W. E. H. Sobes and J. Evarts," 14 May, No Year [probably 1874], Papers of the Patrons of Husbandry, Ypsilanti Grange, No. 56, 1873–1959, Bentley Historical Library, University of Michigan, Ann Arbor, Michigan.

2. See, for example, Hon. Fielder S. Snow, "Address Delivered Before the Lenawee County Fair, October 7, 1852," *Transactions of the State Agricultural Society with Reports of the County Agricultural Societies for 1852* (Lansing: Geo. W. Peck, 1853), 353; and F. L., "Agricultural Societies—Their Benefits," *Michigan Farmer* 4 (18 January 1862): 21.

3. "Report of the Jackson County Agricultural Society," *Transactions of the State Agricultural Society with Reports of the County Agricultural Societies for 1854* (Lansing: Geo. W. Peck, 1855), 428; and "Horticultural Societies and the Horticultural Community," *Gardener's Monthly* 4, no. 3 (March 1862): 84.

4. See Sally McMurry, *Families and Farmhouses in Nineteenth-Century America*, 4. McMurry uses the term "cultural mediator" to describe the progressive farm families who initiated cultural and material changes in the mid-nineteenth-century rural north. For commentary on additional rural institutions which acted as cultural mediators, see Thomas J. Schlereth, "Country Stores, County Fairs, and Mail-Order Catalogues: Consumption in Rural America," in *Consuming Visions, Accumulation and Display of Goods in America, 1880–1920*, ed. Simon J. Bronner (New York: W. W. Norton and Co., 1989), 339–375.

5. For several examples of travel experiences related in the agricultural press, see Warren Isham, "The Great Exhibition," *Michigan Farmer* 10 (January 1852):

20–21; A. C. H. "Massachusetts Horticultural Society," *Michigan Farmer* 1 (24 September 1859): 307; and "Editorial Correspondence," *Prairie Farmer* 8 (October 1848): 298–300. For commentary on the agricultural press and its use of travel and local descriptions to further its aims, see Danhof, *Change in Agriculture*, 53 and 58.

6. S. B. Noble, "Rambles—No. 2," *Michigan Farmer* 11 (September 1853): 277–280; "The Residence of Wm. Burnett Esq.—Apple Trees Dying—A New Variety of Wheat," *Michigan Farmer* 14 (August 1856): 234; and "The Farm of D. L. Quirk, Esq. of Ypsilanti," *Michigan Farmer* 2 (1 April 1871): 100.

7. "An Excursion—Ann Arbor," *Michigan Farmer* 1 (1 August 1843): 92; and S. B. Noble, "City of Ann Arbor," *Michigan Farmer* 12 (June 1854): 169. For an interpretation of the advocacy of nature as a way to overcome the negative elements of urbanization and industrialization, see Thomas Bender, *Toward an Urban Vision*; and James L. Machor, *Pastoral Cities*.

8. Examples of comments in support of associations and their many activities include "Howell Farmer's Club," *Michigan Farmer* 12 (June 1854): 181, "Town Horticultural Societies," *American Agriculturist* 17 (May 1858): 149; "County Fairs," *Jackson Weekly Citizen*, 23 September 1873, 5; and Henry Colman, *Agriculture and Rural Economy from Personal Observation*, vol. 1 (Boston: Phillips, Sampson and Co., 1856), 31.

9. For information on formation, membership, and visiting committees of the Ypsilanti Grange, see Papers of the Patrons of Husbandry, Ypsilanti Grange, No. 56, 1873–1959, Bentley Historical Library, University of Michigan, Ann Arbor, Michigan. All examples of grange visits are drawn from Papers of the Patrons of Husbandry. For general commentary on the early history of the Patrons of Husbandry and its growth in 1873 due to widespread economic depression, see David B. Danbom, *Born in the Country, A History of Rural America* (Baltimore: Johns Hopkins University Press, 1995): 154–6; and R. Douglas Hurt, *American Agriculture, A Brief History*, rev. ed. (West Lafayette, IN: Purdue University Press, 2002), 203–207; D. Sven Nordin, *Rich Harvest: A History of the Grange, 1867–1900* (Jackson: University of Mississippi Press, 1974); Fred Trump, *The Grange in Michigan, An Agricultural History of Michigan Over the Past 90 Years* (Grand Rapids, MI: Published by the Author, Dean-Hick Co., 1963), 1–26; and Charles M. Gardner, *The Grange—Friend of the Farmer* (Washington, DC: National Grange, 1949).

10. See "Visit to the Farm of Brother Wm. Watling," 12 June 1875; and "Visit to the Farm of Brother Benjamin Miller by Committee of W. E. H. Sobes and J. Evarts," 14 May, No Year.

11. See "Visit to the Farm of B. D. Loomis," 17 June 1875; "Visit to the Home of Brother Randall," 11 August 1877; and "Visit to the Farm of Henry Preston," 21 May 1875.

12. J. C. Holmes, "Detroit Horticultural Society," *Michigan Farmer* 6 (1 May 1848): 129–130; and "The Coming Fair," *Jackson Daily Citizen*, 27 September 1873. A number of scholars have documented the rise of agricultural societies and exhibitions in the United States. See, for example, Danhof, *Change in Agriculture*, 60–64; Kenyon L. Butterfield, "Farmers' Social Organizations," in *Cyclopedia of*

American Agriculture, vol. 4, ed. Liberty Hyde Bailey (New York: MacMillan Co., 1909), 289–297; Wayne Caldwell Neely, *The Agricultural Fair* (New York: Columbia University Press, 1935); Marti, *To Improve the Soil and the Mind, Agricultural Societies, Journals and Schools in the Northeastern States, 1791–1865*; Donald B. Marti, *Historical Directory of American Agricultural Fairs* (New York: Greenwood Press, 1986); Mark A. Mastromarino, "Elkanah Watson and Early Agricultural Fairs, 1790–1860," *Historical Journal of Massachusetts* 17 (Summer 1989): 105–118; Chris A. Rasmussen, "State Fair: Culture and Agriculture in Iowa, 1854–1941" (Ph.D. diss., Rutgers University, 1992); and Leslie Prosterman, *Ordinary Life, Festival Days: Aesthetics in the Midwestern County Fair* (Washington, DC: Smithsonian Institution Press, 1995), 42–54. For a concise account of the history of several Michigan county fairs and agricultural societies, see Julie Ann Avery, "An Exploration of Several Early Michigan County Fairs as Community Arts Organizations of the 1850s, 1860s, and 1870s" (Ph.D. diss., Michigan State University, 1992), especially 67–75.

13. For a history of the Lenawee County Agricultural Society, see "Lenawee County Agricultural Society," *Transactions of the Michigan State Agricultural Society with Reports of County Agricultural Societies for 1849* (Lansing: R. W. Ingalls, 1850), 191–200; "Report of the Lenawee County Agricultural Society," *Transactions of the Michigan State Agricultural Society with Reports of County Agricultural Societies for 1851* (Lansing: Ingalls, Hedges and Co., 1852), 395–7; "Report of the Lenawee County Agricultural Society," *Transactions of the Michigan State Agricultural Society with Reports of County Agricultural Societies for 1853* (Lansing: G. W. Peck, 1854), 507–9; "Lenawee County," *Transactions of the Michigan State Agricultural Society with Reports of County Agricultural Societies for 1854* (Lansing: Hosmer and Fitch, 1855), 580. See also Richard I. Bonner, ed., *Memoirs of Lenawee County, Michigan*, vol. 1 (Madison, WI: Western Historical Association, 1909), 673–4.

14. For an account of a Washtenaw County Agricultural Society Fair in 1843, see "Washtenaw County Agricultural Society," *Michigan Farmer* 1 (15 August 1843): 99. For details of the official formation of the society, see "Washtenaw Agricultural Society," *Transactions of the Michigan State Agricultural Society with Reports of County Agricultural Societies for 1849* (Lansing: R. W. Ingalls, 1850), 231. For a description of Washtenaw County Agricultural Society activities and transactions over the years, see *History of Washtenaw County, Michigan*, 536–546. For Finley's list of Agricultural Society members, see William Finley Record Book, 1849–1873, Bentley Historical Library, University of Michigan, Ann Arbor, Michigan. See also "Life Members of the Washtenaw County Agricultural Society," in *Washtenaw County Agricultural and Horticultural Society List of Premiums and Regulations, 20th Annual Fair, October 7–9, 1868* (Ann Arbor: Dr. Chase's Steam Printing House, 1868), 22–24, Michigan Collection, Michigan Library and Historical Center, Lansing, Michigan.

15. See "Report of the Jackson County Agricultural Society," *Transactions of the Michigan State Agricultural Society with Reports of County Agricultural Societies for 1853* (Lansing: G. W. Peck, 1854), 427–8; *History of Jackson County, Michigan*, 433–443; and "Societies for Promoting Agricultural, State Boards, Etc., Re-

port of the Commissioner of Patents for the Year 1858, Agriculture (Washington, DC: James B. Steedman, 1859), 150.

16. For commentary on the improvement of fairgrounds in Lenawee County, see "Lenawee County," *Transactions of the Michigan State Agricultural Society with Reports of County Agricultural Societies for 1854* (Lansing: Hosmer and Fitch, 1855), 580; "Lenawee County," *Fourth Annual Report of the State Board of Agriculture of the State of Michigan* (Lansing: John A. Kerr and Co., 1865), 288; and "Our County Fair," *Adrian Times and Expositor*, 30 September 1869. For commentary on improvements in Washtenaw County, see *History of Washtenaw County, Michigan*, 557; and "The Washtenaw County Agricultural Society," *Michigan Farmer* 2 (22 July 1871): 228. For comments on Jackson County, see "Jackson County," *Transactions of the State Agricultural Society of Michigan with Reports of the County Agricultural Societies for the Year 1857* (Lansing: Hosmer and Kerr, 1859), 388–389; and "Jackson County Agricultural Society—What Remains to Be Done," *Jackson Daily Citizen*, 12 October 1869, 8.

17. "Lenawee County," *Michigan Farmer* 14 (May 1856): 155; and "A New Agricultural Society," *Michigan Farmer* 1 (10 December 1859): 397; and "The Washtenaw and Wayne Union Agricultural Society," *Michigan Farmer* 1 (6 August 1859): 252. For commentary on the Farmers' Social Club in Tecumseh and other farmers' clubs in the region, see Clara Waldron, *One Hundred Years a County Town, The Village of Tecumseh, Michigan, 1824–1924* (n. p.: Thomas A. Riordan, 1968), 128–9. For a nineteenth-century commentary on the rise and purpose of farmers' clubs, see Jonathan Periam, *The Groundswell* (Cincinnati: E. Hannaford and Co., 1874).

18. See "Hillsdale and Lenawee Union Fair," *Michigan Farmer* 2 (September 1863): 102; and "The Fair at Tecumseh," *Michigan Farmer* 14 (October 1856): 314.

19. See J. C. Holmes, "Early History of the State Agricultural Society," *Fourth Annual Report of the Secretary of the State Pomological Society of Michigan, 1874* (Lansing: W. S. George and Co., 1875), 381–386; "Societies for Promoting Agricultural, State Boards, Etc., *Report of the Commissioner of Patents for the Year 1858, Agriculture* (Washington, DC: James B. Steedman, 1859), 147; and S. B. McCracken, ed., *Michigan and the Centennial* (Detroit: Printed for the Publisher S. B. McCracken at the Office of the Detroit Free Press, 1876), 637–8. Also see Lyon, *History of Michigan Horticulture*, 193–197. Residents of the three-county area were often listed as officers or executive committee members in the annual reports of the State Agricultural Society. See, for example, "The Fifth Annual Fair," *Transactions of the State Agricultural Society with Reports of the County Agricultural Societies for 1853* (Lansing: G. W. Peck, 1854), n. p.; and "Report of the Secretary of the State Agricultural Society of Michigan," *Seventh Annual Report of the Secretary of the State Board of Agriculture of the State of Michigan for the Year 1868* (Lansing: John A. Kerr and Co., 1868), 336.

20. E. D. Lay, Washtenaw County, to J. C. Holmes, Detroit, Michigan, 23 August 1849, J. C. Holmes Papers, Folder 2, Burton Historical Collection, Detroit Public Library, Detroit, Michigan.

21. See McCracken, *Michigan and the Centennial*, 637–8.

22. For a general account of the history of the Detroit Horticultural Society, see Holmes, "The Early History of Horticulture in Michigan," 75–95. The Society's Preamble and Constitution were reprinted along with a description of the founding by J. C. Holmes in "May Meeting," *Second Annual Report of the Secretary of the State Pomological Society of Michigan, 1872* (Lansing: W. S. George and Co., 1873), 96–103. For a listing of Detroit Horticultural Society members, see "List of Members and Accounts April 16, 1846–December 28, 1852," Detroit Horticultural Society Account Book, University Archives and Historical Collection, Michigan State University, East Lansing, Michigan.

23. The *Michigan Farmer* regularly reported on activities of the Detroit Horticultural Society. See, for example, "Horticultural Society in Detroit," *Michigan Farmer* 4 (April 1846): 17; "Detroit Horticultural Society," *Michigan Farmer* 5 (August 1847): 75; "Detroit Horticultural Society," *Michigan Farmer* 6 (15 April 1848): 113; "Horticultural Exhibition," *Michigan Farmer* 6 (1 June 1848): 172; "Horticultural Exhibition," *Michigan Farmer* 6 (1 July 1848): 205; "Detroit Horticultural Society," *Michigan Farmer* 7 (1 April 1849): 103; "Detroit Horticultural Society's Exhibition," *Michigan Farmer* 7 (1 June 1849): 166; "Detroit Horticultural Society," *Michigan Farmer* 7 (1 July 1849): 199; "Detroit Horticultural Society," *Michigan Farmer* 10 (February 1852): 47; and "Detroit Horticultural Society," *Michigan Farmer* 10 (April 1852): 111.

24. For an account of the Association, see Holmes, "The Early History of Horticulture in Michigan," 79–83; and "Constitution of the Michigan Nurseryman's and Fruit Growers' Association," J. C. Holmes Papers, 1849–1854, Burton Historical Collection, Detroit Public Library, Detroit, Michigan.

25. For an account of the Society, see Holmes, "The Early History of Horticulture in Michigan," 83–84. Many calls for the new organization appeared in the *Michigan Farmer*. See, for example, "To Nurserymen and Pomologist—Shall We Have a State Horticultural Society?," *Michigan Farmer* 15 (June 1857): 178; "A State Horticultural and Pomological Society, " *Michigan Farmer* 15 (July 1857): 209–10; "The Meeting of the State Horticultural Society," *Michigan Farmer* 16 (January 1858): 47; "State Horticultural Society," *Michigan Farmer* 16 (January 1858): 48; and "The State Horticultural Society," *Michigan Farmer* 16 (August 1858): 240–1.

26. See Lyon, *History of Michigan Horticulture*, 16–80. Lyon also described the state law relating to horticultural societies. See *History of Michigan Horticulture*, 187–8. Michigan lagged behind some other Midwestern states in successfully establishing a statewide horticultural society. The Indiana Horticultural Society, for example, was established in 1860, the Illinois Horticultural Society in the mid-1850s, and the Ohio State Horticultural Society in 1847. See Bailey, *The Standard Cyclopedia of Horticulture*, 1553–1559.

27. See Lyon, *History of Michigan Horticulture*, 48–49 and 57.

28. For a summary of local horticultural societies, their date of founding, and their membership numbers, see Charles W. Garfield, "A Glimpse of Michigan Horticulture" (no date or place of publication), 12–13, Michigan Collection, Michigan Library and Historical Center, Lansing, Michigan. Once local societies became auxiliaries of the State Society, their reports appeared in the annual trans-

actions. See, for example, R. T. McNaughton, "Jackson County Horticultural Society," *Tenth Annual Report of the Secretary of the State Horticultural Society of Michigan, 1880* (Lansing: W. S. George and Co., 1881), 246–251.

29. See "100th Anniversary, 1851–1951, Lenawee County Horticultural Society," Lenawee County Horticultural Society Records, 1877–1959, University Archives and Historical Collection, Michigan State University, East Lansing, Michigan; and Woodland Owen, "Adrian Horticultural Society," *Eleventh Annual Report of the Michigan State Horticultural Society, 1881* (Lansing: W. S. George and Co., 1882), 288. For commentary on the organization's library, see "Adrian Horticultural Society," *Adrian Times and Expositor*, 27 January 1870; and "Adrian Horticultural Society," *Seventh Annual Report of the Secretary of the State Pomological Society of Michigan, 1877* (Lansing: W. S. George and Co., 1878), 331–333. For an example of an advertisement for and account of the exhibition, see "Horticultural Exhibition," [advertisement], *Adrian Times and Expositor*, 23 June 1869; and "Exhibition of the Adrian Horticultural Society," *Adrian Times and Expositor*, 26 June 1869.

30. "Michigan State Fair," *Michigan Farmer* 7 (1 October 1849): 290; "Our County Fair," *Adrian Times and Expositor*, 1 October 1869; and "The County Fair," *Adrian Times and Expositor*, 28 September 1871.

31. "The County Fair," *Jackson Daily Citizen*, 3 October 1873; and "The State Fair," *Adrian Times and Expositor*, 21 September, 1870. The population for the city of Jackson in 1870 was 11,447, and Jackson County had a total of 35,620 residents. See Walker, *The Statistics of the Population of the United States*, 38 and 171.

32. For examples of commentary on the railroads, see "The State Fair," *Adrian Times and Expositor*, 25 September 1869; "The State Fair," *Jackson Weekly Citizen*, 27 September 1865; "The County Fair," *Adrian Times and Expositor*, 15 October 1872.

33. Ann Arbor prided itself on tree-lined streets. See, for example, S. B. Noble, "City of Ann Arbor," *Michigan Farmer* 12 (June 1854): 169. The agricultural press had long discussed the benefits of incorporating landscape gardening and rural scenes into a new kind of urban cemetery, and Jackson's City Cemetery was an example of newly developed "rural cemeteries" that sprang up in a number of cities at mid-century. For typical comments on rural cemeteries, see "Rambles Around Greenwood Cemetery," *Michigan Farmer* 10 (January 1852): 30; and "Rural Cemeteries," *American Agriculturist* 15 (October 1856): 293. For a complete historical analysis of rural cemeteries, see Blanche Linden-Ward, *Silent City on a Hill, Landscapes of Memory and Boston's Mount Auburn Cemetery* (Columbus: Ohio State University Press, 1989). See also David Schuyler, "The Evolution of the Anglo-American Rural Cemetery: Landscape Architecture as Social and Cultural History," *Journal of Garden History* 4, no. 3 (1984): 291–304.

34. For examples of these types of horticultural premiums, see *Constitution and By-laws of the Lenawee County Agricultural and Horticultural Society, Also, The List of Judges and Premiums and the Regulations Adopted for the Annual Fair for the Year 1851* (Adrian: R. W. Ingals, 1851), 16, Michigan Library and Historical Center, Lansing, Michigan; "Lenawee County," *Transactions of the State Agricultural Society: With Reports of County Agricultural Societies for 1854* (Lansing:

Hosmer and Fitch, 1855), 538–9; "Report of the Lenawee County Agricultural Society," *Transactions of the State Agricultural Society: With Reports of County Agricultural Societies for 1853* (Lansing: G. W. Peck, 1854), 515; "County Fair," *Adrian Times and Expositor*, 26 September 1872; and "Premiums for Orchards, Vineyards, and Nurseries," *Fourth Annual Report of the Secretary of the State Pomological Society of Michigan, 1874* (Lansing: W. S. George and Co., 1875), 116–118.

35. See "Flowers," *Transactions of the State Agricultural Society: With Reports of County Agricultural Societies for 1850* (Lansing: R. W. Ingalls, 1851), 82–3; and "Fruits, Flowers, and Vegetables," and "Division G.—Fruits, Flowers and Vegetables-Class I Flowers," *Transactions of the State Agricultural Society: With Reports of County Agricultural Societies for 1855* (Lansing: Hosmer and Fitch, 1856), 70–79 and 129–30. See also "Report of the Lenawee County Agricultural Society for 1852," *Transactions of the State Agricultural Society: With Reports of County Agricultural Societies for 1852* (Lansing: George W. Peck, 1853), 337.

36. "Flowers," *Transactions of the State Agricultural Society: With Reports of County Agricultural Societies for 1850* (Lansing: R. W. Ingalls, 1851), 82–3.

37. *Constitution and By-Laws of the Lenawee County Agricultural and Horticultural Society, Also the List of Judges and Premiums and the Regulations Adopted for the Annual Fair for the Year 1851*, 15. In contrast to many discretionary premiums, Washtenaw County awards seemed to involve recognition but no monetary prize. See "Floral Department, 1871 and 1875," Washtenaw County Fair Records, 107 and 175–6, Bentley Historical Library, University of Michigan, Ann Arbor, Michigan.

38. "Plants in Pots," *Ninth Annual Report of the Secretary of the State Pomological Society of Michigan, 1879* (Lansing: W. S. George and Co., 1880), 115–6.

39. "Horticultural Exhibition," *Michigan Farmer* 6 (1 June 1848): 172; and "Horticultural Exhibition," *Michigan Farmer* 6 (1 July 1848): 205. The Detroit Horticultural Society did not offer premiums for these early exhibitions.

40. "The County Fair," *Adrian Times and Expositor*, 30 September 1870; "The State Fair," *Adrian Times and Expositor*, 21 September 1872; and "Michigan State Fair," *Jackson Weekly Citizen*, 26 September 1871, 4.

41. For descriptions of horticultural decorations in floral halls, see "The County Fairs," *Michigan Farmer* 12 (November 1854): 336–7; "Washtenaw County Fair," *Michigan Farmer* 14 (November 1856): 336; S. B. Noble, "Washtenaw County Fair," *Michigan Farmer* 14 (January 1856): 5–6; "Washtenaw County Fair," *Michigan Farmer* 11 (November 1853): 326; and "Remarks about the Fair," *Michigan Farmer* 2 (13 October 1860): 326.

42. "The State Fair for 1863," *Michigan Farmer* 2 (September 1863): 121. For additional accounts of fountains in floral halls at the State Fair, see "The State Fair," *Michigan Farmer* 2 (August 1863): 73; "The State Fair," *Michigan Farmer* 2 (16 September 1871): 292; and "The State Fair," *Adrian Times and Expositor*, 25 September 1869. Several descriptions of the Washtenaw County Fair noted canaries and a parrot gracing floral hall. See "Washtenaw County Fair," *Michigan Farmer* 11 (November 1853): 326; and S. B. Noble, "Washtenaw County Fair," *Michigan Farmer* 14 (January 1856): 5.

43. For a description of the goods accommodated in floral halls, see "The State Fair," *Michigan Farmer* 2 (16 September 1871): 292; "The County Fair," *Jackson Weekly Citizen*, 18 October 1870, 5; "The County Fair," *Adrian Times and Expositor*, 2 October 1868; "The State Fair," *Michigan Farmer* 2 (August 1863): 73–4; and "The State Fair of 1863," *Michigan Farmer* 2 (October 1863): 171.

44. Michigan fruit grower T. T. Lyon voiced these concerns when commenting on the combined display of horticultural products and decorative arts at the 1859 State Fair. See T. T. Lyon, "Horticultural Arrangements at the Recent State Fair," *Michigan Farmer* 1 (29 October 1859): 347. Historian Thomas Schlereth has pointed out a significant paradox, suggesting that county fairs "frequently fostered traditional lifeways yet championed technologies that altered rural life forever." See Schlereth, "Country Stores, County Fairs, and Mail-Order Catalogues: Consumption in Rural America," 373. Warren J. Gates makes a similar argument in his study of a Pennsylvania Grange Fair, noting that an element in the inclusion of commercial displays was promotion of the idea that "change is good and technology its handmaiden." See Warren J. Gates, "Modernization as a Function of an Agricultural Fair, The Great Grangers' Picnic Exhibition at Williams Grove, Pennsylvania, 1873–1916," *Agricultural History* 58 (July 1984): 267.

45. William H. Scott, "A Few Hints on Farmer's Houses," *Horticulturist* 8 (1 June 1853): 269; Robert Allyn, "Out-Door Pictures, or Landscape Gardening: How to Beautify our Villages and Country Homes," *Transactions of the Illinois State Agricultural Society, with Reports from County and District Agricultural Societies*, vol. 8-1869–1870 (Springfield: Illinois Journal Printing Office, 1871), 170; and "Nudge Your Neighbor," *Prairie Farmer* 2 (19 August 1858): 118.

46. For an image of the Starkweather homegrounds, see *Combination Atlas Map of Washtenaw County*, 71. John Starkweather moved to Ypsilanti Township in 1841 as a young man of thirty-three, and immediately began the systematic improvement of his "nearly worn out" farm. By 1855, J. C. Holmes, Secretary of the Michigan State Agricultural Society, described Starkweather as one of the state's best agriculturists. See J. C. Holmes, Detroit, Michigan to John C. Kennicott, The Grove, Illinois, 27 July 1855, KBP, 54. For comments on the Starkweather farm, see S. B. Noble, "Rambles—No. 2," *Michigan Farmer* 11 (September 1853): 278; and "Some Farms at Ypsilanti," *Michigan Farmer* 14 (October 1856): 303. For commentary on Starkweather's orchard, see "Report of Special Committees," *Sixth Annual Report of the Secretary of the State Board of Agriculture of the State of Michigan for the Year 1867* (Lansing: John A. Kerr, 1867), 367. Starkweather's progressive farm practices led to economic success. The 1870 Federal Population Census listed Starkweather's combined real and personal worth at $35,000. See the 1870 Federal Population Census, Washtenaw County, Michigan, Microfilm Copy, M-593, roll 708, 456. For examples of Mrs. Starkweather's fair entries, see "Division G," *Transactions of the State Agricultural Society of Michigan with Reports of the County Agricultural Societies for the Year 1858* (Lansing: Hosmer and Kerr, 1860), 186; and "Floral Department," Washtenaw County Fair Records, 1871–1880.

47. For a visual image of the Boieses' home, see *Combination Atlas Map of Lenawee County*, 85. Mr. Boies's combined real and personal property was valued at

Notes to Chapter 6

$81,000 in 1870. See 1870 Federal Population Census, Lenawee County, Michigan, Microfilm Copy, M-593, roll 685, 291. J. K. Boies typified successful small town businessmen in the region. Born in Massachusetts in 1828, Boies attended Oberlin College for a short time, but in 1845 decided to join his brother Henry as a merchant in Hudson, Michigan. The Boies brothers operated a dry goods store, but also dealt in grain, wool, and other regional agricultural products. In 1855, Boies, along with his brother, organized the first bank in Hudson. While his business and bank prospered, John Boies found time to serve as President of the Village of Hudson, as a member of both the State House and Senate, and on the State Board of Control of Railroads. For details of J. K. Boies's life, see *Portrait and Biographical Album of Lenawee County, Michigan*, 748–751; and *Combination Atlas Map of Lenawee County*, 19.

48. For an image of the O'Donnell home see *Combination Atlas Map of Jackson County*, 116. Born in Connecticut in 1840, James O'Donnell moved to Michigan with his parents in 1848, and at fourteen apprenticed himself to a Jackson printer and publisher of the local *Jackson Weekly Citizen*. After serving in the Civil War, O'Donnell returned to Jackson, and took on the position of assistant newspaper editor. He purchased the *Weekly Citizen* in 1864, and a year later established the *Jackson Daily Citizen*. While the position of newspaper publisher was influential, it was not particularly lucrative, and O'Donnell reported a total economic worth of $12,000 by 1870. For details of O'Donnell's life, see *Combination Atlas Map of Jackson County*, 22; and 1870 Federal Population Census, Jackson County, Michigan, Microfilm Copy, M-593, roll 678, 197.

49. Whitney and Bonner, *History and Biographical Record of Lenawee County, Michigan*, 503–5. For commentary on his political activities, see "Woodland Owen," *Adrian Times and Expositor*, 5 April 1873; and "City Ticket," *Adrian Times and Expositor*, 8 April 1873.

50. Woodland Owen was frequently mentioned as an officer or participant in society activities. See, for example, "Adrian Horticultural Society," *Adrian Times and Expositor*, 5 January 1872. For additional comments on his role in the Adrian Horticultural Society, see *Adrian City Directory, 1874–5, Also Complete Business Directories of Tecumseh, Blissfield, Morenci and Monroe City* (Detroit: Burch, Montgomery and Co., 1874), 144.

51. See *Combination Atlas Map of Lenawee County*, 75. For examples of the Owens' fair entries, see "The County Fair," *Adrian Times and Expositor*, 30 September 1870; "County Fair," *Adrian Times and Expositor*, 27 September 1872; "Exhibition of the Adrian Horticultural Society," *Adrian Times and Expositor*, 26 June 1869; and "Adrian Horticultural Society," *Ninth Annual Report of the Secretary of the State Pomological Society of Michigan, 1879* (Lansing: W. S. George and Co., 1880), 244–5.

52. For commentary on Mrs. Owen's and Henry's horticultural interest, see "The County Fair," *Adrian Times and Expositor*, 29 September 1871; and "Cut Flowers and Bedding Plants," *Ninth Annual Report of the Secretary of the State Pomological Society of Michigan, 1879* (Lansing: W. S. George and Co., 1880), 116–7. Henry eventually graduated from Michigan Agricultural College and held a position as teacher of horticulture at the college. See Whitney and Bonner, *His-*

tory and Biographical Record of Lenawee County, Michigan, 505. The Owen family had several boarders including, in 1870, a bootmaker, his wife and three children, and a salesman. Mrs. Owen had help with household chores, but there was no evidence of additional hired help to assist with horticultural tasks. For details of the Owens' economic status and household composition, see 1870 Federal Population Census, Lenawee County, Michigan, Microfilm Copy, M-593, roll 685, 123. Dr. Owen's economic level represented the combined value of his real and personal estate as reported in the 1870 Federal Population Census. Of the 731 homes pictured in the combination atlas maps for the three counties, comparable economic data was available for 663 families. The average economic worth for all those families, both city and rural, was $14,644. While well below the average for this selected group, Dr. Owen was far from impoverished. Historian Lee Soltow has calculated the arithmetic mean of total worth for males in the northern United States in 1870 at $2,921, and for adult males in Wisconsin in 1870 as $2,369. See Soltow, *Patterns of Wealthholding in Wisconsin Since 1850*, 80; and Soltow, *Men and Wealth in the United States 1850–1870*, 65.

53. For details of Powell's career, living circumstances, and garden development, see Bonner, *Memoirs of Lenawee County, Michigan*, 502–3; 1870 Federal Population Census, Lenawee County, Michigan, Microfilm Copy, M-593, roll 685, 90; Enthusiast, Adrian, Michigan, "Clergymen Gardeners," *Gardener's Monthly* 8, no. 2 (February 1866): 40–41; Charles E. Brown, compiler, *Brown's City Directory of Adrian, Michigan* (Adrian, MI: Adrian Times and Expositor Steam Printing Office, 1870), 122; and "For Sale," [advertisement] *Adrian Times and Expositor*, 3 May 1871. Rev. Powell placed several orders with Ellwanger and Barry including a $20 order in December 1862 and a second order for $7.75 in April 1864. See Ellwanger and Barry Cash Books, vol. 58, 2 December 1861–30 November 1864, Ellwanger and Barry Company Papers, Department of Rare Books and Special Collections, Rush Rhees Library, University of Rochester, Rochester, New York.

54. See Enthusiast, Adrian, Michigan, "Room Plants," *Gardener's Monthly* 6, no. 1 (January 1864): 4–5.

55. For examples of Powell's entries, see "Exhibition of the Adrian Horticultural Society," *Adrian Times and Expositor*, 26 June 1869; "Our County Fair," *Adrian Times and Expositor*, 30 September 1869; "The State Fair," *Jackson Daily Citizen*, 27 September 1865; "Division I," *Seventh Annual Report of the Secretary of the State Board of Agriculture of the State of Michigan for the Year 1868* (Lansing: John A. Kerr and Co., 1868), 358; and *Eighth Annual Report of the Secretary of the State Board of Agriculture of the State of Michigan for the Year 1869* (Lansing: W. S. George and Co., 1869), 335.

56. See "A Floral Tribute," *Adrian Times and Expositor*, 21 June 1869; and Enthusiast, Adrian, Michigan, "Clergymen Gardeners," *Gardener's Monthly* 8, no. 2 (February 1866): 40–1; and "Personal," *Adrian Times and Expositor*, 3 June 1871.

57. Powell also supported local florists and nurserymen. See, for example, E. P. Powell, "Concerning Nursery Agents," *Adrian Times and Expositor*, 4 January 1870. In commenting on the nursery agent threat to local horticulturists, Powell urged readers to buy from well-known firms and local florists like Loud and Trask.

58. See, for example, E. T. Powell [sic], "How to Make Production Popular," *Gardener's Monthly* 13, no. 6 (June 1871): 163–5. Powell signed other articles "Enthusiast," Adrian, Michigan," but the content clearly linked the article to him. See "Clergymen Gardeners," *Gardener's Monthly* 8, no. 2 (February 1866): 40–1. For comments from the *New York Independent* quoted in the *Adrian Times and Expositor*, see "Personal," *Adrian Times and Expositor*, 3 June 1871.

59. In 1870, E. P. Powell reported total resources of $7,000, $5,000 in real property and an additional $2,000 in personal property. See 1870 Federal Population Census, Lenawee County, Michigan, Microfilm Copy, M-593, roll 685, 90.

60. See "Personal," *Adrian Times and Expositor*, 3 June 187; and "For Sale," [advertisement] *Adrian Times and Expositor*, 3 May 1871.

61. The term "improving spirit" appeared frequently in the mid-nineteenth-century agricultural and horticultural press. See, for example, "Mr. Downing and the Horticulturist," 396.

Chapter Seven: Bergamont Balm and Verbenas

1. See Breck, *The Flower Garden*, 15; and "Attractions of Home," *Michigan Farmer* 6 (November 1848): 329.

2. See Daniel McCarthy, Davenport, Iowa, to John Kennicott, The Grove, Illinois, 5 February 1861, KBP, 948; H. Rowell, Joliet, Illinois, 12 April 1862, KBP, 1245; Morilla Cates, Garden Prairie, Illinois, 2 May 1862, KBP, 1332; L. L. Klemin, Chicago, Illinois, 9 May 1855, KBP, 51; Margaret B. Carle, Urbana, Illinois, 25 October 1858, KBP, 464; and A. W. Arnold, Cortland, Illinois, 10 March 1862, KBP, 1144.

3. See B. G. Buell, "Diary," Barber G. Buell Papers, 1763–1934; Esther Copley Lawrence, "Diary;" Charlotte Copley, "Diary;" and Frank W. Copley, ed., "Charlotte's Diary, The Account of an Overland Trip from Dayton, Ohio to Little Prairie Ronde, Michigan Territory in June, 1833," Copley Family Papers, Bentley Historical Library, University of Michigan, Ann Arbor, Michigan. Both a copy of the original and typescript copy of Esther Copley Lawrence's diaries are available at the University Archives and Regional History Collections, Western Michigan University, and were used in this analysis. See Esther Copley Lawrence, "Diary," in Mrs. Eda Kelly and Mrs. Esther Pellowe Papers, University Archives and Regional History Collections, Western Michigan University, Kalamazoo, Michigan. In succeeding entries, diary entries are abbreviated as follows: ECL-Esther Copley Lawrence; BGB-Barber G. Buell; and CC-Charlotte Copley. For an example of an historical study examining women's lives in Michigan based on first-person documents, see Marilyn Ferris Motz, *True Sisterhood, Michigan Women and Their Kin 1820–1920* (Albany: State University of New York Press, 1983). For general commentary on the use and interpretation of diaries as historical documentation, see Suzanne L. Bunkers, "Diaries: Public and Private Records of Women's Lives," *Legacy* 7 (Fall 1990): 17–26; Suzanne L. Bunkers, "'Faithful Friend': Nineteenth-Century Midwestern American Women's Unpublished Diaries," *Women's Studies International Forum* 10 (1987): 7–17; Margo

Culley, ed., *A Day at a Time: The Diary Literature of American Women from 1764 to the Present* (New York: Feminist Press, 1985); Elizabeth Hampsten, *Read This Only to Yourself: The Private Writings of Midwestern Women, 1880–1920* (Bloomington: Indiana University Press, 1982); Leonore Hoffman and Margo Culley, eds., *Women's Personal Narratives: Essays in Criticism and Pedagogy* (New York: Modern Language Association of America, 1985); and Marilyn Ferris Motz, "Folk Expression of Time and Place: 19th-Century Midwestern Rural Diaries," *Journal of American Folklore* 100 (April–June 1987): 131–47.

4. See Helen Hathaway, "Alexander Bennett Copley," *Decatur Republican*, 22 February 1945; 1 March 1945; 8 March 1945; and 15 March 1945, for a detailed account of the early history of the Copley family. See also L.C.A. Copley, "Autobiography," [handwritten copy], 7–8; Frank W. Copley, "A Sentimental Journey" [typescript], II; and Frank W. Copley, ed., "Charlotte's Diary, The Account of an Overland Trip from Dayton, Ohio to Little Prairie Ronde, Michigan Territory in June, 1833," 3–4, Copley Family Papers, Bentley Historical Library, University of Michigan, Ann Arbor, Michigan.

5. *History of Cass County, Michigan* (Chicago: Waterman, Watkins and Co., 1882), 296–8 and 289–91.

6. See Helen Hathaway, "Alexander Bennett Copley," *Decatur Republican*, 22 February 1945, for an account of the Lawrence family's arrival in Michigan. See Miscellaneous Papers, Mrs. Eda Kelly and Mrs. Esther Pellowe Papers, University Archives and Regional History Collections, Western Michigan University, Kalamazoo, Michigan, for a listing of Lawrence family members and their place of birth. See also *History of Cass County, Michigan* (Chicago: Waterman, Watkins and Co., 1882), 288, for an account of Levi Lawrence and his mechanical skills. L. H. Glover, *A Twentieth Century History of Cass County, Michigan* (Chicago: Lewis Publishing Co., 1906), 109–110 and 138, also mentions Lawrence's blacksmithing skills and his role in platting the village of Volinia.

7. See Helen Hathaway, "Alexander Bennett Copley," *Decatur Republican*, 8 March 1845, for an account of the Buell family's arrival in Michigan. See also *History of Cass County, Michigan*, 288. For general descriptions of Nicholsville and the Volinia Township area, see *History of Cass County, Michigan*, 283–295; Cass County Historical Commission, *Historical Reflections of Cass County* (n. p.: Cass County Historical Commission, 1981), 19–20; and Glover, *A Twentieth Century History of Cass County, Michigan*, 109–110 and 138–9.

8. Cass County was organized in 1829. For comparison of Cass County population figures, see Walker, "Population By Counties—1790–1870," *The Statistics of the Population of the United States, Ninth Census*, 38. See also "Table II," *Census of the State of Michigan, 1874* (Lansing: W. S. George and Co., 1875), 160. For commentary on the agricultural history of the county, see Glover, *A Twentieth Century History of Cass County, Michigan*, 198–204. Also see Dunbar, *Michigan, A History of the Wolverine State*, 201 and 288.

9. See Dunbar, *Michigan, A History of the Wolverine State*, 433; and Cass County Historical Commission, *Historical Reflections of Cass County*, 15–17.

10. Glover, *A Twentieth Century History of Cass County, Michigan*, 204–6; and *History of Cass County, Michigan*, 293–4.

11. The diaries often noted paying for a journal, binding issues of magazines, or books that family members were reading. See, for example, ECL, 23 March 1856, 3; 3 December 1865, 47; 11 January 1866, 48; 6 March 1873, 80; and 30 March 1877, 106. See also BGB, 3 November 1869; 6 November 1869; 24 January 1870; 21 February 1870; 2 March 1870; 13 March 1870; 21 January 1871; and 9 April 1873. For Buell's comments on newspaper subscriptions, see BGB, 30 January 1871; 29 May 1872; and 3 July 1872. Before the marriage of Hattie and B. G. Buell, Esther Lawrence noted that Hattie came and exchanged "magazines, books, and papers," and it seems safe to presume that the exchange between the two families continued after the marriage. See ECL, 25 November 1855, 2. For comments on Johnstone's visit see ECL, 5–6 March 1873, 80; and BGB, 5–6 March 1873. For comments on Mr. Wood's visit, see ECL, 27 January 1879, 123.

12. For examples of attendance and comments on lectures, see BGB, 25 October 1869; 14 March 1870. See also ECL, 25 February 1869, 61; 7 December 1869, 65; and 17 October 1874, 89. For commentary on Farmers' Institutes, see BGB, 13 January 1876.

13. Diary entries for both Esther Copley Lawrence and B. G. Buell record attendance at farmers' clubs or home culture society meetings. See especially BGB, 27 September 1869; 20 November 1869; 5 January 1870; 18 September 1870; 18 August 1875; 14 January 1876; and 15 April 1878. For commentary on grange attendance, see BGB, 14 March 1874; 25 March 1874; 28 January 1876; 27 February 1876; and 24 March 1877. For commentary on the importance and activism of the Volinia Farmers' Club, see *History of Cass County, Michigan*, 293–4; Glover, *A Twentieth Century History of Cass County, Michigan*, 205–6; Lyon, *History of Michigan Horticulture*, 234; and McCracken, *Michigan and the Centennial*, 497–8.

14. See, for example, ECL 17 June 1875, 93; 21 August 1875, 94; and ECL, 2 December 1878, 116. Also see BGB, 19 September 1872; and 11 March 1874.

15. See ECL, 11 June 1878, 113, for comments on her son's experimentation with a telephone; and 29 June 1878, 126, for their work with a new Woods Twine Binder for wheat. She described using a new sweeper on 8 November 1878, 115. See also BGB, 10 October 1869; 16 March 1870; and 3 September 1877, for descriptions of inspecting a neighbor's vineyard, for commentary on examining a new microscope, and for comments on a neighbor's use of a shoe drill for sowing wheat.

16. Lyon, *History of Michigan Horticulture*, 233; and BGB, 22 August 1869; and 30 August 1869. For comments on selling trees and shrubs, see BGB, 10 February 1870; 7 May 1870; and 9 May 1870. For comments on selling fruit, see BGB, 3 June 1872; 3 September 1872; and 15 September 1872.

17. For record of the families' economic advance and farm production, see U.S. Federal Census, Productions of Agriculture for 1850, 1860, 1870, and 1880, Cass County, Michigan, Microfilm Copy, T-1164, roll 1 (1850), T-1164, roll 7 (1860), T-1164, roll 17 (1870), and T-1164, roll 33 (1880). See also 1870 Federal Population Census, Cass County, Michigan, Microfilm Copy, M-593, roll 668, 265. For a record of the construction of the new home, see ECL, 25 October 1872, 78; and 23 January 1873, 79. The actual record of the construction of the house ex-

tends from 19 March 1874 through 12 March 1875. See ECL, 27 May 1874, 86; 24 and 25 August 1874, 88; and 30 September 1874, 88, for comments on the furnace, roofing, eaves, and walkway.

18. BGB, 1 June 1872; and 15 June 1872. Croquet became a very popular lawn game beginning in the mid-1860s.

19. See BGB, 7 May 1870; 19 May 1870; and 21 May 1870; and "Essay on Orchards," handwritten copy, Barber G. Buell Papers, Bentley Historical Library, University of Michigan, Ann Arbor, Michigan. For examples of the wide range of comments on evergreen trees and the particular virtues of the Norway spruce, see "A Chapter on Evergreens," *American Agriculturist* 16 (May 1857): 108; George Jacques, "Evergreens," *Magazine of Horticulture* 31, no. 4 (April 1865): 109; and George Taylor, "When and How to Plant Evergreens," *Fifth Annual Report of the Secretary of the State Pomological Society of Michigan, 1875* (Lansing: W. S. George and Co., 1876), 54.

20. For comments on coleus, see ECL, 20 September 1877, 108; and 27 May 1878, 113. For examples of Esther Lawrence's many references to the type of plants she cultivated, see 12 September 1858, 15; 3 May 1859, 18; 20 March 1859, 18; 26 September 1861, 28; 29 April 1871, 68; 1 January 1876, 96; and 6 September 1877, 108. For examples of commentary on fashionable flowers and their arrangement, see "Planning the Flower Garden," *American Agriculturist* 19 (May 1860): 147; "Landscape Gardening—Of Planting the Flower Garden," *Godey's Ladies' Book* 45 (December 1852): 528; and Edward S. Rand, Jr., "Flowers in Masses," *Horticulturist* 20 (June 1865): 165. For a typical comment on the use of vines in home embellishment, see Anna Hope, "My Madeira Vine," *American Agriculturist* 13 (25 October 1854): 105.

21. While some commentators, along with nursery and seedsmen, were promoting bedding plants and colorful annuals, other horticultural advisers were urging caution and suggesting that perennial gardens were far more appropriate, especially for busy farm families. For commentary on the benefits of perennial gardens, see John A. Kennicott, "Theory and Practice of Horticulture on a Farm," *Prairie Farmer* 3 (25 April 1859): 263; and "Biennials and Perennials," *Michigan Farmer* 11 (May 1853): 148.

22. See BGB, 21 April 1870; and 22 April 1870. For examples of commentary on roadside planting, see Mrs. E. T. Lyon, "Roadside Improvements," *Horticulturist* 24 (June 1869): 176–7; and "Village and Country Road-Side," *Horticulturist* 20 (November 1865): 341–4. For information on Michigan laws regarding tree planting, see Thomas M. Cooley, compiler, *The Compiled Laws of the State of Michigan* (Lansing: Hosmer and Kerr, 1857), 373–4; and *Acts of the Legislature of the State of Michigan Passed at the Regular Session of 1867*, vol. 1 (Lansing: John A. Kerr and Co., 1867), 188–9. For information on premiums for tree planting, see "Shade Trees," *Michigan Farmer* 10 (July 1852): 209.

23. See BGB, 12 April 1871; 14 April 1871; and 29 April 1871. Also see "Flowers in the School Yard," *American Agriculturist* 18 (August 1859): 229. For additional comments, see "Why Is Not Horticulture Taught in Our Common Schools?" *Michigan Farmer* 1 (4 June 1859): 179. Popular culture scholar Fred E. H. Schroeder also offers some historical analysis of changing concerns with

schoolhouse architecture and appearance. See Fred E. H. Schroeder, "The Little Red Schoolhouse," in *Icons of America*, ed. Ray B. Browne and Marshall Fishwick (Bowling Green, OH: Bowling Green State University Popular Press, 1978), 139–160.

24. ECL, 15 April 1876, 98. For the text of the governor's address, see S. B. McCracken, *Michigan and the Centennial*, 174–5.

25. For commentary on Benjamin Hathaway and his nursery, see BGB, 17 October 1869; and "Premium Farms," *Fifth Annual Report of the Secretary of the State Board of Agriculture of the State of Michigan for the Year 1866* (Lansing: John A. Kerr and Co., 1866), 283–5. George R. Fox, "Benjamin Hathaway—Unrecognized Genius," *Dowagiac Daily News*, 20 July 1942; and 21 July 1942. George R. Fox, "Hathaway's Writing Done at Plow, Bench," *Dowagiac Daily News*, 22 September 1942, also details Hathaway's life and accomplishments.

26. See ECL, 25 January 1861, 25; 30 January 1876, 97; and 8 February 1876, 97; and BGB, 6 February 1870; 13 February 1870; 8 March 1870; 11 March 1870; 10 April 1872; and 15 April 1872.

27. See ECL, 22 April 1861, 22; and BGB, 11 May 1870; 16 May 1870; and 17 April 1871.

28. See ECL, 10 November 1858, 16; and 21 September 1877, 108.

29. ECL, 22 March 1866, 49; and 11 April 1866, 49. See also *Vick's Illustrated Catalogue of Seeds and Guide to the Flower Garden* (1865), 3. The diary entry does not indicate where the ladies ordered their seeds, but some seedsmen like James Vick would send club orders to individual homes.

30. BGB, 10 April 1870. For a brief history of government seed distribution programs, see Baker, Rasmussen, Wiser, and Porter, *Century of Service*, 4–18; and Klose, *America's Crop Heritage*, 39–44 and 56–62.

31. For examples of these visits and the exchange of plant materials, see ECL, 31 March 1858, 13; 10 March 1859, 18; 5 May 1860, 22; 10 November 1861, 29; 20 June 1862, 31; 14 April 1862, 31; 10 April 1863, 35; 12 October 1875, 95; 2 November 1875, 96; 22 April 1877, 106; 21 September 1877, 108; 3 July 1878, 114; 15 April 1879, 118; 5 September 1879, 121; and 15 April 1880, 124. For a study of patterns of visiting documented in rural diaries, see Jane Marie Pederson, "The Country Visitors: Patterns of Hospitality in Rural Wisconsin, 1880–1925," *Agricultural History* 58 (July 1984): 347–65. Pederson argues that visiting among neighbors and kin was an important means of cementing ties and was particularly important among rural women.

32. BGB, 21 May 1870; and 12 April 1870.

33. See, for example, ECL, 31 March 1858, 13; 12 April 1879, 118; and 22 April 1863, 35. Also see *History of Cass County, Michigan*, 287. See also George R. Fox, "Benjamin Hathaway—Unrecognized Genius," *Dowagiac Daily News*, 20 July 1942. In 1870, Hathaway owned 120 acres, worth $9,500. The Finch, Huyck, and Morris families' total properties ranged in value from $5,400 to $10,500. See U. S. Federal Census, Productions of Agriculture for 1870, Cass County, Michigan, Microfilm Copy, T-1164, roll 17. Also see 1870 Federal Population Census, Cass County, Michigan, Microfilm Copy, M-593, roll 668, 262r and 263 (Goodspeeds), 264 r (Huycks), 265 (Morris), and 264 r (Finch).

34. ECL, 13 April 1878, 111. These village families had rather modest holdings in 1870, with the Thorps valued at $8,500 and the Bitelys at $3,500. See 1870 Federal Population Census, Cass County, Michigan, Microfilm Copy, M-593, roll 668, 261 r (Lyon), 263 (Thorp), and 263 r (Bitely); and U. S. Federal Census, Productions of Agriculture for 1870, Cass County, Michigan, Microfilm Copy, T-1164, roll 17. Also see D. J. Lake, *Atlas of Cass County, Michigan* (Philadelphia: C. O. Titus, 1872), 41.

35. Grant McCracken's study of the diffusion of cultural meaning offers a model for this repeated transferal and transformation of significance. McCracken suggests that cultural tastemakers "unhitch" cultural meaning and then transfer it to cultural goods. Consumers then "fetch" the meaning out of the goods, and use it for their own purposes. See Grant McCracken, *Culture and Consumption, New Approaches to the Symbolic Character of Consumer Goods and Activities* (Bloomington: Indiana University Press, 1988), xiv.

36. The Lawrence and Buell families had many other opportunities to learn about plant culture and horticultural advances through literature. In 1873, B. G. Buell joined the Michigan State Pomological Society, and through his participation had access to their voluminous annual reports discussing a variety of horticultural topics. County agricultural societies and organizations like the Volinia Farmers' Club also traded reports and information with other groups, opening up vast new sources of information. Both families also subscribed to a range of journals like the *Michigan Farmer* or the *American Agriculturist* that regularly published articles on gardening, floriculture, greenhouse management, and many other topics. On April 15, 1878, B. G. Buell confirmed his access to Pomological Society materials, noting "I got a box of books at Depot from Secretary Garfield of Pomological Society." As was common practice, Buell probably distributed the books to other Farmers' Club members. See BGB, 15 April 1878.

37. For general information on Hathaway and his horticultural activities, see Fox, "Benjamin Hathaway-Unrecognized Genius," and "Hathaway's Writing Done at Plow, Bench." Hathaway's orders are documented in Ellwanger and Barry Cash Books, vol. 58, December 2, 1861–November 30, 1864; and vol. 59, November 18, 1864–October 29, 1867, Ellwanger and Barry Company Papers, Department of Rare Books and Special Collections, Rush Rhees Library, University of Rochester, Rochester, New York. For examples of Hathaway's writings, see B. Hathaway, "The Northern Illinois Horticultural Society," *Michigan Farmer* 3 (22 February 1872): 59. For an account of Hathaway's involvement in horticultural societies, see "State Horticultural Society," *Michigan Farmer* 16 (January 1858): 48; Holmes, "The Early History of Horticulture in Michigan," 83; and Lyon, *History of Michigan Horticulture*, 39.

38. See "Premium Farms," *Fifth Annual Report of the Secretary of the State Board of Agriculture of the State of Michigan for the Year 1866* (Lansing: John A. Kerr and Co., 1866), 283–5. Hathaway's farm fell into a category for farms less than 160 acres.

39. See, for example, ECL, 29 September 1878, 115; BGB, 17 October 1869; and 18 July 1869.

40. BGB, 14 August 1872; and 15 August 1872.

41. ECL, 21 August 1875, 95.

42. See, for example, BGB, October 5–6, 1869; September 29–October 1, 1869; and 28–9 September 1870. See also, ECL, 10 October 1867, 55.

43. ECL, 23 September 1863, 37. The *Michigan Farmer* offered a further description of what Esther saw. The floral hall, the *Farmer* declared, was a "bower of beauty" with a beautiful fountain filled with native fish. See "The State Fair," *Michigan Farmer* 2 (August 1863): 73; and "The State Fair for 1863," *Michigan Farmer* 2 (September 1863): 121–2.

44. For a detailed description of the floral hall, see "The State Fair," *Jackson Weekly Citizen*, 23 September 1873; and "The State Fair," *Adrian Times and Expositor*, 25 September 1869.

45. BGB, 20–22 September 1869. For a description of the Jackson City Cemetery, see *History of Jackson County, Michigan*, 549–550. Also see "Mount Evergreen Cemetery," [Map and Interpretive Brochure] (Jackson, MI: City of Jackson, n. d.), Ella Sharpe Museum, Jackson, Michigan.

46. For commentary on the Volinia entries, see McCracken, *Michigan and the Centennial*, 486. Esther's account of her trip extends from September 11 to September 26, 1876, 101–103. Levi Lawrence also attended, but began his journey after Esther had returned, a pattern the family often followed when they traveled, presumably so that one of them was at home to manage the farm. See ECL, 10 October 1876, 103.

47. See ECL, 15–19 September 1876, 101.

48. For a detailed account of the Exhibition and Horticultural Hall, see James D. McCabe, *The Illustrated History of the Centennial Exhibition* (Philadelphia: National Publishing Co., 1877), 564–576. Also see J. S. Ingram, *Centennial Exposition Described and Illustrated, being a Concise and Graphic Description of this Grand Enterprise Commemorative of the First Centenary of American Independence* (Philadelphia: Hubbard Brothers, 1876). For historical analysis of the Exhibition, Thomas J. Schlereth, "The Philadelphia Centennial as a Teaching Model," *Hayes Historical Journal* 1, no. 3 (1977): 201–210; Robert C. Post, *1876: A Centennial Exhibition* (Washington, DC: Smithsonian Institution Press, 1976); Dee Brown, *The Year of the Century: 1876* (New York: Charles Scribner's Sons, 1966); and John Maass, *The Glorious Enterprise, The Centennial Exhibition of 1876 and H. J. Schwartzmann, Architect-in-Chief* (Watkins Glen, NY: American Life Foundation, 1973).

49. ECL, 22 September 1876, 102. For an inclusive history of Central Park and the changes that occurred in the early 1870s with William "Boss" Tweed's restructuring of the Park administration, see Roy Rosenzweig and Elizabeth Blackmar, *The Park and the People, A History of Central Park* (Ithaca: Cornell University Press, 1992), 263–283. See also Frederick Law Olmsted, *Creating Central Park, 1857–1861, The Papers of Frederick Law Olmsted*, vol. 3, ed. Charles E. Beveridge and David Schuyler (Baltimore: Johns Hopkins University Press, 1983); and Henry Hope Reed, *Central Park: A History and A Guide* (New York: C. N. Potter, 1967). For examples of commentary on Central Park in the agricultural and horticultural press, see "New York Central Park," *Magazine of Horticulture* 26, no. 12 (December 1869): 529–534; "Public Parks," *Gardener's Monthly* 15,

no. 1 (January 1873): 16–17; and T. T. Lyon, "Central Park, New York," *Michigan Farmer* 1 (29 January 1859): 35.

50. For comments on lawn mowing, see BGB, 3 June 1869; 16 June 1869; 2 August 1869; 5 August 1869; 12 August 1871; 4 June 1872; and 8 June 1872. For comments on establishing the lawn, see ECL, 15 August 1874, 88; 25 April 1876, 98; 25 May 1876, 99; and 30 May 1876, 99. For commentary on tree planting and care, see BGB, 28 June 1869; 5 August 1869; 21 April 1870; 22 April 1870; 7 May 1870; 9 May 1870; and 19 May 1870. The mid-nineteenth horticultural and agricultural press routinely noted the gendered division of home embellishment tasks. They suggested that flower gardening was a task best suited to women, and one that women seemed to prefer. See, for example, Breck, *The Flower Garden*, 18.

51. For comments on the yearly cycle of plant work, see ECL, 18 April 1860, 22; 28 March 1878, 111 [replanting]; 7 April 1863, 35; 23 March 1858, 13; and 12–13 May 1865, 45. For comments on the help with the gardens and the planting of flower seeds, see ECL, 15 April 1865, 44; 17 April 1877, 106, 10 May 1878, 115; and 7 May 1859, 17 May 1859, and May 20, 1859, 18. Almost every May, Esther Lawrence noted that she planted flower seeds in the garden. For commentary on setting out tender plants, see ECL, 3 May 1859, 18; 9 May 1863, 35; 29 April 1871, 81; 27 May 1878, 113; and 25 April 1879, 118. For summer gardening chores, see ECL, 3 August 1872, 76; 10 June 1876, 99; and 4 July 1877, 107.

52. For commentary on fall chores, see ECL, 30 October 1878, 115; 26 September 1861, 28; 6 September 1877, 108; and 20 September 1877, 108; 23 October 1862, 33; and 21 November 1858, 16; 11 September 1861, 28; 18 October 1861, 28; and 22 October 1874, 89. For commentary on winter chores, see ECL, 25 January 1863, 38; 7 March 1878, 111; 8 November 1879, 122; and 18 November 1879, 122.

53. See ECL, 21 November 1858, 16; and ECL, 11 November 1858, 16.

54. For commentary on roles in establishing the lawns, see ECL, 15 August 1874, 88; 25 April 1876, 98; 25 May 1876, 99; and 30 May 1876, 99. See also ECL, 4 May 1876, 98; 1 June 1876, 99; and 15 April 1880, 124, for comments on helping "Pa" with gardening chores. For comments on vegetable gardening, see BGB, 22 June 1869; 28 June 1869; 5 July 1869; and 13 August 1869. The Buell family appeared to follow a similar division of labor. Buell himself, or his sons, perhaps because of their age, did not seem to prepare flower beds for their mother, but Buell did note when one of the hired men performed the task. See, for examples, BGB, 25 April 1871; and 21 March 1874.

55. Esther also used flowers as a special gift to sisters she saw with less regularity. See ECL, 13 June 1858, 14, for an example of gathering flowers as a gift for sister Phim.

56. See CC, 4 March 1832; 20 March 1832; 24 March 1832; 4 February 1833; and 24 February 1833.

57. Euphemia was born on March 29, 1832.

58. CC, 14 March 1832; 18 March 1832; and 20 March 1832.

59. See ECL, 3 March 1861, 26; 22 April 1878, 112; 6 May 1878, 113; and ECL, 9 May 1878, 113. For an account of Almon Copley's westward movement, see L.C.A. Copley, "Autobiography," [handwritten copy], 7–8, Copley Family Papers, Bentley Historical Library, University of Michigan, Ann Arbor, Michigan. See also McCracken, *Culture and Consumption*, 137. McCracken describes goods as a way to ameliorate social change through their seeming permanence and concreteness.

60. See ECL, 11 September 1865, 41; and BGB, 11 May 1870; 16 May 1870; and 21 May 1870.

61. ECL, 30 October 1856, 6; 1 March 1858, 7; 26 September 1858, 15; and ECL, 26 May 1861, 27.

62. BGB, 23 April 1871.

63. ECL, 31 May 1857, 9.

64. See ECL, 8 May 1858, 14; and 30 June 1876, 99. Also see *History of Cass County, Michigan*, 287, for an account of the early connection between the Copleys and the Goodspeeds.

65. BGB, 19 January 1876. There were several indications of the Buell family's support of the temperance cause. While in New York State, Buell had been a member of the Sons of Temperance, and his own sons Frank and Lincoln attended temperance lectures in Nicholsville. See BGB, 5 April 1877; and "Miscellaneous Papers," Barber G. Buell Papers, Bentley Historical Library, University of Michigan, Ann Arbor, Michigan.

66. ECL, 15 April 1879, 118. For comment on the medicinal uses of bergamot, see William H. Hylton, ed., *The Rodale Herb Book* (Emmaus, PA: Rodale Book Division, 1976), 363–366; and Charles F. Millspaugh, *Medicinal Plants* (Philadelphia: John C. Yorston and Co., 1892; reprint edition, New York: Dover Books, 1974), 453–69 (page references to reprint edition).

67. ECL, 15 April 1879, 118.

68. See ECL, 4 September 1858, 15; 7 September 1858, 15; 12 September 1858, 15; and 3 October 1858, 15. Esther and other family members had been suffering from recurring bouts of ague at the time of Napoleon's death, so her comments on not feeling well reflected both her emotional and physical state. See also ECL, 21 November 1877; and 23 November 1877, 109.

69. See Frank W. Copley, "A Sentimental Journey" [typescript], II; and Frank W. Copley, ed., "Charlotte's Diary, The Account of an Overland Trip from Dayton, Ohio to Little Prairie Ronde, Michigan Territory in June, 1833," 1–4, for an account of preparations for the trip.

70. CC, 12 June 1833; 13 June 1833; CC, 22 June 1833; and 24 June 1833.

71. CC, 4 March 1832. Charlotte herself experienced four moves before the family settled in Ohio. ECL 20 January 1877, 105.

72. ECL 20 January 1877, 105.

73. ECL, 21 January 1877, 105. For commentary on Arthur and his writings, see Mott, *A History of American Magazines, 1850–1865*, 416–418.

74. Esther often commented on her closeness to Archie as a young child. See, for example, ECL, 21 March 1857, 8; and 23 March 1857, 9.

75. ECL, 13–14 December 1874, 90.

76. ECL, 12 February 1876, 97; and 16 April 1877, 106. On February 12, 1876, B. G. Buell commented that "Archie Lawrence returned home yesterday a married man." Esther never mentioned a marriage and does not comment about a wife when Archie finally returned in 1877. Whatever actually happened in 1874, Archie married Matilda VanRiper of Lawton, Michigan, in 1880.

77. ECL, 27 January 1877, 105.

Chapter Eight: Conclusion

1. Henry Ward Beecher, *Plain and Pleasant Talk about Fruits, Flowers and Farming* (New York: Derby and Jackson, 1859),10–11.

2. J. P. Kirtland, "Professor Kirtland's Essay," *Transactions of the Illinois State Agricultural Society*, vol. 1, 1853–54 (Springfield, IL: Lanpheir and Walker, 1855), 456.

3. "Tree Peddlers," *Horticulturist* 25 (May 1870): 151.

4. "Purchasing Fruit Trees," *Moore's Rural New Yorker* 14 (21 November 1863): 375.

5. Beecher, *Plain and Pleasant Talk about Fruits, Flowers and Farming*, 10–11.

Selected Bibliography

Unpublished Primary Sources

Manuscript and Visual Image Collections

Allmendinger Family Photograph Series. Bentley Historical Library, University of Michigan, Ann Arbor, Michigan.

Barber G. Buell Papers, 1763–1934. Bentley Historical Library, University of Michigan, Ann Arbor, Michigan.

Copley Family Papers. Bentley Historical Library, University of Michigan, Ann Arbor, Michigan.

Detroit Horticultural Society Account Book, 1846–1857. University Archives and Historical Collection, Michigan State University, East Lansing, Michigan.

Ellwanger and Barry Company Papers. Department of Rare Books and Special Collections, Rush Rhees Library, University of Rochester, Rochester, New York.

Ferry Family Papers. Bentley Historical Library, University of Michigan, Ann Arbor, Michigan.

William Finley Papers. Bentley Historical Library, University of Michigan, Ann Arbor, Michigan.

J. C. Holmes Papers, 1849–1854. Burton Historical Collection, Detroit Public Library, Detroit, Michigan.

Mrs. Eda Kelly and Mrs. Esther Pellowe Papers. Diary of Esther Copley Lawrence. University Archives and Regional History Collections, Western Michigan University, Kalamazoo, Michigan.

John Kennicott Business Correspondence and Papers. The Grove National Historic Landmark, Glenview, Illinois. Microfilm Copy, Rolls 13–17, Illinois State Historical Library, Springfield, Illinois.

Lenawee County Horticultural Society Records. University Archives and Historical Collection, Michigan State University, East Lansing, Michigan.

T. T. Lyon Letters, 1874–1880. University Archives and Historical Collection, Michigan State University, East Lansing, Michigan.

Papers of the Patrons of Husbandry, Ypsilanti Grange, No. 56, 1873–1959. Bentley Historical Library, University of Michigan, Ann Arbor, Michigan.

Postcard Collection. Bentley Historical Library, University of Michigan, Ann Arbor, Michigan.

Prince Family Manuscript Collection. National Agricultural Library, United States Department of Agriculture, Beltsville, Maryland.

Smithe Family Photograph Series. Bentley Historical Library, University of Michigan, Ann Arbor, Michigan.

Stereograph Collection, Adrian, Tecumseh, Ann Arbor, and Ypsilanti, Michigan. Clarke Historical Library, Central Michigan University, Mount Pleasant, Michigan.

United States Post Office Records, South Lyon, Michigan, 1853–1865. Bentley Historical Library, University of Michigan, Ann Arbor, Michigan.

Clara Waldron Collection, 1821–1899. University Archives and Historical Collection, Michigan State University, East Lansing, Michigan.

Washtenaw County Fair Records, 1871–1880. Bentley Historical Library, University of Michigan, Ann Arbor, Michigan.

Manuscript Census Records

United States Federal Census, Productions of Agriculture, 1850, 1860, 1870, and 1880, Cass County, Michigan. Microfilm Copy, T-1164, Roll 1 (1850); Roll 7 (1860); Roll 17 (1870); and Roll 33 (1880).

United States Federal Census, Productions of Agriculture, 1870, Jackson County, Michigan. Microfilm Copy, T-1164, Roll 19.

United States Federal Census, Productions of Agriculture, 1870, Lenawee County, Michigan. Microfilm Copy, T-1164, Roll 20.

United States Federal Census, Productions of Agriculture, 1870, Washtenaw County, Michigan. Microfilm Copy, T-1164, Roll 23.

United States Federal Population Census, 1870, Cass County, Michigan. Microfilm Copy, M-593, Roll 668.

United States Federal Population Census, 1870, Jackson County, Michigan. Microfilm Copy, M-593, Roll 678 and 679.

United States Federal Population Census, 1850–1870, Lenawee County, Michigan. Microfilm Copy, M-432, Roll 355 (1850); M-653, Roll 551 (1860); and M-593, Roll 685 and 686 (1870).

United States Federal Population Census, 1870, Washtenaw County, Michigan. Microfilm Copy, M-593, Roll 707 and 708.

Published Primary Sources

Miscellaneous Publications and Collections

Constitution and By-laws of the Lenawee County Agricultural and Horticultural Society, Also, The List of Judges and Premiums and the Regulations Adopted for the Annual Fair for the Year 1851. Adrian, MI: R. W. Ingals, 1851.

Michigan Collection, Michigan Library and Historical Center, Lansing, Michigan.

Garfield, Charles W. *A Glimpse of Michigan Horticulture.* No date or place of publication. Michigan Collection, Michigan Library and Historical Center, Lansing, Michigan.

Washtenaw County Agricultural and Horticultural Society List of Premiums and Regulations, 20th Annual Fair, October 7–9, 1868. Ann Arbor, MI: Dr. Chase's Steam Printing House, 1868. Michigan Collection, Michigan Library and Historical Center, Lansing, Michigan.

Government Documents

Acts of the Legislature of the State of Michigan Passed at the Regular Session of 1867, Vol. 1. Lansing: John A. Kerr and Co., 1867.

"Annual Report of the Postmaster General, 1860." In *Message from the President of the United States to the Two Houses of Congress at the Commencement of the Second Session of the 36th Congress.* Vol. 3. Washington: George W. Bowman, 1860. Serial Set 1080. 36th Congress, 2nd Session, December 3, 1860–March 2, 1861. Senate Executive Document, Document I, Part 3, 400–499. (Readex microprint).

Census and Statistics of the State of Michigan, 1864. Lansing: John A. Kerr and Co., 1865.

Census of the State of Michigan, 1874. Lansing: W. S. George and Co., 1875.

Compendium of the Tenth Census, 1880, Part II. Washington, DC: Government Printing Office, 1885. Microfilm Copy, 1880.6–1880.7-2, Roll 3.

Cooley, Thomas M., compiler. *The Compiled Laws of the State of Michigan.* Lansing: Hosmer and Kerr, 1857.

Kennedy, Joseph C. G., compiler. *Population of the United States in 1860; Compiled from the Original Returns of the Eighth Census.* Washington, DC: Government Printing Office, 1864.

Manufactures of the United States in 1860, Compiled from the Original Returns of the Eighth Census, 1860. Washington, DC: Government Printing Office, 1865.

Statistical View of the United States, being a Compendium of the Seventh Census, 1850. Microfilm Copy, 1850.1–1850.3, Roll 1.

Statistics of the State of Michigan Collected for the Ninth Census of the United States, June 1, 1870. Lansing: W. S. George and Co., 1873.

Walker, Francis A., compiler. *The Statistics of the Population of the United States, Embracing the Tables of Race, Nationality, Sex, Selected Ages, and Occupations. To Which Are Added the Statistics of School Attendance and Illiteracy, of Schools, Libraries, Newspapers and Periodicals, Churches, Pauperism and Crime, and of Areas, Families and Dwellings Compiled from the Original Returns of the Ninth Census, June 1, 1870.* Vol. 1. Washington, DC: Government Printing Office, 1872.

Walker, Francis A., compiler. *The Statistics of the Wealth and Industry of the United States, Embracing The Tables of Wealth, Taxation, and Public Indebtedness; Of Agriculture; Manufactures; Mining; and the Fisheries. With Which Are Repro-*

duced, from the Volume of Population, The Major Tables of Occupations. Compiled from the Original Returns of the Ninth Census, June 1, 1870. Vol. 3. Washington, DC: Government Printing Office, 1872.

Waring, George E. Jr., compiler. *Report on the Social Statistics of Cities. Part II. The Southern and the Western States.* Washington, DC: Government Printing Office, 1887.

Weeks, Jos. D., compiler. *Report on the Statistics of Wages in Manufacturing Industries with Supplementary Reports on the Average Retail Prices of Necessaries of Life and on Trade Societies, and Strikes and Lockouts.* Washington, DC: Government Printing Office, 1886.

Newspapers and Magazines

Adrian Times and Expositor, Adrian, Michigan, 1867–1875.
American Agriculturist, 1850–1880.
Decatur Republican, Decatur, Michigan, 1945.
Dexter Leader, Dexter, Michigan, 1869–1875.
Dowagiac Daily New, Dowagiac, Michigan, 1942.
Gardener's Monthly, 1859–1875.
Hearth and Home, 1874–1876.
Horticulturist and Journal of Rural Art and Rural Taste, 1846–1875.
Jackson Daily Citizen, Jackson, Michigan, 1865–1875.
Jackson Weekly Citizen, Jackson, Michigan, 1865–1875.
Ladies' Floral Cabinet and Pictorial Home Companion, 1874–1880.
Magazine of Horticulture, Botany, and All Useful Discoveries and Improvement in Rural Affairs, 1835–1868.
Michigan Farmer, 1843–1880.
Michigan State Journal, Ann Arbor, Michigan, 1839–1845.
Moore's Rural New Yorker, 1860–1880.
Prairie Farmer, 1849–1875.
Rural Annual and Horticultural Directory, 1856–1863.
Tecumseh Herald, Tecumseh, Michigan, 1863–1866.
Tilton's Journal of Horticulture and Floral Magazine, 1868–1875.
Vick's Illustrated Monthly Magazine, 1878–1881.
Ypsilanti Commercial, Ypsilanti, Michigan, 1869–1875.

Agricultural and Horticultural Society and Department Reports

Annual Report of the Secretary of the State Board of Agriculture of the State of Michigan, 1862–1880.
Annual Report of the Secretary of the State Pomological Society of Michigan, 1871–1881.
Report of the United States Commissioner of Agriculture, 1862–1874.
Report of the United States Commissioner of Patents, 1850–1861.

Transactions of the Illinois State Agricultural Society with Reports from County Agricultural Societies and from Kindred Associations, 1853–1870.
Transactions of the Illinois State Horticultural Society, 1868–1875.
Transactions of the Indiana Horticultural Society, 1870–1880.
Transactions of the Michigan State Agricultural Society, 1849–1861.
Transactions of the Minnesota State Horticultural Society, 1866–1879.

Nursery and Seed Catalogues

William Adair and Co. Catalogues. 1871, 1872. Clarke Historical Library, Central Michigan University, Mount Pleasant, Michigan.
M. B. Bateham and Co. Columbus Nursery Catalogue. 1859. John Kennicott Business Correspondence and Papers. The Grove National Historic Landmark, Glenview, Illinois.
Benjamin K. Bliss Co. Catalogues. 1860, 1867. National Agricultural Library, United States Department of Agriculture, Beltsville, Maryland.
Benjamin K. Bliss Co. Catalogue. 1872. Special Collections, Michigan State University Library, East Lansing, Michigan.
Robert Buist, Rosedale Nurseries Catalogue. 1859–60. National Agricultural Library, United States Department of Agriculture, Beltsville, Maryland.
Detroit and Oakland Horticultural Gardens Co. Catalogue. 1846. Michigan Collection, Michigan Library and Historical Center, Lansing, Michigan.
Detroit Nursery Co. Catalogue. 1852. Michigan Collection, Michigan Library and Historical Center, Lansing, Michigan.
H. A. Dreer Co. Catalogues. 1860, 1868, 1869. National Agricultural Library, United States Department of Agriculture, Beltsville, Maryland.
Ellwanger and Barry Co. Catalogues. 1850–51, 1860, 1870. Ellwanger and Barry Company Papers, Department of Rare Books and Special Collections, Rush Rhees Library, University of Rochester, Rochester, New York.
D. M. Ferry Co. Catalogues. 1877 and 1881. Clarke Historical Library, Central Michigan University, Mount Pleasant, Michigan.
D. M. Ferry Co. Catalogue. 1881. Special Collections, Michigan State University Library, East Lansing, Michigan.
Hubbard and Davis Co. Catalogues. 1853 and 1871. Clarke Historical Library, Central Michigan University, Mount Pleasant, Michigan.
I. E. Ilgenfritz and Co. Monroe Nursery Catalogue. 1874. University Archives and Historical Collection, Michigan State University, East Lansing, Michigan.
E. G. Mixer Co. Catalogue. 1852. Clarke Historical Library, Central Michigan University, Mount Pleasant, Michigan.
F. K. Phoenix. Bloomington Nursery Catalogue. 1859. National Agricultural Library, United States Department of Agriculture, Beltsville, Maryland.
F. K. Phoenix. Bloomington Nursery Catalogue. 1861. John Kennicott Business Correspondence and Papers, The Grove National Historic Landmark, Glenview, Illinois.

William Prince. *Catalogue of Fruit and Ornamental Trees and Plants, Bulbous Flower Roots, Green-House Plants etc. etc. Cultivated at the Linnaean Botanic Garden*. New York: T. and J. Swords, 1823.

William R. Prince and Co. Catalogues. 1847 and 1860–1. National Agricultural Library, United States Department of Agriculture, Beltsville, Maryland.

J. M. Thorburn and Co. Catalogues. 1856 and 1870. National Agricultural Library, United States Department of Agriculture, Beltsville, Maryland.

James Vick and Co. Catalogues. 1865, 1868, 1871, 1873, 1875, 1876, 1877. Department of Rare Books and Special Collections, Rush Rhees Library, University of Rochester, Rochester, New York.

Washburn and Co.'s Amateur Cultivator's Guide to the Flower and Kitchen Garden. Boston: Washburn and Co., 1868.

Books

Adrian City Directory, 1874–5, Also Complete Business Directories of Tecumseh, Blissfield, Morenci and Monroe City. Detroit: Burch, Montgomery and Co., 1874.

Allen, Lewis. *Rural Architecture, Being a Complete Description of Farm Houses, Cottages, and Out Buildings Together with Lawns, Pleasure Grounds, and Parks*. New York: C. M. Saxton, 1863.

Beecher, Catharine, and Harriet Beecher Stowe. *The American Woman's Home; or Principles of Domestic Science; Being a Guide to the Formation and Maintenance of Economical, Healthful, Beautiful and Christian Homes*. New York: J. B. Ford, 1869.

Beecher, Henry Ward. *Plain and Pleasant Talk About Fruit, Flowers and Farming*. New York: Derby and Jackson, 1859.

Bolles, Albert S. *Industrial History of the United States*. Norwich, CT: Henry Bill Publishing Co., 1879.

Breck, Joseph. *The Flower Garden; or, Breck's Book of Flowers*. Boston: John P. Jewett and Co., 1851.

———. *New Book of Flowers*. New York: Orange Judd and Co., 1866.

Bremer, Fredrika. *The Homes of the New World; Impressions of America*. Vols. 1 and 2. Translated by Mary Howitt. New York: Harper and Brothers, 1853.

Brown, Charles E., compiler. *Brown's City Directory of Adrian, Michigan*. Adrian, MI: Adrian Times and Expositor Steam Printing Office, 1870.

Buel, Jesse. *The Farmer's Companion; or, Essays on the Principles and Practice of American Husbandry*. New York: Harper and Brothers, 1847.

Bushnell, Horace. *Christian Nurture*. New York: C. Scribner, 1863.

Cleveland, H. W. *Landscape Architecture As Applied to the Wants of the West*. Chicago: Jansen, McClurg and Co., 1873; reprint, Pittsburgh: University of Pittsburgh Press, 1965.

Cleveland, H. W., William Backus, and Samuel D. Backus. *The Requirements of American Village Homes Considered and Suggested; with Designs for Such Houses of Moderate Cost*. New York: D. Appleton and Co., 1856.

Colman, Henry. *Agriculture and Rural Economy from Personal Observation.* Vol. 1. 5th ed. Boston: Phillips, Sampson and Co., 1856.

Combination Atlas Map of Jackson County, Michigan, Compiled, Drawn, and Published from Personal Examinations and Surveys. Chicago: Everts and Stewart, 1874.

Combination Atlas Map of Lenawee County, Michigan, Compiled, Drawn, and Published from Personal Examinations and Surveys. Chicago: Everts and Stewart, 1874.

Combination Atlas Map of Macomb County, Michigan, Compiled, Drawn, and Published from Personal Examinations and Surveys. Philadelphia: D. J. Stewart, 1875.

Combination Atlas Map of St. Clair County, Michigan, Compiled, Drawn, and Published from Personal Examinations and Surveys. Philadelphia: Everts and Stewart, 1876.

Combination Atlas Map of Washtenaw County, Michigan, Compiled, Drawn, and Published from Personal Examinations and Surveys. Chicago: Everts and Stewart, 1874.

Dewey, D. M. *The Tree Agents' Private Guide. A Manual for the Use of Agents and Dealers Containing Suggestions and Directions for Successful Work in Canvassing for the Sale of Nursery Stock and also a Brief Pronouncing Dictionary of Leading Horticultural Terms, Names of Plants, Flowers etc.* Rochester, NY: D. M. Dewey Publishing Co., 1875.

Downing, Andrew Jackson. *The Architecture of Country Houses Including Designs for Cottages, and Farm-Houses, and Villas, with Remarks on Interiors, Furniture, and the Best Modes of Warming and Ventilating.* New York: D. Appleton and Co., 1850; reprint, New York: Dover Publications, 1969.

———. *Cottage Residences or, A Series of Designs for Rural Cottages and Cottage Villas, and their Gardens and Grounds. Adapted to North America.* Ed. George E. Harney. New York: John Wiley and Son, 1873; reprint, New York: Dover Publications, 1981.

———. *Rural Essays.* New York: Leavitt and Allen, 1854.

———. *A Treatise on the Theory and Practice of Landscape Gardening Adapted to North America; With a View to the Improvement of Country Residences.* 7th edition with a supplement by Henry Winthrop Sargent. New York: Orange Judd and Co., 1865; reprint, New York: Dover Publications, 1991.

Elder, Walter. *The Cottage Garden of America Containing Practical Directions for the Culture of Flowers, Fruits and Vegetables, The Nature and Improvement of Soils, Manures and Their Application, Wounds, Diseases, Cures, Monthly Calendar, Insects, etc.* 2nd ed. Philadelphia: Moss and Brother, 1856.

Elliott, F. R. *Hand Book of Practical Landscape Gardening.* Rochester, NY: D. M. Dewey Publishing Co., 1877.

———. *Popular Deciduous and Evergreen Trees and Shrubs for Planting in Parks, Gardens, Cemeteries, etc., etc.* New York: George R. Woodward, 1870.

Fessenden, Thomas G. *The New American Gardener.* Boston: Odiorne and Co., 1833.

Harrington, Bates. *How 'Tis Done, A Thorough Ventilation of the Numerous Schemes Conducted by Wandering Canvassers together with the Various Advertising Dodges for the Swindling of the Public.* Chicago: Fidelity Publishing Co., 1879.

Henderson, Peter. *Gardening for Pleasure, A Guide to the Amateur in the Fruit, Vegetable, and Flower Garden with Full Directions for the Greenhouse, Conservatory, and Window Garden.* New York: Orange Judd Co., 1885.

———. *Practical Floriculture, A Guide to the Successful Cultivation of Florists' Plants, for the Amateur and Professional Florist.* New York: Orange Judd and Co., 1869.

Ingram, J. S. *Centennial Exposition Described and Illustrated, being a Concise and Graphic Description of this Grand Enterprise Commemorative of the First Centenary of American Independence.* Philadelphia: Hubbard Brothers, 1876.

Jackson City Directory, 1873–4. Jackson, MI: Polk, Murphy and Co., 1873.

Jackson City Directory, 1876. Jackson, MI: R. L. Polk and Co., 1876.

Jackson City Directory, 1883. Detroit: R. L. Polk and Co., 1883.

Johnson, Edwin A. *Winter Greeneries at Home.* 2nd ed. New York: Orange Judd Co., 1878.

Johnson, Mrs. S. O. *Every Woman Her Own Flower Gardener. A Handy Manual of Flowering Gardening for Ladies.* New York: Henry T. Williams, 1871.

Kern, G. H. *Practical Landscape Gardening with References to the Improvement of Rural Residences Giving General Principles to the Art with Full Directions for Planting Shade Trees, Shrubbery and Flowers, and Laying Out Grounds.* Cincinnati: Moore, Wilstach, Keys and Co., 1855.

Kirkland, Caroline. *A Book for the Home Circle or, Familiar Thoughts on Various Topics, Literary, Moral and Social, A Companion for the Evening Book.* New York: Charles Scribner, 1856.

———. *A New Home—Who'll Follow? or Glimpses of Western Life.* Ed. Sandra A. Zagarell. New York: C. S. Francis, 1840; reprint, New Brunswick, NJ: Rutgers University Press, 1990.

Lake, D. J. *Atlas of Cass County, Michigan.* Philadelphia: C. O. Titus, 1872.

"List of the Principal Nurserymen, Florists, and Seedsmen." In *American Horticultural Annual, 1867,* 147–150. New York: Orange Judd and Co., 1867.

Lyon, T. T. *History of Michigan Horticulture: Being a Part of the Seventeenth Annual Report of the Secretary of the State Horticultural Society of Michigan.* Lansing, MI: Thorp and Godfrey, 1887.

McCabe, James D. *The Illustrated History of the Centennial Exhibition.* Philadelphia: National Publishing Co., 1877.

McCracken, S. B., ed. *Michigan and the Centennial, Being a Memorial Record Appropriate to the Centennial Year.* Detroit: Printed for the Publisher S. B. McCracken at the Office of the Detroit Free Press, 1876.

"Nurserymen, Florists, Seedsmen, and Dealers in Horticultural Stock." In *American Horticultural Annual, 1871,* 140–152. New York: Orange Judd and Co., 1871.

Olmsted, Frederick Law. *Creating Central Park, 1857–1861.* The Papers of Frederick Law Olmsted. Vol. 3. Ed. Charles E. Beveridge and David Schuyler. Baltimore: Johns Hopkins University Press, 1983.

Periam, Jonathan. *The Groundswell. A History of the Origin, Aims, and Progress of the Farmers' Movement.* Cincinnati: E. Hannaford and Co., 1874.

Rand, Edward Sprague, Jr. *Flowers for the Parlor and Garden.* Boston: J. E. Tilton and Co., 1869.

Robinson, Solon. *Facts for Farmers; Also for the Family Circle.* Vol. 2. New York: A. J. Johnson, 1866.

Scott, Frank J. *The Art of Beautifying Suburban Home Grounds of Small Extent.* New York: D. Appleton, 1870; facsimile edition, Watkins Glen, NY: Library of Victorian Culture, American Life Foundation, 1982.

Stowe, Harriet Beecher. *House and Home Papers.* Boston: Fields, Osgood, and Co., 1869.

Thomas, J. J. "A Complete Country Residence." In *The Illustrated Register of Rural Affairs and Cultivators Almanac for the Year 1858,* 22–23. Albany, NY: Luther Tucker and Sons, 1858.

Vaux, Calvert. *Villas and Cottages, A Series of Designs Prepared for Execution in the United States.* New York: Harper Brothers, 1864.

Warder, John A. *Hedges and Evergreens, A Complete Manual for the Cultivation, Pruning and Management of all Plants Suitable for American Hedging; especially for Maclura or Osage Orange. To Which is Added, A Treatise on Evergreens.* New York: Orange Judd and Co., 1858.

Ware, John F. W. *Home Life: What It Is and What It Needs.* Chicago: R. M. Morris, 1864.

Waring, George E., Jr. *Village Improvements and Farm Villages.* Boston: James R. Osgood and Co., 1877.

Watson, Alexander. *The American Home Garden Being Principles and Rules for the Culture of Vegetables, Fruits, Flowers and Shrubbery.* New York: Harper and Brothers, 1870.

Weidenmann, Jacob. *Beautifying Country Homes; A Handbook of Landscape Gardening.* New York: Orange Judd and Co., 1870.

Wheeler, Gervase. *Homes for the People in Suburb and Country, The Villa, the Mansion, and the Cottage Adapted to American Climate and Wants. With Examples Showing How To Alter and Remodel Old Buildings. In a Series of One Hundred Original Designs.* New York: Charles Scribner, 1858.

Williams, Henry T., ed. *Window Gardening Devoted Specially to the Culture of Flowers and Ornamental Plants for In Door Use and Parlor Decoration.* New York: Henry T. Williams, 1872.

Williams, Henry T. and Mrs. C. S. Jones. *Beautiful Homes; or, Hints in House Furnishing.* New York: Henry T. Williams, 1878.

Unpublished Secondary Sources

Dissertations

Avery, Julie Ann. "An Exploration of Several Early Michigan County Fairs as Community Arts Organizations of the 1850s, 1860s, and 1870s." Ph.D. diss., Michigan State University, 1992.

Lawrence, Henry W. "The Geography of the U.S. Nursery Industry: Locational Change and Regional Specialization in the Production of Woody Ornamental Plants." Ph.D. diss., University of Oregon, 1985.

Prendergast, Norma. "The Sense of Home: Nineteenth-Century Domestic Architectural Reform." Ph.D. diss., Cornell University, 1981.

Rasmussen, Chris A. "State Fair: Culture and Agriculture in Iowa, 1854–1941." Ph.D. diss., Rutgers University, 1992.

Watson, Daryl G. "Shade and Ornamental Trees in the Nineteenth-Century Northeastern United States." Ph.D. diss., University of Illinois, 1978.

Weyeneth, Robert Richardi. "Moral Spaces: Reforming the Landscape of Leisure in Urban America, 1850–1920." Ph.D. diss., University of California, Berkeley, 1984.

Published Secondary Sources

Articles

Aeberli, William A., and Margaret Becket. "Joseph Harris-Captain of the Rochester Seed Industry." *University of Rochester Library Bulletin* 35 (1982): 69–83.

Ames, Kenneth L. "Material Culture as Non-Verbal Communication: A Historical Case Study." *Journal of American Culture* 3, no. 4 (Winter 1980): 619–641.

Ball, Franklin. "Just Express It." *Railroad Magazine* 47, no. 1 (October 1948): 94–98.

Bell, Susan Groag. "Women Create Gardens in Male Landscapes: A Revisionist Approach to Eighteenth-Century English Garden History." *Feminist Studies* 16, no. 3 (Fall 1990): 471–491.

Blinn, Thomas W. "Early Michigan Postal History." *American Philatelic Congress* 13 (1947): 5–24.

Bloomberg, Susan E., Mary Frank Fox, Robert M. Warner, and Sam Bass Warner, Jr. "A Census Probe into Nineteenth-Century Family History: Southern Michigan, 1850–1880." *Journal of Social History* 5 (1971): 26–45.

Boorstin, Daniel J. "Welcome to the Consumption Community." *Fortune* 76 (1 September 1967): 118–20, 131–2, 134–8.

Brewer, L. G., T. W. Hodler, and H. A. Rapp, "The Presettlement Vegetation of Southwestern Michigan." *Michigan Botanist* 23 (1984): 153–6.

Brown, Richard D. "Modernization and the Modern Personality in Early America, 1600–1865: A Sketch of a Synthesis." *Journal of Interdisciplinary History* 2 (Winter 1972): 201–28.

———. "Modernization: A Victorian Climax." *American Quarterly* 27 (December 1975): 533–548.

Bullion, Brenda. "Early American Farming and Gardening Literature: 'Adapted to the Climates and Seasons of the United States.'" *Journal of Garden History* 12, no. 1 (1992): 29–51.

Bunkers, Suzanne L. "Diaries: Public and Private Records of Women's Lives." *Legacy* 7 (Fall 1990): 17–26.

———. "'Faithful Friend': Nineteenth-Century Midwestern American Women's Unpublished Diaries." *Women's Studies International Forum* 10 (1987): 7–17.

Carson, Gerald. "Get the Prospect Seated ... And Keep Talking." *American Heritage* 9 (August 1958): 38–41 and 77–80.

Clark, Clifford E., Jr. "Domestic Architecture as an Index to Social History: The Romantic Revival and the Cult of Domesticity in America, 1840–1870." *Journal of Interdisciplinary History* 7, no. 1 (Summer 1976): 33–56.

Constantine, Stephen. "Amateur Gardening and Popular Recreation in the Nineteenth and Twentieth Centuries." *Journal of Social History* 14 (1981): 387–406.

Conzen, Michael P. "The County Landownership Map in America, Its Commercial Development and Social Transformation 1814–1939." *Imago Mundi* 36 (1984): 9–31.

———. "Landownership Maps and County Atlases." *Agricultural History* 58 (1984): 118–122.

———. "Woodland Clearances." *Geographical Magazine* 52 (April 1980): 483–491.

Crawford, Pleasance. "Some Early Ontario Nurserymen." *Canadian Horticultural History* 1, no. 1 (1985): 28–64.

Dandekar, Hemalata C., and Mary Bockstahler. "The Changing Farmscape, A Case Study of German Farmers in Southeast Michigan." *Michigan History* 74, no. 1 (1990): 42–7.

Danton, Robert. "The Symbolic Element in History." *Journal of Modern History* 58 (March 1986): 218–34.

Davis, Rodney O. "Coming to Terms with County Histories." *Western Illinois Regional Studies* 2, no. 2 (1979): 144–155.

Demaree, Albert L. "The Farm Journals, Their Editors, and Their Public, 1830–1860." *Agricultural History* 15 (October 1941): 182–188.

Desmond, Ray. "Technical Problems in Transporting Living Plants in the Age of Sail." *Canadian Horticultural History* 1, no. 2 (1986): 74–90.

Dick, W. B. "A Study of the Original Vegetation of Wayne County, Michigan." *Papers of the Michigan Academy of Science* 22 (1937): 329–334.

Dodge, Sheridan L. "Presettlement Forest of South-Central Michigan." *Michigan Botanist* 26 (1987): 139–152.

Doell, M. Christine Klim. "Verdant Frames: Plant Portraits from the Ellwanger and Barry Collection." *University of Rochester Library Bulletin* 39 (1986): 11–23.

Drepperd, Carl W. "The Tree, Fruit and Flower Prints of D. M. Dewey, Rochester, New York from 1844." *Spinning Wheel* 12, no. 5 (May 1956): 12–15, 46.

Duncan, James P. "Landscape Taste as a Symbol of Group Identity: A Westchester Co. Village." *Geographical Review* 63 (July 1973): 334–55.

Elkins, James. "On the Conceptual Analysis of Gardens." *Journal of Garden History* 13, no. 4 (1993): 189–198.

Ernst, Erik A. "John A. Kennicott of The Grove. Physician, Horticulturist, and Journalist in Nineteenth-Century Illinois." *Illinois State Historical Society Journal* 74 (Summer 1981): 109–118.

Favretti, Rudy. "The Ornamentation of New England Towns: 1750–1850." *Journal of Garden History* 2, no. 4 (October–November 1983): 325–342.

———. "Thomas Jefferson's 'Ferme Ornée' at Monticello." *Proceedings of the American Antiquarian Society* 103, no. 1 (1993): 17–29.

Friedman, Lee M. "The Drummer in Early American Merchandise Distribution." *Bulletin of the Business Historical Society* 21 (April 1947): 39–44.

Gates, Jane. "Old Nursery Catalogs." *Pacific Horticulture* 42, no. 2 (1981): 9–11.

Gates, Warren J. "Modernization as a Function of an Agricultural Fair, The Great Grangers' Picnic Exhibition at Williams Grove, Pennsylvania, 1873–1916." *Agricultural History* 58 (July 1984): 262–279.

Gordon, Jean, and Jan McArthur. "American Women and Domestic Consumption, 1800–1920: Four Interpretive Themes." *Journal of American Culture* 8, no. 3 (Fall 1985): 35–46.

Grosso, Diane Holahan. "From the Genesee to the World." *University of Rochester Library Bulletin* 35 (1982): 3–24.

Groth, Paul. "Lot, Yard, and Garden. American Distinctions." *Landscape* 30, no. 3 (1990): 29–35.

Harvey, P. D. A. "The Documents of Landscape History: Snares and Delusions." *Landscape History* 13 (1991): 47–52.

Hayter, Earl W. "Horticultural Humbuggery Among the Western Farmers, 1850–1890." *Indiana Magazine of History* 43, no. 3 (September 1947): 205–224.

Hayward, D. Geoffrey. "Home as an Environmental and Psychological Concept." *Landscape* 20, no. 1 (October 1975): 2–9.

Hodder, Ian. "Economic and Social Stress and Material Culture Patterning." *American Antiquity* 44, no. 3 (1979): 446–454.

Hollander, Stanley C. "Nineteenth Century Anti-Drummer Legislation in the United States." *Business History Review* 38 (Winter 1964): 479–500.

Horsfall, Frank Jr. "Horticulture in Eighteenth-Century America." *Agricultural History* 43 (January 1969): 159–167.

Howe, Daniel Walker. "American Victorianism as a Culture." *American Quarterly* 27 (December 1975): 507–532.

Hugill, Peter J. "English Landscape Tastes in the United States." *Geographical Review* 76 (1990): 408–423.
Jaffee, David. "Peddlers of Progress and the Transformation of the Rural North, 1760–1860." *Journal of American History* 78 (September 1991): 511–535.
Jeffrey, Kirk. "The Family as Utopian Retreat from the City: The Nineteenth-Century Contribution." *Soundings* 55 (1972): 21–41.
Jenks, William L. "A Michigan Family of Mapmakers." *Michigan History Magazine* 11, no. 2 (1927): 242–250.
Kabelac, Karl Sanford. "Advice for Gardeners: Vick's Monthly Magazine (The First Series, 1878–1891)." *University of Rochester Library Bulletin* 39 (1986): 24–35.
———. "Nineteenth-Century Rochester Fruit and Flower Plates." *University of Rochester Library Bulletin* 35 (1982): 93–113.
Keir, R. Malcolm. "The Tin Peddler." *Journal of Political Economy* 21 (1913): 255–8.
Kline, Priscilla Carrington. "New Light on the Yankee Peddler." *The New England Quarterly* 12 (March 1939): 80–98.
Kwolek-Folland, Angel. "The Elegant Dugout. Domesticity and Moveable Culture in the United States, 1870–1900." *American Studies* 25, no. 2 (1984): 21–37.
Leach, William R. "Transformation in a Culture of Consumption: Women and Department Stores, 1890–1925." *Journal of American History* 71, no. 2 (September 1984): 319–342.
Lees, Carlton B. "The Golden Age of Horticulture." *Historic Preservation* 24–5 (October–December 1972): 32–7.
Leone, Mark P. "The Relationship Between Archaeological Data and the Documentary Record: Eighteenth-Century Gardens in Annapolis, Maryland." *Historical Archaeology* 22, no. 1 (1988): 29–35.
Lowen, Sara. "The Tyranny of the Lawn." *American Heritage* 42, no. 5 (September 1991): 45–55.
Lux, Karen D. "A Folkloric Approach to Nineteenth-Century County Histories." *New York Folklore* 8, no. 1–2 (1982): 25–34.
Maclean, Jayne T. "Nursery and Seed Trade Catalogs." *Journal of NAL Associates* 5, no. 3–4 (1980): 88–92.
Manning, Warren. "The History of Village Improvement in the United States." *Craftsman* 5 (February 1904): 423–435.
Marsh, Margaret. "From Separation to Togetherness: The Social Construction of Domestic Space in American Suburbs, 1840–1915." *Journal of American History* 76 (June–September 1989): 506–527.
Marti, Donald B. "Agricultural Journalism and the Diffusion of Knowledge: The First Half Century in America." *Agricultural History* 54 (January 1980): 28–37.
———. "Sisters of the Grange: Rural Feminism in the Late Nineteenth Century." *Agricultural History* 58 (July 1984): 247–261.

Marti, Donald B. "Woman's Work in the Grange: Mary Ann Mayo of Michigan 1882–1903." *Agricultural History* 56 (April 1982): 439–452.

Mastromarino, Mark A. "Elkanah Watson and Early Agricultural Fairs, 1790–1860." *Historical Journal of Massachusetts* 17 (Summer 1989): 105–118.

McClaugherty, Martha Crabill. "Household Art, Creating the Artistic Home, 1868–1893." *Winterthur Portfolio* 18, no. 1 (1983): 1–26.

McClelland, Linda Flint. "Imagery, Ideals, and Social Values: The Interpretation and Documentation of Cultural Landscapes." *Public Historian* 13, no. 2 (Spring 1991): 107–124.

McDermott, John J. "Nature Nostalgia and the City: An American Dilemma." *Soundings* 55 (1972): 1–20.

McMurry, Sally. "City Parlor, Country Sitting Room, Rural Vernacular Design and the American Parlor, 1840–1900." *Winterthur Portfolio* 20, no. 4 (1985): 261–280.

———. "Who Read the Agricultural Journals? Evidence from Chenango County, New York." *Agricultural History* 63 (1989): 1–18.

———. "Women in the American Vernacular Landscape." *Material Culture* 20, no. 1 (1988): 33–49.

Merish, Lori. "'The Hand of Refined Taste' in the Frontier Landscape: Caroline Kirkland's *A New Home, Who'll Follow?* and the Feminization of American Consumerism." *American Quarterly* 45, no. 4 (December 1993): 485–523.

Miller, George J. "Some Geographic Influences in the Settlement of Michigan and in the Distribution of its Population." *Bulletin of the American Geographical Society* 45, no. 5 (1913): 321–348.

Moore, Edward N. "Early Mail Service on the Ohio and Erie Canal." *American Philatelic Congress* 11 (1945): 23–28.

Motz, Marilyn Ferris. "Folk Expressions of Time and Place: 19th-Century Midwestern Rural Diaries." *Journal of American Folklore* 110 (April–June 1987): 131–147.

Mugerauer, Robert. "Midwestern Yards." *Places* 2 (1985): 31–9.

O'Malley, Therese. "Appropriation and Adaptation: Early Gardening Literature in America." *Huntington Library Quarterly* 55 (1992): 401–431.

Onion, Margaret Kent. "Drummers Accommodated, A Nineteenth-Century Salesman in Minnesota." *Minnesota History* 46, no. 2 (1978): 59–65.

Parks, Dan. "The Cultivation of Flower City." *Rochester History* 45, no. 3–4 (1983): 25–45.

Pederson, Jane Marie. "The Country Visitor: Patterns of Hospitality in Rural Wisconsin, 1880–1925." *Agricultural History* 58 (July 1984): 347–65.

Ristow, Walter W. "Nineteenth-Century Cadastral Maps in Ohio." *Papers of the Bibliographic Society of America* 59, no. 3 (1965): 306–315.

Rose, Gregory S. "The County Origins of Southern Michigan's Settlers: 1800–1850." *East Lakes Geographer* 22 (1987): 74–87.

Rowntree, Lester B., and Margaret W. Conkey. "Symbolism and the Cultural Landscape." *Annals of the Association of American Geographers* 70, no. 4 (December 1980): 459–474.

Sarudy, Barbara W. "Genteel and Necessary Amusements: Public Pleasure Gardens in Eighteenth-Century Maryland." *Journal of Garden History* 9, no. 3 (1989): 118–124.

———. "Nurserymen and Seed Dealers in the Eighteenth-Century Chesapeake." *Journal of Garden History* 9, no. 3 (1989): 111–117.

———. "Writings About Pleasure and Kitchen Gardening Available in Eighteenth-Century Maryland." *Journal of Garden History* 9, no. 3 (1989): 153–9.

Schlereth, Thomas J. "Mail-Order Catalogs as Resources in American Culture Studies." *Prospects* 7 (1982): 141–161.

———. "The New England Presence on the Midwest Landscape." *Old Northwest* 9 (Spring 1983/Winter 1984): 125–142.

———. "The Philadelphia Centennial as a Teaching Model." *Hayes Historical Journal* 1, no. 3 (1977): 201–210.

———. "Plants Past: An Historian's Use of Vegetation as Material Cultural Evidence." *Environmental Review* 4, no. 1 (1980): 20–28.

Schmiechen, James A. "The Victorians, the Historians, and the Idea of Modernism." *American Historical Review* 93 (1988): 287–316.

Schuyler, David. "The Evolution of the Anglo-American Rural Cemetery: Landscape Architecture as Social and Cultural History." *Journal of Garden History* 4, no. 3 (1984): 291–304.

Senn, T. L. "Farm and Garden: Landscape Architecture and Horticulture in Eighteenth-Century America." *Agricultural History* 43 (January 1969): 149–157.

Shepherd, Jack. "Seeds of the Presidency, The Capitol Schemes of John Quincy Adams." *Horticulture* 61 (January–June 1983): 38–47.

Skinner, Helen Ross. "With a Lilac by the Door. Some Research into Early Gardens in Ontario." *APT Bulletin* 15, no. 4 (1983): 35–37.

Skocpol, Theda. "The Tocqueville Problem, Civic Engagement in American Democracy." *Social Science History* 21, no. 4 (Winter 1997): 455–479.

Snyder, Ellen Marie. "Victory over Nature. Victorian Cast-Iron Seating Furniture." *Winterthur Portfolio* 20, no. 4 (1985): 221–242.

Spahn, Betty, and Raymond Spahn. "Wesley Raymond Brink, History Huckster." *Journal of the Illinois State Historical Society* 58 (Summer 1965): 117–38.

Stetson, Sarah P. "The Traffic in Seeds and Plants from England's Colonies in North America." *Agricultural History* 23 (1949): 45–56.

Swenson, Russell G. "Illustrations of Material Culture in Nineteenth-Century County and State Atlases." *Pioneer America Society Transactions* 5 (1982): 63–70.

Taber, Morris C. "The New England Influence in South Central Michigan." *Michigan History* 45 (December 1961): 305–336.

Thornton, Tamara Plakins. "The Moral Dimensions of Horticulture in Antebellum America." *New England Quarterly* 57 (1984): 3–24.

Thrower, Norman J. W. "The County Atlas of the United States." *Surveying and Mapping* 21 (1961): 365–373.

Tishler, William H., and Virginia S. Luckhardt. "H. W. S. Cleveland. Pioneer Landscape Architect to the Upper Midwest." *Minnesota History* 49 (1984–5): 281–290.

Traub, Hamilton. "Tendencies in the Development of American Horticultural Associations." *National Horticultural Magazine* 9 (January 1930): 18–26.

Turner, Victor. "Symbolic Studies." *Annual Review of Anthropology* 4 (1975): 145–161.

Upton, Dell. "Pattern Books and Professionalism, Aspects of the Transformation of Domestic Architecture in America, 1800–1860." *Winterthur Portfolio* 19 (1984): 107–150.

Van DeWetering, Maxine. "The Popular Concept of 'Home' in Nineteenth-Century America." *Journal of American Studies* 18 (April 1984): 5–28.

Van Ravenswaay, Charles. "Drawn and Colored from Nature. Painted Nurserymen's Plates." *Magazine Antiques* 123 (January–March 1983): 594–599.

Volkman, Nancy J. "Landscape Architecture on the Prairie: The Work of H. W. S. Cleveland." *Kansas History* 10, no. 2 (1987): 89–110.

Wolschke-Bulmahn, Joachim. "Ethics and Morality. Questions in the History of Garden and Landscape Design: A Preliminary Essay." *Journal of Garden History* 14, no. 3 (1994): 140–146.

Wott, John A. "A Short History of Consumer Horticulture." *HortScience* 17, no. 3 (June 1982): 313–316.

Books

Adams, John N. "Early History of Lenawee County." In *Pioneer Collections: Report of the Pioneer Society of the State of Michigan together with Reports of County, Town, and District Societies.* Vol. 2, 380–381. Detroit: Wm. Graham's Presses, 1880.

Allen, Jessica H., and Thomas J. Schlereth. *Sense of Place, American Regional Cultures.* Lexington: University Press of Kentucky, 1990.

Aston, Michael. *Interpreting the Landscape. Landscape Archaeology in Local Studies.* London: B. T. Batsford, 1985.

Bailey, Liberty Hyde, ed. *Cyclopedia of American Agriculture, A Popular Survey of Agricultural Conditions, Practices and Ideals in the United States and Canada.* Vol. 4—Farm and Community. New York: Macmillan Co., 1909.

Bailey, Liberty Hyde. *The Standard Cyclopedia of Horticulture.* Vols. 1–3. New York: Macmillan Co., 1937.

Baker, Gladys L., Wayne D. Rassmussen, Vivian Wiser, and Jane M. Porter. *Century of Service, The First 100 Years of the United States Department of Agriculture.* Washington, DC: Centennial Committee, U.S.D.A., 1963.

Ball, Michael S., and Gregory W. H. Smith. *Analyzing Visual Data. Qualitative Research Methods.* Vol. 24. London: Sage Publications, 1992.

Barnes, Burton V., and Warren H. Wagner, Jr. *Michigan Trees, A Guide to the Trees of Michigan and the Great Lakes Region.* Ann Arbor: University of Michigan Press, 1981.

Barron, Hal S. *Mixed Harvest, The Second Great Transformation in the Rural North, 1870–1930.* Chapel Hill: The University of North Carolina Press, 1997.

Bender, Thomas. *Community and Social Change in America.* New Brunswick, NJ: Rutgers University Press, 1978.

———. *Toward an Urban Vision, Ideas and Institutions in Nineteenth-Century America.* Lexington: University Press of Kentucky, 1975.

Benes, Peter, ed. *Itinerancy in New England and New York. Annual Proceedings of the Dublin Seminar for New England Folklife, 1984.* Boston: Boston University, 1986.

Benson, Albert Emerson. *History of the Massachusetts Horticultural Society.* Boston: Massachusetts Horticultural Society, 1929.

Beveridge, Charles E. *Frederick Law Olmsted; Designing the American Landscape.* New York: Rizzoli, 1995.

Bledstein, Burton. *The Culture of Professionalism, The Middle Class and the Development of Higher Education in America.* New York: Norton, 1976.

Blumin, Stuart M. *The Emergence of the Middle Class, Social Experience in the American City, 1760–1900.* New York: Cambridge University Press, 1989.

Bonner, Richard I., ed. *Memoirs of Lenawee County, Michigan.* Vol. 1. Madison, WI: Western Historical Association, 1909.

Borchert, John. *America's Northern Heartland: An Economic and Historical Geography of the Upper Midwest.* Minneapolis: University of Minnesota Press, 1987.

Boorstin, Daniel J. *The Americans: The Democratic Experience.* New York: Random House, 1973; reprint, New York: Vintage Books, 1974.

Bournes, Russell. *Americans on the Move: The History of Waterways, Railways, and Highways.* Golden, CO: Fulcrum Publications, 1995.

Boyd, James. *A History of the Pennsylvania Horticultural Society, 1827–1927.* Philadelphia: Pennsylvania Horticultural Society, 1929.

Boydston, Jeanne. *Home and Work, Housework, Wages, and the Ideology of Labor in the Early Republic.* New York: Oxford University Press, 1990.

Boyer, Paul S. *Urban Masses and Moral Order in America, 1820–1920.* Cambridge: Harvard University Press, 1978.

Bronner, Simon J. *Consuming Visions, Accumulation and Display of Goods in America 1880–1920.* New York: W. W. Norton and Co. for the Henry Francis duPont Winterthur Museum, 1989.

Brown, Dee. *The Year of the Century: 1876.* New York: Charles Scribner's Sons, 1966.

Brown, Richard D. *Modernization: The Transformation of American Life, 1600–1865.* New York: Hill and Wang, 1976.

Bullion, Brenda. "The Agricultural Press: 'To Improve the Soil and the Mind'." In *The Farm, Annual Proceedings of the Dublin Seminar for New England Folklife, 1986.* Ed. Peter Benes, 74–94. Boston: Boston University Press, 1988.

Burrill, Dr. T. J. "Advance of Ornamental Horticulture in the State." In *Transactions of the Illinois Horticultural Society,* 160–171. Springfield: Illinois State Horticultural Society, 1906.

Bushman, Richard L. *The Refinement of America, Persons, Houses, Cities.* New York: Alfred A. Knopf, 1992; reprint, New York: Vintage Books, Random House, 1993.

Campbell, Colin. *The Romantic Ethic and the Spirit of Modern Consumerism.* New York: Basil Blackwell, 1987.

Cayton, Andrew R. L., and Susan E. Gray, eds. *The American Midwest, Essays on Regional History.* Bloomington: Indiana University Press, 2001.

Cayton, Andrew R. L., and Peter S. Onuf. *The Midwest and the Nation, Rethinking the History of an American Region.* Bloomington: Indiana University Press, 1990.

Clark, Christopher. *The Roots of Rural Capitalism, Western Massachusetts, 1760–1860.* Ithaca: Cornell University Press, 1990.

Clark, Clifford Edward, Jr. *The American Family Home, 1800–1960.* Chapel Hill: University of North Carolina Press, 1986.

Clifford, Derek. *A History of Garden Design.* Rev. ed. New York: Frederick A. Praeger, 1966.

Cohn, Jan. *The Palace or the Poorhouse, The American House as a Cultural Symbol.* East Lansing: Michigan State University Press, 1979.

Comton, F. E. *Subscription Books.* New York: New York Public Library, 1939.

Conzen, Michael P., ed. *Chicago Mapmakers: Essays on the Rise of the City's Map Trade.* Chicago: Chicago Historical Society, 1984.

———. *The Making of the American Landscape.* London: Harper Collins Academic, 1990.

Cosgrove, Denis, and Stephen Daniels, eds. *The Iconography of Landscape, Essays on the Symbolic Representation, Design, and Use of Past Environments.* New York: Cambridge University Press, 1988.

Cott, Nancy. *The Bonds of Womanhood, Woman's Sphere in New England, 1780–1835.* New Haven: Yale University Press, 1977.

Csikszentmihalyi, Mihaly, and Eugene Rochberg-Halton. *The Meaning of Things, Domestic Symbols and the Self.* Cambridge: Cambridge University Press, 1981.

Culley, Margo, ed. *A Day at a Time: The Diary Literature of American Women from 1764 to the Present.* New York: Feminist Press, 1985.

Cushing, Marshall Henry. *The Story of Our Post Office: The Greatest Government Department in All Its Phases.* Boston: A. M. Thayer and Co., 1893.

Danbom, David B. *Born in the Country, A History of Rural America.* Baltimore: Johns Hopkins University Press, 1995.

Danhof, Clarence H. *Change in Agriculture: The Northern United States, 1820–1870.* Cambridge: Harvard University Press, 1969.

Demaree, Albert L. *The American Agricultural Press, 1819–1860.* New York: Columbia University Press, 1941.

Deming, Brian. *Jackson, An Illustrated History.* Woodland Hills, CA: Windsor Publications, Inc., 1984.

Depew, Chauncey M., ed. *One Hundred Years of American Commerce.* Vol. I. New York: D. O. Haynes and Co., 1895.

Doell, M. Christine Klim. *Gardens of the Gilded Age, Nineteenth-Century Gardens and Homegrounds of New York State.* Syracuse, NY: Syracuse University Press, 1986.

Dolan, J. R. *The Yankee Peddlers of Early America.* New York: Clarkson N. Potter, 1964.
Dorr, John A., Jr., and Donald F. Eschman. *Geology of Michigan.* Ann Arbor: University of Michigan Press, 1970.
Douglas, Mary, and Baron Isherwood. *The World of Goods.* New York: Basic Books, 1979.
Dunbar, Willis F. Rev. ed. by George S. May. *Michigan, A History of the Wolverine State.* Grand Rapids, MI: William B. Eerdmans Publishing Co., 1980.
Edgar, Neal L. *A History and Bibliography of American Magazines, 1810–1820.* Metuchen, NJ: Scarecrow Press, 1975.
Eighty Years of Growing, 1856–1936. Detroit: Ferry-Morse Seed Co., 1936.
Endres, Kathleen L., and Therese L. Lueck. *Women's Periodicals in the United States, Consumer Magazines.* Westport, CT: Greenwood Press, 1995.
Evernden, Neil. *The Social Creation of Nature.* Baltimore: Johns Hopkins University Press, 1992.
Ewen, Stuart. *Captains of Consciousness, Advertising and the Social Roots of the Consumer Culture.* New York: McGraw-Hill Book Co., 1976.
Fackler, P. Mark, and Charles H. Lippy. *Popular Religious Magazines of the United States.* Westport, CT: Greenwood Press, 1995.
Faragher, John Mack. *Sugar Creek, Life on the Illinois Prairie.* New Haven: Yale University Press, 1986.
Favretti, Rudy J., and Joy Putman Favretti. *Landscapes and Gardens for Historic Buildings, A Handbook for Reproducing and Creating Authentic Landscape Settings.* Nashville, TN: American Association for State and Local History, 1978.
Firth, Raymond. *Symbols, Public and Private.* Ithaca: Cornell University Press, 1973.
Fleming, E. McClung. "Artifact Study: A Proposed Model." In *Material Culture Studies in America.* Ed. Thomas J. Schlereth, 162–173. Nashville, TN: American Association for State and Local History, 1986.
Fox, Stephen. *The Mirror Makers, A History of American Advertising and Its Creators.* New York: William Morrow and Co., 1984.
Francis, Mark, and Randolph T. Hester, Jr., eds. *The Meaning of Gardens, Idea, Place, and Action.* Cambridge: MIT Press, 1990.
Francis, Mark, and Margarita Hill. *Gardens in the Mind and in the Heart: Some Meanings of the Norwegian Garden.* Davis: University of California Center for Design Research, 1989.
Fraser, W. Hamish. *The Coming of the Mass Market, 1850–1914.* Hamden, CT: Archon Books, Imprint of Shoe String Press, 1981.
Fuller, George N., ed. *Historic Michigan, Land of the Great Lakes.* Vol. 1. Dayton, OH: National Historical Association, 1924.
Fuller, George N. *Michigan, A Centennial History of the State and Its People.* Vol. 1. Chicago: Lewis Publishing Co., 1939.
Fuller, Wayne E. *The American Mail: Enlarger of the Common Life.* Chicago: University of Chicago Press, 1972.

The Garden of the Genesee. Rochester, NY: Rochester Historical Society, 1940.

Gardner, Charles M. *The Grange—Friend of the Farmer.* Washington, DC: National Grange, 1949.

Garkovitch, Lorraine. *Population and Community in Rural America.* New York: Praeger, 1989.

Garrett, Elisabeth Donaghy. *At Home. The American Family 1750–1870.* New York: Harry N. Abrams, 1989.

Garrison, J. Ritchie. *Landscape and Material Life in Franklin County, Massachusetts, 1770–1860.* Knoxville: University of Tennessee Press, 1991.

Garvey, Ellen Gruber. *The Adman in the Parlor: Magazines and the Gendering of Consumer Culture, 1880s to 1910s.* New York: Oxford University Press, 1996.

Gates, Paul W. *Agriculture and the Civil War.* New York: Alfred D. Knopf, 1965.

———. *The Farmer's Age, Agriculture 1815–1860.* New York: Holt, Rinehart and Winston, 1960.

Geertz, Clifford. "Ideology as a Cultural System." In *Ideology and Discontent.* Ed. David E. Apter, 46–76. New York: Free Press of Glencoe, 1964.

———. *The Interpretation of Cultures, Selected Essays.* New York: Basic Books, 1973.

Giddens, Anthony. *Class Structure of the Advanced Societies.* New York: Harper and Row, 1975.

Gjerde, Jon. *The Minds of the West: Ethnocultural Evolution in the Rural Middle West, 1830–1917.* Chapel Hill: University of North Carolina Press, 1997.

Glover, L. H. *A Twentieth-Century History of Cass County, Michigan.* Chicago: Lewis Publishing Co., 1906.

Goodrum, Charles A., and Helen Dalrymple. *Advertising in America, The First 200 Years.* New York: Harry N. Abrams, 1990.

Gray, Susan E. *The Yankee West, Community Life on the Michigan Frontier.* Chapel Hill: University of North Carolina Press, 1996.

Grier, Katherine C. *Culture and Comfort, People, Parlors and Upholstery 1850–1930.* Rochester, NY: Strong Museum, 1988.

Groth, Paul, ed., *Visions, Culture and Landscape. Working Papers from the Berkeley Symposium on Cultural Landscape Interpretation.* Berkeley: Department of Landscape Architecture, University of California, 1990.

Hafen, LeRoy R. *The Overland Mail 1849–1869, Promoter of Settlement, Precursor of Railroads.* Cleveland, OH: Arthur H. Clark Co., 1926.

Hahn, Steven, and Jonathan Prude, eds. *The Countryside in the Age of Capitalist Transformation, Essays in the Social History of Rural America.* Chapel Hill: University of North Carolina Press, 1985.

Hall, Peter Dobkin. *The Organization of American Culture, 1700–1900: Private Institutions, Elites, and the Origins of American Nationality.* New York: New York University Press, 1982.

Halttunen, Karen. *Confidence Men and Painted Women, A Study of Middle-Class Culture in America, 1830–1870.* New Haven: Yale University Press, 1982.
Hampsten, Elizabeth. *Read This Only to Yourself: The Private Writings of Midwestern Women, 1880–1920.* Bloomington: Indiana University Press, 1982.
Handlin, David P. *The American Home: Architecture and Society, 1815–1915.* Boston: Little, Brown and Co., 1979.
Harvey, John. *Early Gardening Catalogues: with Complete Reprints of Lists and Accounts of the 16th–19th Centuries.* London: Phillimore, 1972.
Hawthorn, Leslie R., and Leonard H. Pollard. *Vegetable and Flower Seed Production.* New York: Blakiston Co., 1954.
Hayden, Delores. *The Grand Domestic Revolution: A History of Feminist Designs for American Homes, Neighborhoods, and Cities.* Cambridge: MIT Press, 1981.
———. *Seven American Utopias, The Architecture of American Socialism, 1790–1975.* Cambridge: MIT Press, 1976.
Hedrick, U.P. *A History of Horticulture in America to 1860.* New York: Oxford University Press, 1950.
Herman, Bernard L. *The Stolen House.* Charlottesville: University Press of Virginia, 1992.
Higham, John. *From Boundlessness to Consolidation, The Transformation of American Culture 1848–1860.* Ann Arbor: William Clements Library, 1969.
Hill, May Brawley. *Grandmother's Garden. The Old-Fashioned American Garden 1865–1915.* New York: Harry N. Abrams, 1995.
Historical Reflections of Cass County. n. p.: Cass County Historical Commission, 1981.
History of Cass County, Michigan. Chicago: Waterman, Watkins and Co., 1882.
History of Jackson County, Michigan. Vol. 1. Chicago: Inter-State Publishing Co., 1881.
History of Washtenaw County, Michigan. Vol. 1. Chicago: Chas. C. Chapman and Co., 1881.
Hix, John. *The Glass House.* Cambridge: MIT Press, 1974.
Hoffman, Leonore, and Margo Culley, eds. *Women's Personal Narratives: Essays in Criticism and Pedagogy.* New York: Modern Language Association of America, 1985.
Holbrook, Stewart. *The Yankee Exodus, An Account of Migration from New England.* New York: Macmillan, 1950.
Holmes, J. C. "The Early History of Horticulture in Michigan." In *Collections and Researches Made by the Pioneer Society of the State of Michigan.* Vol. 10, 69–84. Lansing: Wynkoop Hallenbeck Crawford Co., 1908.
Horowitz, Daniel. *The Morality of Spending, Attitudes toward the Consumer Society in America, 1875–1940.* Baltimore: Johns Hopkins University Press, 1985.
Horwitz, Howard. *By the Law of Nature, Form and Value in Nineteenth-Century America.* New York: Oxford University Press, 1991.

Howe, Daniel Walker. *The Political Culture of the American Whigs.* Chicago: University of Chicago Press, 1979.

Hubka, Thomas C. *Big House, Little House, Back House, Barn, The Connected Farm Buildings of New England.* Hanover, NH: University Press of New England, 1985.

Hunt, John Dixon. *Gardens and the Picturesque. Studies in the History of Landscape Architecture.* Cambridge: MIT Press, 1992.

Hunt, John Dixon, and Joachim Wolschke-Bulmahn, eds. *The Vernacular Garden.* Washington, DC: Dumbarton Oaks Research Library and Collection, 1993.

Hurt, R. Douglas. *American Agriculture, A Brief History.* Revised edition. West Lafayette, IN: Purdue University Press, 2002.

Hylton, William H., ed. *The Rodale Herb Book, How to Use, Grow, and Buy Nature's Miracle Plants.* Emmaus, PA: Rodale Press Book Division, 1976.

Jackson, John Brinkerhoff. *American Space, The Centennial Years 1865–1876.* New York: W. W. Norton and Co., 1972.

———. *Discovering the Vernacular Landscape.* New Haven: Yale University Press, 1984.

———. "A New Kind of Space." In *Changing Rural Landscapes.* Ed. Ervin H. Zube and Margaret J. Zube, 66–73. Amherst: University of Massachusetts Press, 1977.

———. "The Past and Present of the Vernacular Garden." In *The Vernacular Garden.* Ed. John Dixon Hunt and Joachim Wolschke-Bulmahn, 11–17. Washington, DC: Dumbarton Oaks Research Library and Collection, 1993.

Jackson, Kenneth T. *Crabgrass Frontier, The Suburbanization of the United States.* New York: Oxford University Press, 1985.

Jay, Robert. *The Trade Card in Nineteenth-Century America.* Columbia: University of Missouri Press, 1987.

Jenkins, Virginia Scott. *The Lawn, A History of An American Obsession.* Washington, DC: Smithsonian Institution Press, 1994.

Jensen, Joan M. *Loosening the Bonds, Mid-Atlantic Farm Women, 1750–1850.* New Haven: Yale University Press, 1986.

John, Richard R. *Spreading the News, The American Postal System from Franklin to Morse.* Cambridge: Harvard University Press, 1995.

Johnson, Paul E. *A Shopkeeper's Millennium, Society and Revivals in Rochester, New York, 1815–1920.* New York: Hill and Wang, 1978.

Jones, Fred Mitchell. *Middlemen in the Domestic Trade of the United States 1800–1860.* Illinois Studies in the Social Sciences 21, no. 3. Urbana: University of Illinois, 1937.

Kassarjian, Harold H., and Thomas S. Robertson. *Perspectives in Consumer Behavior.* Englewood Cliffs, NJ: Prentice Hall, 1991.

Kielbowicz, Richard B. *News in the Mail: The Press, Post Office, and Public Information, 1700–1860s.* New York: Greenwood Press, 1989.

Klein, Maury. *The Flowering of the Third America. The Making of an Organizational Society, 1850–1920.* Chicago: Ivan R. Dee, 1993.

Kloppenburg, Jack Ralph, Jr. *First the Seed, The Political Economy of Plant Biotechnology 1492–2000.* New York: Cambridge University Press, 1988.
Klose, Nelson. *America's Crop Heritage. The History of Foreign Plant Introduction by the Federal Government.* Ames: Iowa State College Press, 1950.
Kohlmaier, Georg, and Barna Van Sartory. *Houses of Glass, A Nineteenth Century Building Type.* Cambridge: MIT Press, 1986.
Koppelkamm, Stefan. *Glasshouses and Wintergardens of the Nineteenth Century.* New York: Rizzoli International Publications, 1981.
Lasch, Christopher. *Haven in a Heartless World, The Family Besieged.* New York: Basic Books, 1977.
Lawrence, George H. M., ed. *America's Garden Legacy, A Taste for Pleasure.* Philadelphia: Pennsylvania Horticultural Society, 1978.
Lears, T. Jackson. *Fables of Abundance, A Cultural History of Advertising in America.* New York: Basic Books, 1994.
———. *No Place of Grace, Antimodernism and the Transformation of American Culture 1880–1920.* New York: Pantheon Books, 1981.
Leighton, Ann. *American Gardens in the Eighteenth Century: "For Use or for Delight."* Boston: Houghton Mifflin, 1976.
———. *American Gardens of the Nineteenth Century: "For Comfort and Affluence."* Amherst: University of Massachusetts Press, 1987.
———. *Early American Gardens: "For Meate or Medicine."* Boston: Houghton Mifflin, 1970.
Leone, Mark P. "Interpreting Ideology in Historical Archaeology: Using the Rules of Perspective in the William Paca Garden in Annapolis, Maryland." In *Ideology, Power and Prehistory.* Ed. Daniel Miller and Christopher Tilley, 24–35. New York: Cambridge University Press, 1984.
Levine, Lawrence W. *Highbrow/Lowbrow, The Emergence of Cultural Hierarchy in America.* Cambridge: Harvard University Press, 1988.
Linden-Ward, Blanche. *Silent City on a Hill. Landscapes of Memory and Boston's Mount Auburn Cemetery.* Columbus: Ohio State University Press, 1989.
Lindquist, Charles N. *Lenawee County, A Harvest of Pride and Promise.* Chatsworth, CA: Windsor Publications, 1990.
———. *Lenawee Reflections, September 1988–August 1989.* Adrian, MI: Lenawee County Historical Society, 1992.
Lockwood, Alice G. B., ed. *Gardens of Colony and State: Gardens and Gardeners of the American Colonies and of the Republic Before 1840.* New York: Scribners, 1931.
Long, Clarence D. *Wages and Earnings in the United States 1860–1890.* Princeton: Princeton University Press, 1960.
Lynes, Russell. *The Taste-Makers.* New York: Harper and Brothers, 1954.
Maass, John. *The Glorious Enterprise, The Centennial Exhibition of 1876 and H. J. Schwarzmann, Architect-in-Chief.* Watkins Glen, NY: American Life Foundation, 1973.
Machor, James L. *Pastoral Cities. Urban Ideals and the Symbolic Landscape of America.* Madison: University of Wisconsin Press, 1987.

Major, Judith K. *To Live in the New World: A. J. Downing and American Landscape Gardening.* Cambridge: MIT Press, 1997.

Marti, Donald B. *Historical Directory of American Agricultural Fairs.* New York: Greenwood Press, 1986.

———. *To Improve the Soil and the Mind, Agricultural Societies, Journals, and Schools in the Northeastern States, 1791–1865.* Ann Arbor: University Microfilms International, 1979.

Martin, Edgar W. *The Standard of Living in 1860.* Chicago: University of Chicago Press, 1942.

Marwil, Jonathan L. *A History of Ann Arbor.* Ann Arbor: Ann Arbor Observer Co., 1987.

Marzio, Peter C. *The Democratic Art, Chromolithography 1840–1900, Pictures for a Nineteenth-Century America.* Boston: David R. Godine, 1979.

Mathews, Glenna. *"Just a Housewife," The Rise and Fall of Domesticity in America.* New York: Oxford University Press, 1987.

McCracken, Grant. *Culture and Consumption, New Approaches to the Symbolic Character of Consumer Goods and Activities.* Bloomington: Indiana University Press, 1988.

McDannell, Colleen. *The Christian Home in Victorian America 1840–1900.* Bloomington: Indiana University Press, 1986.

McGuire, Diane K. *Gardens of America: Three Centuries of Design.* Charlottesville, VA: Thomasson-Grant, 1989.

McKelvey, Blake. "The Flower City: Center of Nurseries and Fruit Orchards." In *Rochester Historical Society Publication Fund Series.* Vol. 18, 121–169. Rochester, NY: Rochester Historical Society, 1923.

———. *Rochester: The Flower City 1855–1890.* Cambridge: Harvard University Press, 1949.

McMurry, Sally. *Families and Farmhouses in Nineteenth-Century America, Vernacular Design and Social Change.* New York: Oxford University Press, 1988.

———. *Transforming Rural Life, Dairying Families and Agricultural Change, 1820–1885.* Baltimore: Johns Hopkins University Press, 1995.

Meinig, D. W. *The Interpretation of Ordinary Landscapes: Geographical Essays.* New York: Oxford University Press, 1979.

Millspaugh, Charles F. *Medicinal Plants.* Philadelphia: John C. Yorston, 1892; reprint edition, New York: Dover Publications, 1974.

Morgan, Joan and Alison Richards. *A Paradise Out of a Common Field.* New York: Harper and Row, 1990.

Mott, Frank Luther. *A History of American Magazines, 1850–1865.* Vol. 2. Cambridge: Harvard University Press, 1938.

———. *A History of American Magazines, 1865–1885.* Vol. 3. Cambridge: Harvard University Press, 1938.

Motz, Marilyn Ferris. *True Sisterhood, Michigan Women and Their Kin, 1820–1920.* Albany: State University of New York Press, 1983.

Neely, Wayne Caldwell. *The Agricultural Fair.* New York: Columbia University Press, 1935.

Newton, Norman T. *Design on the Land, The Development of Landscape Architecture.* Cambridge: Belknap Press of Harvard University Press, 1971.
Noble, Allen G., ed. *To Build in a New Land, Ethnic Landscapes in North America.* Baltimore: Johns Hopkins University Press, 1992.
Nordin, D. Sven. *Rich Harvest: A History of the Grange, 1867–1900.* Jackson: University of Mississippi Press, 1974.
Norris, James D. *Advertising and the Transformation of American Society, 1865–1920.* New York: Greenwood Press, 1990.
Nourie, Alan, and Barbara Nourie, eds. *American Mass-Market Magazines.* Westport, CT: Greenwood Press, 1990.
Novack, Barbara. *Nature and Culture, American Landscape and Painting, 1825–1875.* New York: Oxford University Press, 1980.
O'Malley, Therese, and Marc Treib, eds. *Regional Garden Design in the United States.* Washington, DC: Dumbarton Oaks Research and Collection, 1995.
Ortner, Sherry B. "Is Female to Male as Nature Is to Culture?" In *Women, Culture, and Society.* Ed. Michelle Zimbalist Rosaldo and Louise Lamphere, 67–87. Stanford, Stanford University Press, 1974.
Osterud, Nancy Grey. *Bonds of Community, The Lives of Farm Women in Nineteenth-Century New York.* Ithaca: Cornell University Press, 1991.
Pope, Daniel. *The Making of Modern Advertising.* New York: Basic Books, 1983.
Porter, Glenn, and Harold C. Livesay. *Merchants and Manufacturers, Studies in the Changing Structures of Nineteenth-Century Marketing.* Baltimore: Johns Hopkins University Press, 1971.
Portrait and Biographical Album of Lenawee County, Michigan. Chicago: Chapman Brothers, 1888.
Post, Robert C. *1876: A Centennial Exhibition.* Washington, DC: Smithsonian Institution Press, 1976.
Powell, Fred Wilbur. *The Bureau of Plant Industry, Its History, Activities and Organization.* Baltimore: Johns Hopkins University Press, 1927.
Presbrey, Frank. *The History and Development of Advertising.* New York: Greenwood Press, 1968.
Prosterman, Leslie. *Ordinary Life, Festival Days: Aesthetics in the Midwestern County Fair.* Washington, DC: Smithsonian Institution Press, 1995.
Punch, Walter. ed. *Keeping Eden: A History of Gardening in America.* Boston: Brown-Little, 1992.
Rapoport, Amos. *House Form and Culture.* Englewood Cliffs, NJ: Prentice-Hall, 1969.
Rasmussen, Wayne D., and Gladys L. Baker. *The Department of Agriculture.* New York: Praeger Publishers, 1972.
Reed, Lula A., compiler. *The Early History, Settlement and Growth of Jackson, Michigan.* Jackson, MI: Jackson County Historical Society, 1965.
Reps, John W. *Views and Viewmakers of Urban America, Lithographs of Towns and Cities in the United States and Canada, Notes on the Artists and Publishers, and a Union Catalog of Their Work, 1825–1925.* Columbia: University of Missouri Press, 1984.

Ristow, Walter W. *American Maps and Mapmakers, Commercial Cartography in the Nineteenth Century.* Detroit: Wayne State University Press, 1985.

Rose, Anne C. *Voices of the Marketplace, American Thought and Culture 1830–1860.* New York: Twayne Publishers, 1995.

Rosenweig, Roy, and Elizabeth Blackmar. *The Park and the People, A History of Central Park.* Ithaca: Cornell University Press, 1992.

Rothman, David J. *The Discovery of the Asylum, Social Order and Disorder in the New Republic.* Rev. ed. Boston: Little, Brown and Co., 1990.

Ryan, Mary P. *Cradle of the Middle Class, The Family in Oneida County, New York.* New York: Cambridge University Press, 1981.

———. *The Empire of Mother. American Writing About Domesticity 1830–1860.* New York: Copublished by Institute for Research in History and Haworth Press, 1982.

Schapsmeier, Edward L., and Frederick H. Schapsmeier. *Encyclopedia of American Agricultural History.* Westport, CT: Greenwood Press, 1975.

Scheele, Carl H. *A Short History of the Mail Service.* Washington, DC: Smithsonian Institution Press, 1970.

Schlereth, Thomas J. *Cultural History and Material Culture, Everyday Life, Landscapes, Museums.* Ann Arbor: UMI Research Press, 1990.

———. *Victorian America, Transformations in Everyday Life, 1876–1915.* New York: Harper Collins, 1991.

Schroeder, Fred E. H. *Front Yard America: The Evolution and Meanings of a Vernacular Domestic Landscape.* Bowling Green, OH: Bowling Green State University Popular Press, 1993.

———. "The Little Red Schoolhouse." In *Icons of America.* Ed. Ray B. Browne and Marshall Fishwick, 139–160. Bowling Green, OH: Bowling Green State University Popular Press, 1978.

———. "Semi-Annual Installment on the American Dream: The Wishbook as Popular Icon." In *Icons of Popular Culture.* Ed. Marshall Fishwick and Ray B. Browne, 73–86. Bowling Green, OH: Bowling Green University Popular Press, 1970.

Schuyler, David. *Apostle of Taste, Andrew Jackson Downing, 1815–1852.* Baltimore: Johns Hopkins University Press, 1996.

———. *The New Urban Landscape, The Redefinition of City Form in Nineteenth-Century America.* Baltimore: Johns Hopkins University Press, 1986.

Scull, Penrose. *From Peddlers to Merchant Princes, A History of Selling in America.* Chicago: Follett Publishing Co., 1967.

Scully, Vincent J., Jr. *The Shingle Style and The Stick Style. Architectural Theory and Design from Downing to the Origins of Wright.* Rev. ed. New Haven: Yale University Press, 1971.

The Seeds of Tomorrow: Ferry-Morse Seed Co. 1856–1956. Detroit: Ferry-Morse Seed Co., 1956.

Sennett, Richard. *Families Against the City, Middle-Class Homes of Industrial Chicago 1872–1890.* Cambridge: Harvard University Press, 1970.

Shannon, Fred A. *The Farmer's Last Frontier, Agriculture 1860–1897.* New York: Rinehart and Co., 1959.

Shapiro, Stanley J., and Alton F. Doody, compilers. *Readings in the History of American Marketing, Settlement to Civil War.* Homewood, IL: Richard D. Irwin, 1968.

Shoetzow, Mae R., compiler. *A Brief History of Cass County.* Marcellus, MI: Cass County Federation of Women's Clubs, 1935.

Shortridge, James. *The Middle West: Its Meaning in American Culture.* Lawrence: University Press of Kansas, 1989.

Sklar, Kathryn Kish. *Catharine Beecher, A Study of American Domesticity.* New Haven: Yale University Press, 1973.

Smith, H. Shelton, ed. *Horace Bushnell.* New York: Oxford University Press, 1965.

Soltow, Lee. *Men and Wealth in the United States 1850–1870.* New Haven: Yale University Press, 1975.

———. *Patterns of Wealthholding in Wisconsin since 1850.* Madison: University of Wisconsin Press, 1971.

Somer, Margaret Frisbee. *The Shaker Garden Seed Industry.* Old Chatham, NY: Shaker Museum, 1972.

Sommers, Lawrence M., ed., *Atlas of Michigan.* East Lansing: Michigan State University Press, 1977.

Spears, Timothy. *100 Years on the Road, The Traveling Salesman in American Culture.* New Haven: Yale University Press, 1995.

Stephenson, Richard W. *Land Ownership Maps, A Checklist of Nineteenth Century United States County Maps in the Library of Congress.* Washington, DC: Library of Congress, 1967.

Stern, Edward. *History of the "Free Franking" of Mail in the United States.* New York: H. L. Lindquist, 1936.

Stilgoe, John R. *Borderland, Origins of the American Suburb, 1820–1939.* New Haven: Yale University Press, 1988.

———. *Common Landscape of America, 1580–1845.* New Haven: Yale University Press, 1982.

———. *Metropolitan Corridor: Railroads and the American Scene.* New Haven: Yale University Press, 1983.

———. "Smiling Scenes." In *Views and Visions: American Landscape Before 1830.* Ed. Edward J. Nygren, 211–226. Washington, DC: Corcoran Gallery of Art, 1986.

Sutherland, Daniel E. *The Expansion of Everday Life, 1860–1876.* New York: Harper and Row, 1990.

Sweeting, Adam. *Reading Houses and Building Books. Andrew Jackson Downing and the Architecture of Popular Antebellum Literature, 1835–1855.* Hanover, NH: University Press of New England, 1996.

Tanner, Helen Hornbeck, ed. *Atlas of Great Lakes Indian History.* Norman, OK: Published for Newberry Library by University of Oklahoma Press, 1986.

Tatum, George B., and Elisabeth B. MacDougall, eds. *Prophet with Honor: The Career of Andrew Jackson Downing. 1815–1852.* Washington, DC: Dumbarton Oaks, 1989.

Taylor, George Rogers. *The Transportation Revolution, 1815–1860*. New York: Rinehart, 1951.

Tedlow, Richard S. *New and Improved, The Story of Mass Marketing in America*. New York: Basic Books, 1990.

Thacker, Christopher. *The History of Gardens*. Berkeley: University of California Press, 1992.

Thompson, George F., ed. *Landscape in America*. Austin: University of Texas Press, 1995.

Thornton, Tamara Plakins. *Cultivating Gentlemen, The Meaning of Country Life among the Boston Elite, 1785–1860*. New Haven: Yale University Press, 1989.

Tice, Patricia M. *Gardening in America, 1830–1910*. Rochester, NY: Strong Museum, 1984.

Tishler, William, ed. *American Landscape Architecture: Designers and Places*. Washington, DC: Preservation Press, John Wiley and Sons, 1989.

Tobey, G. B. *A History of Landscape Architecture, The Relationship of People to Environment*. New York: American Elsevier Publishing Co., 1973.

Trachtenberg, Alan. *Reading American Photographs, Images as History, Matthew Brady to Walker Evans*. New York: Hill and Wang, 1989.

Trump, Fred. *The Grange in Michigan, An Agricultural History of Michigan Over the Past 90 Years*. Grand Rapids, MI: By the Author, Dean-Hick Co., 1963.

Turner, Victor. *The Forest of Symbols*. Ithaca: Cornell University Press, 1967.

United States Domestic Postage Rates 1789–1956. Washington, DC: Post Office Department, 1956.

Upton, Dell, and John Michael Vlach, eds. *Common Places, Readings in American Vernacular Architecture*. Athens: University of Georgia Press, 1986.

Van Ravenswaay, Charles. *Drawn from Nature. The Botanical Art of Joseph Prestele and His Sons*. Washington, DC: Smithsonian Institution Press, 1984.

———. "Horticultural Heritage—The Influence of the U.S. Nurserymen: A Commentary." In *Proud Heritage, Future Promise: A Bicentennial Symposium*, 143–148. Washington, DC: Associates of the National Agricultural Library, 1977.

———. *A Nineteenth-Century Garden*. New York: Universe Books, Main Street Press, 1977.

Waldron, Clara. *One Hundred Years, A Country Town, The Village of Tecumseh, Michigan, 1824–1924*. n. p.: Thomas A. Riordan, 1968.

Walters, Ronald G. *American Reformers, 1815–1860*. New York: Hill and Wang, 1978.

Warner, Sam Bass, Jr. *To Dwell is to Garden: A History of Boston's Community Gardens*. Boston: Northeastern University Press, 1987.

———. *Streetcar Suburbs, The Process of Growth in Boston, 1870–1900*. Cambridge: Harvard University Press and MIT Press, 1962.

———. *The Urban Wilderness, A History of the American City*. New York: Harper and Row, 1972.

Watts, May T. *Reading the Landscape of America*. New York: Collier/Macmillan, 1975.

Weir, Levi C. "The Express." In *One Hundred Years of American Commerce 1795–1895*. Vol. 1. Ed. Chauncey M. DePew, 137–140. New York: D. O. Haynes and Co., 1895.

Westmacott, Richard. *African-American Gardens and Yards in the Rural South*. Knoxville: University of Tennessee Press, 1992.

White, Richard P. *A Century of Service, A History of the Nursery Industry Association of the United States*. Washington, DC: American Association of Nurserymen, 1975.

Whitney, William A., and R. I. Bonner. *History and Biographical Record of Lenawee County, Michigan*. Vol. 1. Adrian, MI: W. Stearns and Co., 1879.

Williamson, Tom. *Polite Landscapes, Gardens and Society in Eighteenth-Century England*. Baltimore: Johns Hopkins University Press, 1995.

Winkler, Gail Caskey, and Roger W. Moss. *Victorian Interior Decoration, American Interiors 1830–1900*, New York: Henry Holt and Co., 1986.

Wright, Gwendolyn. *Building the Dream: A Social History of Housing in America*. Cambridge: MIT Press, 1983.

———. *Moralism and the Model Home*. Chicago: University of Chicago Press, 1980.

Wright, Richardson. *Hawkers and Walkers in Early America. Strolling Peddlers, Preachers, Lawyers, Doctors, Players, and Others, from the Beginning to the Civil War*. Philadelphia: J. B. Lippincott Co., 1927.

INDEX

(References to figures appear in bold face.)

Accessories, horticultural, 18, 79–81, **80, 84,** 84–86
Adair, William, 22, 100, 121, 144, 157
  Descriptive Catalogue of Choice and Select Flower, Vegetable, and Agricultural Seeds, **101,** 101–102
Adams, H. Dale, 140, 198
Adams, John Quincy, 90
Adams, Mrs. L. B., 44, 92
Adams Express Company, 111
Adornments. *See* Accessories, horticultural
Adrian College, 5
Adrian, Mich., 5, 6, 7, 16, 156
  fairgrounds, 154
Adrian Horticultural Society, 23, 107, 109, 154, 174, 175
  activities of, 159–160
  founding of, 172
Adrian Times and Expositor, 61, 80, 82, 120, 160, 165
  advertisement, **41, 108**
  "Farm and Garden" column, 58
  Lenawee County Fair, 161
  nursery agents, 141
Advertisement, **41,** 109, 123, 142, 143. *See also* Publications: advertising
Advisers, x, 16, 21, 66, 86. *See also* Advocates; Commentators; Critics; Proponents; Reformers
  garden plans, 73–79

  on gender roles, 47–49
  how-to advice, 53–54
  on indoor plants, 81–86
  in newspapers, 58
  on character, 44–45
Advisory Board of Agriculture of the Patent Office, 112
Advocates, xviii–xix, 148, 179–180, 217–218. *See also* Advisers; Commentators; Critics; Proponents; Reformers
  ornamental plants and social status, 43, 43–44
Agents, nursery, 135–136, 144
  critics of, 139–140
  plate books, 137–138, **138**
Agriculture, progressive, 149–150
Allen, Lewis, 31
Allyn, Robert, 169
American Agriculturist, xi, 29, 31–32, 34, 56, 66, 79, 90, 186, 190, 192
  advertising in, 188–119, **119, 122**
  circulation, 61
  garden plan, 67, 69, 75, 76
  houseplants, 83, 84
  seeds, 89, 93
  value of ornamental planting, 40–41
American Express Company, 111
American Farmer, 56
American Flower-Garden Directory (Buist), 63

Index

American Garden, 61. See also *Ladies' Floral Cabinet*
American Gardener, 55
American Gardener's Calendar, 55
American Horticultural Annual (Judd), 94
American Journal of Horticulture, 58
American Philosophical Society for Promoting Useful Knowledge, 54
American Pomological Society, 65
Ann Arbor, Mich., 4, 5, 7, 16, 20
   description of, 150
   fairgrounds, 154
Ann Arbor Garden and Nursery, 105. See also Noble, S. B.
Arnold, R. W., 180
Arthur, Timothy Shay, 211
Arthur's Home Magazine, 186, 211
Art of Beautifying Suburban Home Grounds of Small Extent, The (Scott), 63
Association, 33–34, 47
Atlantic Monthly, 28, 30, 186
Atlas map, combination, 18, 24, 172, 212
   contents of, xv–xvi
   demographics in, 7–9
   homegrounds, 2, 19–22
   ornamentals, 10–17
Atlas of Cass County, Michigan, 195

Bagley, John J., 191
Bailey, Liberty Hyde, 62
Balm, bergamot, 181, 207–208, 212
Baltimore, Md., x, 95
Baptists, 6
Barry, Patrick, 135
Battle Creek, Mich., 32
Barry, Patrick, 23, 59, 62, 69. See also Ellwanger and Barry
Bateham, M. B., 99, 146
Baugher, Rev. H. L., 31
Beals, Kelly, 174
Beecher, Henry Ward, 220
   Plain and Pleasant Talk about Fruits, Flowers and Farming, 218
Bellet, Peter, 95
Berckmans' Nursery, 95
Bitely, Mrs. Jerome, 196

Bliss, B. K., 22, , 96, 128, 132, 145
   Benjamin K. Bliss and Sons' Illustrated Spring Catalogue and Amateurs' Guide to the Flower and Kitchen Garden for 1872, **130**
   catalogue of, 131, 142
Bloomington, Ill., 98
Bloomington Nursery, 98
Boies, J. K., **170**, 170–171, 172
Books, guide, 62–63
Boorstin, Daniel, 132
Boston, Mass., 28, 95
Bowlsby, George W., 93
Braun, John M., 16
Breck, Joseph, 44–45, 96
   Flower-Garden; or Breck's Book of Flowers, The, 63
Briggs and Brothers, 116–117, 120, 165–166
Briggs family, 97
Bremer, Fredrika, 34, 37
Brown, Mrs. Jeremiah, 32, 37
Brown, Lancelot "Capability," 55
Buel, Jesse, x
Buell, Barber Grinnel (B. G.), 181, 186, 187, 190–192, 198, 212
   family ties, **183**, 205, 206, 207
   free seeds, 93, 193
   homegrounds, 188, 189, 210–212
   at Jackson State Fair, 199
   marriage of, 183–184
   plant exchange, 194, 197
Buell, Bertha, **183**
Buell, Emmons, 183
Buell, Eunice, 183
Buell, Everett, **183**
Buell, Flora, **183**, 207
Buell, Frank, **183**
Buell, Hal, **183**
Buell, Hattie (Harriet; Mrs. B. G.; née Copley), **183**, 183, 186, 194, 203, 210
   family ties, 206, 208, 209
   homegrounds, 202
   Kalamazoo State Fair, 199
   seed club, 193
Buell, Jennie, **183**, 184
Buell, Josiah, **183**

Index

Buell, Lincoln, **183**
Buell, Mrs. Lincoln, **183**
Buell family, 206, 208, 212, 217, 220
   gendered roles, 205
   homegrounds, 203
   home map, **195**
   portrait of, **183**
Buffalo Nursery and Horticultural Garden, 120
Buist, Robert, 95
   American Flower-Garden Directory, 63
Burnett, William, 149
Burnett, Mrs., 149–150
Bushnell, Horace
   "Discourses on Christian Nurture," 27

Cabinet Fruit and Flower Plates, **43**, 139
California Farmer, 61
Carle, Margaret, 180
Cass County, Mich., 184–185, 186
Cass County Agricultural Society, 93, 185, 186
   Fair, 198
Cassopolis Vigilant, 186
Catalogues, nursery, 124, 135, 143, 145
   advertising for, 123
   chromolithographs in, 129–131
   illustration in, 126–127
   information in, **125**, 128
   personalized, 131–132, 133, 134
Cates, Morilla, 180
Cayton, Andrew, xiv
Census, Federal 1870, xvi
Centennial Exhibition, 199–201
   Illustrated History of the Centennial Exhibition, The, **200**
Central Park, 201
Chandler, Thomas, 16
   family, **15**
Chautauqua Library Study Course, 187
Chelsea, Mich., 4
Chicago, Ill., 4, 98
Chicago Road, 4
Christian Advocate, 30
Christian Examiner, 28, 38
Chromolithographs, **129**, 129–131

Church, E. F., 102
Cincinnati, Ohio, 99
Cincinnati Horticultural Society, 66
Cleveland, Horace William Shaller, 38
   Requirements of American Village Homes, The, 32
Cole, E. M., 14
Cole, Ezra, 18
Columbus Nursery, 99, 120
Combination Atlas Map, 212. *See also* Atlas map, combination
   of Jackson County, Michigan, **10**, **13**, **14**, **17**, **106**, **171**
   of Lenawee County, Michigan, **15**, **19**, **170**, **173**
Commentators, x–xi, 30–32, 46, 150, 169. *See also* Advisers; Advocates; Critics; Proponents; Reformers
   on advertising, 122–123
   on association, 33–34
   on cut-in gardens, 189
   on homeground plans, 67–72. *See also* Atlas map, combination; Homegrounds
   on ideal home, 35–36
Commissioner of Agriculture, 92
Commissioner of Patents, 92, 112
Committee on Ornamental Trees and Shrubs, 116
Congress (United States), 3–4, 112–113
Cook, Daniel, 158
Cook, family, **13**
Copley, Alexander (father), 181–182, 207, 209–210
Copley, Almon, 204
Copley, Charlotte, 181, 182, 203–204, 209–210, 212, 221
Copley, Esther, 182, 183, 204. *See also* Lawrence, Esther
Copley, Esther (née Nott), 194, 207, 208, 209
Copley, Euphemia (Phim), 204, 205–206, 208, 209
Copley, Flora, 204
Copley, Harriet (Hattie), 182, 183–184, 205. *See also* Buell, Hattie
Copley, Napoleon, 182, 204, 206, 208
Copley, Olivia, 204

Copley family, 181–182, 184, 217
  agricultural activities of, 185–186
  death of Napoleon, 208–209
  family ties in, 203–206
  home map, **195**
  value of ornamental planting to, 212, 220
Cottage Residences (Downing), 66
Country Gentlemen, 186
Critics, xvii, 45, 80. *See also* Advisers; Advocates; Commentators; Proponents; Reformers
Crosman, C. F., 97
Crosman Seeds, 97
Crozier, Mrs. M. P. A., 49
Crystal Palace Exhibition (1851), 84, 85

D. M. Ferry and Company, 103, 120
  catalogue of, 126, 134
  Catalogue of Dutch Bulbs and other Flowering Roots, also Seeds and Plants, **125**
  quality seeds of, 143, 144
D. M. Ferry & Co. Seed Annual, **103**
D. M. Ferry & Co.'s Illustrated, Descriptive and Priced Catalogue of Garden, Flower and Agricultural Seeds, **71**
Davis, Franklin, 95
Dean, Artemus J., 156
Dean, Mrs. Artemus J., 23
Dean family, 23
Delevan, Wis., 98
Delightful Hill, 183, 209, 210. *See also* Copley family
Demaree, Albert L., 29
Deming, Mr., 156
Demographics, 20–21
  for Lanawee, Jackson, and Washtenaw counties, 7–9
Department of Agriculture, (U.S.), 64, 65, 193.
  seed distribution by, 91–94
Department of State (U.S.), 91
Department of the Interior (U.S.), 91
Descriptive Catalogue of Choice and Select Flower, Vegetable, and Agricultural Seeds. See Adair, William

Detroit, Mich., 4, 5, 7, 99
  seed companies in, 102–103, **103**
  state fair, 156
Detroit Advertiser and Tribune, 30, 92
Detroit Free Press, 30
Detroit Horticultural Society, 105, 153, 156–157, 165
Detroit Nursery, 120, 123
Detroit Tribune, 186
Dewey, Dellon Marcus, 144
  The Tree Agent's Private Guide, 137–139
Dexter, Mich., 4, 7
Dexter Leader, 61, 120
"Discourses on Christian Nurture" (Bushnell), 27
Dougall, James, 157
Douglas, Robert, 99, 120
Downing, Andrew Jackson, 31, 37, 39, 49, 57, 59, 68, 69
  family of, 95
  Horticulturist, 58
  Rural Architecture, 66
  Treatise on the Theory and Practice of Landscape Gardening Adapted to North America, A, 62–63, 67
  on trees, 70, 72
Dreer, Henry A., 96, 201, 132–133

East Saginaw, Mich., 7
Edmiston, D. M., 107
Elder, Walter, 40
Elgin Botanic Garden, x
Ellsworth, Henry L., 64
Ellsworth, Oliver, 90, 91
Ellwanger, George, 23, 116, 135
Ellwanger and Barry, 22, 23, 107, 144, 174, 176, 197. *See also* Mount Hope Nursery
  advertising of, 120, 121
  catalogue, 123, 133
  nursery agents of, 135, 141
  plate books, **138**
Elmwood Gardens, 120
Erie Canal, 97
Evangelical Review, 31
Evening Journal (Chicago), 122
Evergreen Place, 205

Everts, L. H., xv, xvi, 2
Everts and Stewart, xvi
Exhibitions, 165–166, 168. *See also* Fairs

F. K. Phoenix, 119, 120
Fahnstock Nursery, 99
Fair, 157, 170, 173, 175, 177. *See also* Exhibitions; Jackson: State Fair; Lenawee County: fair; Michigan State Fair
 advertisement for, **160**
 associations and, 153–156, 158–160
 floral hall, 166–169
 horticultural information, 161–162
 nurserymen at, 165
 premiums, 163–164
Farm, progressive, 149
Farmer's Companion, 30
Farmers' Institute, 186
Federal Population and Agricultural Census, 1870, xvi, 2
Fernery, 85–86. *See also* Wardian Case
Ferry, Dexter Mason, 102. *See also* D. M. Ferry and Company; M. T. Gardner and Company
Fessenden, Thomas, 56, 74
Finch family, 196
Finley, William, 154
Floral hall, 167–168
Florence flask, 85
Flower-Garden; or Breck's Book of Flowers, The (Breck), 63
Flowers. *See* Plants, ornamental
Flowers for the Parlor and Garden (Rand), 46
Flushing, N.Y., 95
Ford, Marian, 34
Frost, A., 97
Furniture, lawn, 81

Gale, G. W., 16
Garden, 18, 74–77, 91. *See also* Plants, ornamental
 beds, 12–16, **77**, 189
 English, x, 54–55, 84–85
 landscape, 68–69
 plan, **76**
 rockery, 16–17, 78–79

Gardening for Profit and Practical Floriculture (Henderson), 63
Gardiner, John, 55
Gardener's Magazine, 55
Gardener's Monthly, xvii, 40, 57, 61, 66, **80**, 115, 126. See also *Horticulturist*
 advice, 58, 70, 82
 commentary of, 113, 148, 175, 176
 merger with *Horticulturist*, 60
Gardenesque, 55
Gardner, M. T., 102. See also *Gardener's Monthly*
Garfield, Charles W., 35, **84**, 159
Gender issues, 47–49, 74, 205
Genesee Farmer, 89, 97
Genesee Valley Nurseries, 97, 120
Gjerde, Jan, xiii
Godley's Lady's Book, 28, 30
Goodspeed, Mary, 208
Goodspeed, Sarah, 194, 207–208
Goodspeed brothers, 194, 196
Grand Rapids, Mich., 7, 156
Grange, 7, 147, 186
 Michigan State, 184
 Ypsilanti, 151–152
Grass Lake, Mich., 7
Gray, Susan, xiv
Great Lakes, 4
Green, William, 88
Greenwood Cemetery, 199
Greenwood Stock Farm, 184, 187
Griffen, J., 118
Grove, The, 98
Guide books, 62–63
Guild, E. C., 32

H. A. Dreer, 126
Harbaugh, Henry, 33, 34
Harnden, William F., 111
Harper's, 28, 30, 186
Harrington, Bates, 140
Harris, Joseph, 98
Harvard, x
Harwood and Dunning, 107. *See also* Jackson Nursery

Index

Hathaway, Benjamin, 190, 191, 194, 197, 199
   family, 196
   nursery, 192, 198
Hathaway, Mrs., 193
Hathaway, Naomi, 194
Hearth and Home, 30, 39, 186
Helme, J. W., 159
Henderson, Peter, 84–85, 120, 124
   Gardening for Profit and Practical Floriculture, 63
Hendrick's Garden, 109
Hepburn, David, 55
Heuisler, Maximillian, 95
Hicks, Herman N., 12
Hicks, Horatio, 10
Higham, John, 26
Hillsdale and Lenawee Union Agricultural Society, 155
Hillsdale County, Mich., 5
Hitchcock family, **17**
Hogg, Thomas J. and Sons, 95
Holmes, J. C., 57, 123, 153, 155, 156, 157
Holt, Joseph, 112
Home, ideal, xvii, 50–51. *See also* Reformers: ideal home
Home Culture Society, 186, 198
Homegrounds, ix. *See also* Atlas map, combination; Commentators: home-ground plans
Horsack, Dr. David, x
Horticulture, 23–24, 25, 49
   democratization of, xi, xiv, 215
   golden age of, xii–xiii
   information on, 54–55, 56, **56**
Horticulture Hall, 200, **200**
Horticulturist, commercial, 115. *See also* Nursery
Horticulturist and Journal of Rural Art and Rural Taste, 49, 57, **59**, 60, 63, 67, 70, 82, 124, 136, 169. See also *Gardener's Monthly*
   circulation of, 62
   on garden plans, **76**
   history of, 59
   on home adornment, 46, 72, 74, 78, 83–84
   as prize, 66

Hours at Home, 28
Houseplants, 81, 82–85
Hovey, Charles M., x–xi, 58–59, 66
Hovey Seed Company, 120
Hubbard and Davis, 22, 132
   advertising of, 120, 121, 123
   catalogue, 99, **100**, 145
Hudson, Mich., 5, 7
Huyck family, 196

Ideology, domestic, 27–30
Ilgenfritz, I. E., 102, 141
Illinois State Agricultural Society, 94
Illinois State Horticultural Society, 31, 65
Illustrated History of the Centennial Exhibition, The, **200**
Immigration, 8, 10, 11, 12, 15, 16, 18, 20
Indiana Farmer, 57
Indiana State Horticultural Society, 65, 117
Industry, horticultural, 87–88

Jackson, Mich., 4, 5, 6, 7, 16, 29
   nurseries, 107
   State Fair, 148, 156, 199
Jackson City Directory, 107
Jackson County, Mich., **3**, 5, 6, 7–9, 167
   Agricultural Society, 154–155, **160**, 161
   atlas map survey, xvi
   history, 2–4
   Horticultural Society, 159
   Pomological Society, 159
Jackson Daily and Weekly Citizen, 171, **171**
Jackson Daily Citizen, 82, 153
Jackson Weekly Citizen, **106**, 120, 165–166
   advertisement in, **160**
Jefferson, Thomas, 54
Johnson, Edwin A.
   Winter Greeneries at Home, 81
Johnstone, H. F., 186
Johnstone, R. F., 57, 156
Judd, Orange
   American Horticultural Annual, 94
Journals. *See* Periodicals

Index   331

Kalamazoo, Mich., 5, 7
   State Fair, 156, 166
Kelley, O. H., 93
Kennicott, John, 88, 104, 140, 111, 118, 180–181, 212
   free seeds, 121–122
   The Grove, 98
Kent County Agricultural Society, 158
Kern, G. M.
   Practical Landscape Gardening, 40, 63
Kinney, Nelson, 11
Kirtland, Dr. Jared P., 219
Klemin, L. L., 180
Knapp, S. O., 78–79, 158, 159

Ladies Companion to the Flower Garden (Loudon, Mrs.), 66
Ladies' Floral Cabinet and Pictorial Home Companion, 35, 48–49, **60**, 60–61, **84**, 88. See also *American Garden*
Ladies' Wreath, 30
Landreth, David, 96
Landreth's Seeds, 120
Lathrop, R., 148
Lawn, 69–71
Lawrence, Archie, 183
Lawrence, Arthur, 211
Lawrence, Austin, 183, 199, 203, 209
Lawrence, Esther (née Copley), 183–184, 186, 187, 221. See also Copley, Esther
   diary, 181, 208, 209, 210, 211, 212
   family ties, 206–207
   gardening of, 188–189, 191–192, 193
Lawrence, Levi (elder), 182
Lawrence, Levi B. (father), 182, 187, 192, 199, 208–209
   Farmers' Club, 186, 191, 194
   homegrounds, 201, 203
   marries Esther Copley, 183
Lawrence, Linneaus, 183, 206, 211
Lawrence family, 204, 217, 220
   gender roles, 205
   homegrounds, 201–203
   home map, **195**
Lay, E. D., 104–105, 107, 156, 157
Lay, Z. K. (brother), 104

Lears, Jackson, 39
Legislature, Illinois, 65
Lenawee County, Mich., xvi. See also Jackson County, Mich.
   Agricultural and Horticultural Society, 153–154, 162–163, 172
   Fair, 23, 107
Lenawee Junction Farmers' Club, 155
Lent, S. Q., 2, 8, 24
   report to the Michigan State Pomological Society, 1, 9, 13, 18
Leroy, Andre, 120
Little Prairie Ronde, 182, 184–185
Locust Grove, 182
Loomis, B. D., 152
Loud and Trask, 107, **108**, 108–109, 141, 165
Loudon, John Claudius, 55
Loudon, Mrs.
   Ladies' Companion to the Flower Garden, 66
Lyon, T. T., 50
Lyon, William, 196
Lyon family, 196

M. T. Gardner and Company, 102
McCarthy, Daniel, 180
M'Mahon, Bernard, 55, 96
McMurry, Sally, xiv
Magazine of Horticulture, Botany, and All Useful Discoveries and Improvements in Rural Affairs, x, 58–59, 66, 75. See also *Tilton's Journal of Horticulture*
Manchester, Mich., 20
Mandeville-King Seed House, 98
Mason, Charles, 92
Masonic Temple, 174
Massachusetts Horticultural Society, x
Maynard, William S., 6
Mead, Peter, 59
Meehan, Thomas, 57, 60, 126, 147
Merritt, John, 158
Methodist, 6
Methodist Quarterly Review, 29, 30, 36
Michigan, xvi, **3**, 181, **184**, 190. See also Jackson County, Mich.
Michigan, Lake, 4, 185
Michigan Anti-Slavery Society, 172

Index

Michigan Farmer and Western Agriculturist, xi, 6, 29, 30, 35, 56, 61, 62, 66, 70, 82, 89, 92, 100, 156, 186, 197
  advertising in, 119–121, 123, 192
  catalogues in, 99, 124, 145
  on fairs, 155, 157, 161
  on homegrounds, 67, 76
  on moral reform, 44, 45, 50
  name change to the *Michigan Farmer and Western Horticulturist*, 57
  nurseries, 101, 102, 105, 107
  progressive agriculture, 149–150
Michigan Farmers' Institute, 47
Michigan Nurseryman's and Fruit Growers' Association, 157–158
Michigan Ornamental Stone Company, 79
Michigan State Agricultural Society, 57, 65, 66, 116–117, 186. *See also* Michigan State Pomological Society; Societies; State Agricultural Society
Michigan State Fair, 116, 155, 158–159, 163–164, 175, 198–199
  Agricultural Fair, 107
  railroads and, 161–162
Michigan State Grange, 184
Michigan State Horticultural Society, 158
Michigan State Journal, 105
Michigan State Pomological Society, 32, 36, 37, 47, 78, 102, 116, 137, 146, 164, 166, 197. *See also* Michigan State Agricultural Society; Societies; State Pomological Society of Michigan
  Annual Report, 1, 65, 66
Miller, 152
Minnesota Farmer and Gardener, 57
Mixer, E. G., 102, 157
Monroe, Mich., 102
Monroe Garden and Nursery (Rochester, N.Y.), 97, 104
Monroe Nursery (Mich.), 102
Monthly Religious Magazine, 28, 39, 45
Moore, D. D. T., 57

Moore's Rural New Yorker, xi, **56**, 66, 89, 149, 186
  catalogues, 136
  circulation of, 29, 30, 56
  on seed agents, 220
Morris, Mr., 199
Morris, Mrs., 199
  family, 196
Mother's Magazine and Family Monitor, 30, 34
Mount Hope Nursery, 97, 116, 133, **138**. *See also* Ellwanger and Barry

Napoleon, Mich., 5
New American Gardener, 74
Newburgh, N.Y., 95
New England Farmer, 48, 56
New Lebanon, N.Y., 96
Newton, Isaac, 91
New York and Erie Railroad, 111
New York City, N.Y., 29, 96, 201
New York Evening Post, 30
New York Horticultural Society, x
New York Independent, 176
New York Mercury, 30
New York Tribune, 30, 186
Nicholsville, Mich., 196
Noble, S. B., 57, 105, 107, 149, 150, 158
Norris, Mark, 157
Norris, Mrs. Mark, 163–164
North American Review, 28, 53
Northern Lenawee County Agricultural Society, 155
Northern Michigan Agricultural and Mechanical Society, 156, 158
Northwest, Old, ix, xii, xiii–xv
Northwest Ordinances, xiii
Northwood Farmers' Club, 93
Nott, Esther, 181, 182. *See also* Copley, Esther
Nurseries, xvii–xviii, 21–23, 114, **133**, 180–181, 216
  advertising of, 118–123, **119**, **122**
  agents of, 135–142, **138**, **139**. *See also* Peddler, plant
  catalogues of, 124–134, **125**, **127**, **129**, **130**, **134**

Index   333

eastern, 94–97
exhibitions of, 165–166
local, 103–110
Michigan, 100–102
Midwestern, 98–99
promotion of, 87–88, 115–117, 142–146, 196
shipping of, 111–113

Oak Grove Nursery, 107
Oakland County, Mich., 29, 99
Odd Fellows Hall, 157
O'Donnell, James, 171, **171**, 172
Ohio Farmer, 57
Old Northwest, ix, xii, xiii–xv
Old Rochester Nursery, 97
Oneida County, N.Y., xiv
Ontario, Lake, 97
Ordway, O., 50
Ornamentation, 34. *See also* Accessories; Plants, ornamental; Trees, ornamental
Owen, Dr. Woodland, 17, 19, 109, 172–174, 175–177
    homegrounds, **173**
Owen, Henry, 174
Owen, Mrs. Woodland, 174

Pacific Guano Company, 200
Painesville, Ohio, 99
Palmer, Mrs. W. A., **10**
Parsons and Company, 95
Patent Office, United States. *See* Department of Agriculture, U.S.
Patent Office Commissioner, 92
Patrons of Husbandry. *See* Grange
Peddler, plant, 122, 219–220. *See also* Nurseries: agents
Peddler, tree. *See* Nurseries: agents
Pennsylvania Horticultural Society, x
Periodicals, 28–30, 60, 64, 66, 67, 125, 151
    advertising in, 118–121, 126
    circulation of, 61–62
    early, 55–56
    free seeds in, 88–90, 122–124
    horticulture in, 57–59
    on progressive agriculture, 149–150

Philadelphia, Pa., x, 96
Phoenix, F. K., 22, 88, 136, 143
Plain and Pleasant Talk about Fruits, Flowers and Farming (Beecher), 218
Plans, garden, 73–79
Plant, miscellaneous
    bean, castor, 75
    bean, marrowful, 199
    corn, dent, 199
    cotton, 173
    moss, sphagnum, 113
    strawberry, 104, 188
    wheat, 149, 199
Plant, ornamental, ix–x, 20. *See also* Tree, ornamental
    arbor vitae, 197; arborvitae, American, 107; abutilon, 83, 175; ageratum, 77, 164; alyssum, 164; amaranth, 164; apple, 194; aster, xi, 117, 164; aster, German, 77, 116, 163, 175; azalea, 164, 200; balsam, xi, 90, 117; beech, American, 4; begonia, 83, 175; bouvardia, 22, 83; cactus, 84; caladium, 175; camellia, 83, 164, 175; cedar, 197; cedar, white, 4, 21; cherry, Chinese, 83; cherry, Jerusalem, 83; chestnut, 192; chestnut, horse, 194; chrysanthemum, 189, 193; chrysanthemum, Chinese, 99; clematis, 35; clover, 199; coleus, 189, 193, 202; cottonwood, 188; croton, 83; cyclamen, 83; cypress vine, 90, 204; dahlia, 22, 96, 99, 101, 105, 107, 118, 163, 164, 168, 175, 189, 194, 202, 217; delphinium, 145; *dianthus*, 145; dianthus, 117; dracaena, 83; dusty miller, 165; eucalyptus, 200; fig, 164; fir, balsam, 205; fuchsia, 83, 99, 124, 145, 173, 175; geranium, 22, 75, 83, 100, 105, 164, 176, 170, 173, 175, 189, 202, 205; geranium, rose, 193; gladiola, 75, 116, 117, 144, 168, 170, 173, 202; grape, 104; grape, Catawba, 196; grape, Delaware, 196; guava, 200; heliotrope, 164, 175; hemlock, 205; hibiscus, 164; hickory, 4; honeysuckle,

35; hollyhock, 210; hoya, 85; hyacinth, 194, 202; iris, 99; ivy, 85; lady's slipper, 209–210; lantanas, 189; lemon, 200; lilac, ix, 50, 210, 165; lilac, white, 194; lily, 105, 194; lily, calla, 83; locust, black, 182; lupine, dwarf, 164; lycopodium, 175; Madeira, 85, 189, 202; maple, 197; maple, sugar, ix, 4, 152, 221; maple, Japanese, 200; mignonette, 175; morning glory, 90; mulberry, 182; oak, 4; oleander, 164; orange, 200; palm, fan, 200; pansy, 76, 117, 163, 173, 175; peach, 42; pear, 42; pelargonium, 165; peony, 99, 104, 165; petunia, 76, 77, 163, 165, 170, 175; phlox, 76, 116, 117, 163, 165, 175, 208; pine, Austrian, 205; pine, Scotch, 107, 205; pine, white, 188, 197; pink, 175; pitcher plant, 200; poinsettia, 95; poplar, Lombardy, 188; portulaccas, 164; pyrethrum, 77; primrose, Chinese, 83, 164; *Prunus,* **73;** quince, 104, 210; rose, 35, 83, 99, 100, 101, 104, 105, 118, 163, 173, 175, 180, 189, 194, 210; rose, China, 105; rubber tree, India, 83, 84, 200; santolina, 77; scabiosa, 164; scarlet runner, xi; spirea, ix; spruce, 205; spruce, black, 4; spruce, Norway, ix, xix, 9, 21, 22, 73, 107, 180, 181, 188, 197, 221; tamarack, 4; tropaeolum, 175; tuberose, 22; tulip, 165, 202; verbena, 22, 75, 76, 77, 99, 107, 118, 168, 164, 168, 173, 175, 181, 189, 202, 212; veronica, 165; weeping tree, 10, **10**, 21, **72;** zinnia, 164, 175

Plymouth Congregational Church, 174, 175

Postal Department (U.S.), 112. *See also* Nurseries: shipping

Potter's American Monthly, 34

Powell, Rev. Edwin P., 109, 141, 174–177

Practical Landscape Gardening (Kern), 40, 63

Prairie Farmer, xi, 29, 57, 66, 98, 124, 169, 197
 on character, 44–47
 regional, 56, 149

Presbyterian Banner, 30

Press. *See also* Publications
 agricultural, 56, 152–153, 190
 horticultural, 117–118

Preston, Henry, 152

Preston, William, 16

Prince Nursery, 95, 120, 135

Professionals, Horticultural, xvii. *See also* Nurseries

Progressive
 household, 29
 farming, 35

Proponents, xv, 113, 215–216. *See also* Advisers; Advocates; Commentators; Critics; Reformers

Quigley, A. A., family, **14**

Quirk, Daniel L., 150

Railroads, 5, 161–162
 "Air Line," 185

Raisin Institute, 5

Raisin Township, (Lenawee County), 23

Rand, Edward Sprague, 63, 85
 Flowers for the Parlor and Garden, 46
 Report of the Commissioner of Agriculture for 1863, 45

Randall, "Brother," 152

Ransom, Governor, 156

Reformers, ix, 41, 42, 44, 47, 145–146. *See also* Advisers; Advocates; Commentators; Critics; Proponents
 on the ideal home, 37–40. *See also* Home, ideal
 on the ideology of home, 25–27
 on moral value of ornamentals, 49–51
 resistance to, 40
 on translocal values, 218–221

Repton, Humphrey, 55

Requirements of American Village Homes, The (Cleveland), 32

Richmond Nurseries, 95

River
  Grand, 4
  Huron, 4
  Kalamazoo, 4
  St. Joseph, 4
Rochester, N.Y., 22–23, 29, 96–97, 98, 99
Rochester Central Nurseries, 120
Rochester Wholesale Nurseries, 97
Rockery. *See* Garden, rockery
Rosedale Nurseries, 95
Rowe, Asa, 97, 104
Rowells, Mary, 180
Rural Architecture (Downing), 66
Rural New Yorker, xi

St. Joseph, Mich., 5
St. Louis, Mo., 28
Saline, Mich., 4
Salyes, Mr., 156
Sanders, Edgar, xi
Sarudy, Barbara W., 135
Saunders, William, 91
Scientific American, 30
Scott, Frank Jessup, 69
  Art of Beautifying Suburban Home Grounds of Small Extent, The, 63
Scott, William H., 169
Seed industry. *See* Nurseries
Share, Hortense, 88
Shrubs 11, 73. *See also* Plants, ornamental
Sigler, Artemus, 159
Siler, Andrew L., 104
Sion Greenhouse, 107–108
Smith, J. Evarts, 147
Smith, J. J., 59
Sobes, W. E. H., 147
Societies, 153, 158, 165. *See also* Michigan State Agricultural Society; Michigan State Horticultural Society; Michigan State Pomological Society
South Lynn, Mich., 29
Springfield, Mass., 96
Starkweather, John, 156, 169–170, 171–172

State Agricultural Society, 23, 158. *See also* Michigan State Agricultural Society
  incorporation of, 155–156
State Fair. *See* Michigan State Fair
State Pomological Society of Michigan, 7, 158–159, 163. *See also* Michigan State Pomological Society
  Second Annual Report of the Secretary of the State Pomological Society of Michigan, **72**, **73**, **84**
  Seventh Annual Report of the Secretary of the State Pomological Society of Michigan, **77**
Steere, Benjamin W., 105–107, 108–109, 158
Steere, David (father), 105–106
Stetson, Sarah P., 110
Stewart, D. J., xvi, 2
Storrs, Harrison and Company, 99, 120, **122**, 123, 192
Swisshelm, Jane, 48
Symbols, cultural, 218
Syracuse Nurseries, 120, 124
Syracuse, N.Y., 120

Tallman, John, 17
Tecumseh, Mich., 5, 7, 155
Tecumseh Herald, 30, 58, 61, 88, 120
Territorial Road, 4
Terry, H. A., 111
Thomas, David, 95
Thomas, H. F., **106**, 107
Thomas, J. J., 36
Thomas J. Hogg and Sons, 95
Thompson, R. O., 104
Thorburn, Grant, 96
Thorburn, J. M., 120
Thorp, Amanda, 196
Thumb, Levi, 140–141
Tilton's Journal of Horticulture and Floral Magazine, 58, 64. *See also Magazine of Horticulture*
Toledo, Ohio, 5
Toledo Nursery, 144
Toms, James, 107–108
Tooker, D. D., 107

Index

Treasury Department (U.S.), 90
Treatise on the Theory and Practice of Landscape Gardening Adapted to North America, A, (Downing), 62–63, 67
Tree, ornamental, 9–11, 70, 72–73. *See also* Plant, ornamental
Tree Agent's Private Guide, The (Dewey), 137–139
Tree bed. *See* garden: beds

Underwood, D. K., 158
Unitarian, 28, 29
United States Department of Agriculture. *See* Department of Agriculture, U.S.
United States Express Company, 111
United States Patent Office. *See* Department of Agriculture, U.S.
University of Michigan, 5
University of South Carolina, x
Updike, Anson
    family of, 13

Van Ravenswaay, Charles, xii
Vick, James, 36, 70, 78, 117, 193
    advertising by, 120, 123
    catalogue of, 124, 126–127, 128
    at fairs, 165, 199
    free seeds, 89, 117
    personalizing business, 116, 131, 132, 133–134
    on plant peddlers, 140
    puffery of, 121
Vick's Floral Guide for 1876, 127, **127**, **133**
Vick's Illustrated Monthly Magazine, **129**
Vick's Quarterly Guide, 127
Vick's Seed Company, 97–98, **134**
Vines, homegrounds adornment, 11–12, 73–74. *See also* Plants, ornamental
Visiting Committee of the Ypsilanti Grange, 142
Volinia Township Farmers' Club, 185, 186, 198
    at Centennial Exhibition, 199
    free seeds, 93, 193–194

Walnut Hill Grapery, 196
Walsh, Alexander, ix, x
Ward, N. B., 85
Ward, Samuel S., 182
Warder, John A., 63–64
Wardian Case, 35, 85–86, 175, 176
Ware, John, 38, 45, 46
Washburn Company, 126, 131, 143, 144
Washtenaw County, xvi, 29. *See also* Jackson County, Mich.
    Agricultural Society, 154, 164
    Fair, 105
Washtenaw County Pomological Society, 159
Water Cure Journal, 30
Watervliet, N.Y., 96
Watling, William, 152
Watts, Frederick, 91
Waukegan, Ill., 99
Wayne County, Mich., 4
Wells, Fargo, and Company, 111
Western Farmer and Gardener, 61, 66
Western Horticultural Review, 61, 63
Western Journal of Agriculture, Manufactures, Mechanic Arts, Internal Improvement, Commerce, and General Literature, 28, 33, 37
Western Literary Cabinet, 30
Western Pomologist, 61
Western Rural, 186
Whitney, Mrs. Dorcas, **19**
William Adair and Company Catalogue, **101**
William H. Reid Company, 98
Williams, Henry T., 57, 59, 60, 62, 63, 83
Williamsburg, Va., 95
Wilson, John L., 122
Window, bay, 18, **84**
Wines, Charles H., 11, 12
Winship Nursery, 95
Winter Greeneries at Home (Johnson), 81
Wisconsin, 88
Wisconsin Farmer, 57
Wisconsin Horticultural Society, 65

Index   337

Women's Advocate, 186
Women's Pavilion, 200
Wood, Mr., 186
Woodman, Jason, 47
Woodward, G. E., 59
Wright County, Minn., 93

Y. M. C. A., Adrian, 66
Yale, x
Ypsilanti, Mich., 4, 5, 6, 7
Ypsilanti Commercial, 58, 108, 124
Ypsilanti Grange, 151–152
Ypsilanti Nursery, 104, 107, 120